ALLELOPATHY IN ECOLOGICAL AGRICULTURE AND FORESTRY

Allelopathy in Ecological Agriculture and Forestry

Proceedings of the III International Congress on Allelopathy in Ecological Agriculture and Forestry, Dharwad, India, 18–21 August 1998

Edited by

S.S. Narwal
R.E. Hoagland
R.H. Dilday

and

M.J. Reigosa

SPRINGER-SCIENCE+BUSINESS MEDIA, B.V.

A C.I.P. Catalogue record for this book is available from the Library of Congress.

ISBN 978-94-010-5817-9 ISBN 978-94-011-4173-4 (eBook)
DOI 10.1007/978-94-011-4173-4

Printed on acid-free paper

Contents

PREFACE

The rapidly growing human population has increased the dependence on fossil fuel based agrochemicals such as fertilizers and pesticides to produce the required agricultural and forestry products. This has exerted a great pressure on the non-renewable fossil fuel resources, which cannot last indefinitely. Besides, indiscriminate use of pesticides for pests (weeds, insects, nematodes, pathogens) control has resulted in serious ecological and environmental problems viz., (A) Increasing incidence of resistance in pest organisms to important pesticides. (B) Shift in pests population, particulary in weeds and insects. In weeds, species that are more closely related to the crops they infest have developed. In insects, scenario is most grim, the predators have been killed and minor insect pests have become major pests and require very heavy doses of highly toxic insecticides for their control. (C) Greater environmental pollution and health hazards (a) particularly from contamination of surface and underground drinking water resources and (b) from their inhalation during handling and application. (D) Toxic residues of pesticides pollute the environment and may prove hazardous to even our future generations. (E) Some agricultural commodities may contain minute quantities of pesticides residues, with long term adverse effects on human and livestock health. Therefore, serious ecological questions about the reliance on pesticides for pests control has been raised.

The use of fertilizers, besides causing environmental problems has also impoverished the soil health and decreased the beneficial soil fauna. For example, in some major crop rotations viz., Rice-wheat, in last 10 years, the yields of both crops has reached plateau and have rather started declining in many east Asian countries including India, despite the application of full dose of plant nutrients. Hence such practices are not sustainable over long periods. In India, crops are being grown since ancient times, without damage to the environment, but the use of agrochemicals during the short span of last fifty years have raised serious doubts about their continuous use. One safe alternative to overcome these problems is the use of allelopathy to sustain development in agriculture and forestry and maintain clean environment for our future generations. Studies have shown a great potential of allelopathy to increase productivity of field crops, vegetables, fruits/forestry trees, fisheries, etc. using allelochemicals as growth regulators, pesticides and using stimulatory allelopathic effects in agriculture. Besides, it aims at reducing environmental pollution and maintaining an ecological balance in fauna and flora through reduced use of agrochemcials, [nitrogenous fertilizers and pesticides (weedicides, nematicides, insecticides and fungicides)] and substituting them with biological nitrogen fixation, natural (plant and microbial) products, allelochemicals and phytochemicals free from all problems arising from the use of agrochemicals. Hence allelopathy may be used in agronomy, plant protection (weed control, entomology, nematology, plant pathology), forestry, horticulture, vegetable crops, soil science, plant breeding, agri-biotechnology, genetics, fisheries, aquaculture, etc.

Therefore, in the 21st century, allelopathy may become the backbone of the strategies for sustainability of Ecological Agriculture and Forestry to develop it further. In many countries, use of such practices have eliminated or drastically reduced the use of agrochemicals to make Agriculture and Forestry sustainable over long periods.

This conference aimed at bringing together active allelopathy scientists from all over the world, to review the progress, identify constraints and plan for the future to provide momentum to applied aspects of this important field of research in agriculture and forestry for the development of technology to make Ecological Agriculture successful for our future generations. The scientific programme of this Conference consisted of six symposia viz., (1) Crop allelopathy, (2) Cropping systems, (3) Agroforestry and Forestry, (4) Allelochemicals, (5) Plant protection and (6) Ecological Agriculture. During the Conference, inaugural lecture was delivered by Prof. R.H. Dilday, "*Father of Rice Allelopathy*", USDA-ARS, National Rice Germplasm Centre, Stuttgart, Arkansas, USA. The Congress was inaugurated by Minister for Cooperation, Government of Karnataka and presided over by Prof. M. Mahadeveppa, Vice-Chancellor, University of Agricultural Sciences, Dharward, Karnataka, India. In the Congress, there were 122 presentations (29 Invited Lectures, 93 Contributory papers). This book contains only Invited Lectures and Contributory papers are being published separately in *Allelopathy Journal*. It is hoped that information contained in these chapters would be beneficial to readers and would help in the development of Ecological Agriculture.

During the short span of eight years, the Indian Society of Allelopathy (ISAL) has strengthened the science of Allelopathy in India. Till now, it has organised three International Conferences viz., I National Symposium on *Allelopathy in Agroecosystems*, February, 1992 at Hisar, Chief guest was Prof. E.L. Rice, USA. II International Symposium on *Allelopathy in Sustainable Agriculture, Forestry and Environment*, September 1996 at New Delhi, Chief Guest was Prof. G.R.Waller, USA. III International Congress on *Allelopathy in Ecological Agriculture and Forestry*, August, 1998 at Dharwad, Chief Guest was Prof. R.H. Dilday, USA. The IV International Congress has been planned for 2001. The ISAL is thankful to Prof. M. Mahadeveppa, Vice-Chancellor, University of Agricultural Sciences, Dharwad for hosting this Conference and to Prof. C.J. Itnal, Director of Research, Organising Secretary and his team of scientists. The financial assistance received from various funding agencies is thankfully acknowledged. We are also thankful to Indian and Foreign participants to participate in this Conference and for revising their manuscripts promptly. SSN is grateful to Profs. R.E. Hoagland, R.H. Dilday and M.J. Reigosa for organising their symposia very well and for editing the manuscripts from their symposia. SSN is also thankful to his family members for encouragement and moral support during the organization of this Congress.

November 1,1999. Editors

LIST OF CONTRIBUTORS

ALAGAWADI, A. R.
Department of Agronomy,
University of Agricultural Sciences,
Dharwad – 580 005, Karnataka, India

AMAL, J. A.
Forest College and Research Institute,
Tamil Nadu Agricultural University,
Mettupalayam 641 301, India

CARTWRIGHT, DAVID
Zeneca Ag Products, 1200 S. 47th St.,
Richmond, CA 94804, USA

CHANNAL, H. T.
Department of Agronomy,
University of Agricultural Sciences,
Dharwad – 580 005, Karnataka, India

CHIN, HSIAO-LING
Zeneca Ag Products, 1200 S. 47th St.,
Richmond, CA 94804, USA

CROMARTIE, THOMAS H.
Zeneca Ag Products, 1200 S. 47th St.,
Richmond, CA 94804, USA

CUTLER, HORACE G.
Natural Products Discovery Group,
Southern School of Pharmacy,
Mercer University,
3001 Mercer University Drive,
Atlanta, Georgia 30341, USA

CUTLER, STEPHEN J.
Natural Products Discovery Group,
Southern School of Pharmacy,
Mercer University,
3001 Mercer University Drive,
Atlanta, Georgia 30341, USA

FACKNATH, S.
Faculty of Agriculture, University of
Mauritius, Reduit, Mauritius.

FRASER, TORQUIL E. M.
Zeneca Ag Products, 1200 S. 47th St.,
Richmond, CA 94804, USA

GILL, A.S.
Division of Crop Production,
Indian Grassland and Fodder Research
Institute, Jhansi-284003, India

GONZALEZ, L.
Department of Plant Biology and Soil Science,
Faculty of Science, University of Vigo, Apdo
874, Vigo-E 36200, Spain

GRAY REED
Zeneca Ag Products, 1200 S. 47th St.,
Richmond, CA 94804, USA

HOAGLAND, ROBERT E.
Natural Products Discovery Group,
Southern School of Pharmacy, Mercer
University, 3001 Mercer University Drive,
Atlanta, Georgia 30341, USA

HOAGLAND, ROBERT E.
USDA-ARS, Southern Weed Science Research
Unit, Stoneville, Mississippi 38776 USA

HUNSHAL, C.S.
Department of Agronomy,
University of Agricultural Sciences,
Dharwad – 580 005, Karnataka, India

JOHN JACOB
Department of Agronomy
College of Agriculture,
Vellayani 695 522,
Thiruvananthapuram, Kerala, India

KNUDSEN, CHRISTOPHER G.
Zeneca Ag Products, 1200 S. 47th St.,
Richmond, CA 94804, USA

LAKE, BYRON H.
Zeneca Ag Products, 1200 S. 47th St.,
Richmond, CA 94804, USA

LALLJEE, B.
Faculty of Agriculture, University of Mauritius, Reduit, Mauritius

LEE, DAVID L.
Zeneca Ag Products, 1200 S. 47th St., Richmond, CA 94804, USA

MICHAELY, WILLIAM J.
Zeneca Ag Products, 1200 S. 47th St., Richmond, CA 94804, USA

NAIR, A.M.
Department of Agronomy
College of Agriculture,
Vellayani 695 522
Thiruvananthapuram, Kerala, India

NARWAL, S. S.
Department of Agronomy,
CCS Haryana Agricultural University,
Hisar – 125 004, India

NGUYEN, NHAN H.
Zeneca Ag Products, 1200 S. 47th St., Richmond, CA 94804, USA

PARAMATHMA, M.
Forest College and Research Institute,
Tamil Nadu Agricultural University,
Mettupalayam 641 301, India

PARIHAR, S. S.
Division of Grassland and Silvipasture,
Indian Grassland and Fodder Research Institute, Jhansi-284 003, India

PASTORIZA, J.E.
Department of Plant Biology and Soil Science,
Faculty of Science, University of Vigo, Apdo 874, Vigo-E 36200, Spain

PATIL, R.H.
Department of Agronomy,
University of Agricultural Sciences,
Dharwad – 580 005, Karnataka, India

PRASAD, J.V.N.S.
Division of Crop Production,
Indian Grassland and Fodder Research Institute, Jhansi-284003, India

RAJKUMAR, M.
Forest College and Research Institute,
Tamil Nadu Agricultural University,
Mettupalayam 641 301, India

REIGOSA, M. J.
Department of Plant Biology and Soil Science,
Faculty of Science, University of Vigo, Apdo 874, Vigo-E 36200, Spain

RUSAY, RONALD J.
Zeneca Ag Products, 1200 S. 47th St., Richmond, CA 94804, USA

SOUTO, X.C.
Department of Plant Biology and Soil Science,
Faculty of Science, University of Vigo, Apdo 874, Vigo-E 36200, Spain

SUGAWARA, F.
Department of Applied Biological Science,
Research Institutes for Science and Technology,
Science University of Tokyo, Noda,
Chiba 278-8510, Japan

VIJAYALAKSHMI MOJUMDER
Division of Nematology,
Indian Agricultural Research Institute,
New Delhi -110012, India

SECTION I

ECOLOGICAL AGRICULTURE

1

Proceedings of congress

The proceedings of III International Congress on Allelopathy in Ecological agriculture and Forestry held on August 18-21, 1998 at University of Agriculture Sciences, Dharwad, Karantaka, India are as under:

SYMPOSIA 1 : CROP ALLELOPATHY

Date	18.8.1998
Time	11.30 am onwards
Chairman	Prof. R. H. Dilday, USA
Co-Chairman	Prof. M. Olofsdotter, Phillipines
Rapporteur	Dr. V. K. Garg, India

In this symposia, a total of six lectures were delivered. The objective of all the lectures was to develop allelopathic varieties of crops which can control the weeds. The presented papers dealt mainly with allelopathic rice varieties.

Lecture 1. The Symposia began with the key note lecture presented by Dr. R. H. Dilday, USA. He gave examples and explained the concept of allelopathy. He stated that phototoxicity was first mentioned by Ancient Philosophers and more recently by E.L.Rice (1984). He expaliend several evaluation experiments that were carried out on rice acessions under his guidance in the fields at the National Rice Germplasm Evaluation and Enhancement Centre, Stuttgart, Arkansas, USA since 1985. Success has been achieved in the control of ducksalad, red stem, barnyardnass and rice flatsedge with allelopathic rice accessions. In his experiments, he found that due to the allelopathic properties of some varieties, the number of ducksalad weed plants was reduced upto 80% under field conditions and PI 338046 (an allelopathic rice variety) reduced the *Echinochloa crus-galli* upto 35%. He suggested that if Breeders, Chemists, Geneticists, Microbiologists, Physiologists and Weed Scientists work together, then the use of herbicides can be reduced significantly by developing allelopathic varieties.

Lecture II. Ms. K. Ebana, from Japan presented her work done at the National Institute of Agrobiological resources, Kannodai, Tsukuba, Ibaraki, Japan. She has developed a method to evaluate allelopathic activity of rice from waster soluble extracts. She found that the orders of allelopathic activity of different plant parts of PI 312777 against ducksalad was leaf > stem > root. She further concluded that extracts from the 6th leaf stage of the allelopathic rice variety PI 312777 were most effective against ducksalad.

Lecture III. Dr. S. M. Hassan, Rice Research Training Centre, Sakha, Kafr EL Sheikh, Egypt, research aimed to improve weed management in rice through the development of

S.S. Narwal et al. (eds.), Allelopathy in Ecological Agriculture and Forestry, 1–9.

allelopathic rice varieties. According to Dr. Hassan, the most important weeds of rice in Egypt are *E. crus-galli*, *Cypeus difformis* and *Dinebra retroflexa*. The work dealing with identification of allelopathic rice varieties in Egypt started in 1993 and presently they have identified many rice varieties which control *E. crus-galli* (50 to 80%), *C. difformis* (30 to 70%) and *D.retroflexa* (40 to 50%) through the release of allelochemicals. The best control of weeds was observed at the 3-4 leaf stage. He concluded that the rice yield depended upon the proper seedling establishment of the allelopathic varieties. With such varieties only 50% of the present day herbicides would be needed to give 100% control of *E. crus-galli*. GIZA 159 and GIZA 1368-5-4, two allelopathic rice varieties from Egypt, are able to control 50-70% of *E. crus-galli* at 1/2 to 2 leaf stage. The allelochemicals released by these varieties significantly reduced the seed bank of *E. crusgalli*.

Lecture IV. Dr. J. D. Mattic, Department of Agronomy, University of Arkansas, Fayetteville AR-72704, USA presented research conducted during 1997. The objectives of the research were to determine the allelochemicals which controlled ducksalad through the use of GCMS and HPLC etc. The target compounds were phenolic and carboxylic acids. He concluded that carboxylic and phenolic acids of PI 312777 were not in high enough concentration to be responsible for allelopathic effects.

Lecture V. Dr. T. L. Lavy from the same institute presented a lecture on plant allelochemicals and pesticide residues. He also presented data on the uptake of herbicide by soybean.

Lecture VI. In the plenary lecture, Dr .K.A.Moldenhauer, Plant Breeder, Rice Research and Extension Cetnre, Stuttgart, Arkansas, USA explained that the plant breeder was only one member of the variety development team in Arkansas especially to develop allelopathic varieties. She has made several crosses between allelopathic rice accessions and the US cultivars Katy, Alan and Newbonnet. She has developed 14 lines from these crosses that have acceptable plant type.

Discussion: At the end of symposia, Dr. Dilday, Chairman remarked on the work carried out on allelopathic rice varieties. During the open discussion, Dr. Saxena (Kenya) suggested that the identification of allelochemicals is important. Dr. S. S. Narwal, while commenting on the lecture of Dr. Dilday, emphasized the need to increase the gossypol content of cotton to solve the *Heliothis* problem in cotton. Dr. Mallik from USA said that plant breeders have done an outstanding job in developing resistant varieties to insects or diseases and now we have the opportunity to develop varieties that also have allelopathic activity against weeds.

RECOMMENDATIONS

1. Breeding of crop varieties for weed control is a major allelopathic strategy and work should continue in this direction.

2. Maximum screening work among the plants for allelopathy has been done on rice. Such research work should be restarted in those crops, where only initial screening work has shown potential for weed control.
3. Prof. Dilday's group and the Group at International Rice research Institute, Philippines should strive to evolve transgenic lines of rice to control weeds.
4. The work of rice allelopathy to control weeds should be started in major rice growing regions of the world.
5. Quick and reliable methods for screening of crop varieties for allelopathic potential need to be developed.

SYMPOSIA 2 : CROPPING SYSTEMS

Date	August 18, 1999
Time	17.30- pm onwards
Chairman	Prof. S. M. Hassan (Egypt)
Co-Chairman	Prof. G. Kazinczi (Hungary)
Rapporteur	Prof. M. K. Porwal (India)

Of the 9 papers proposed for presentation, 8 were presented by 6 speakers.

Lecture 1. Dr. S. S. Narwal in his keynote address highlighted the importance of cropping sequence/rotations, intercropping and crop mixtures in relation to allelopathic effects. He indicated the detrimental effects of monocropping including that of ratooning in certain crops. He also suggested the future areas of allelopathic research in cropping system (a) Survey of all crops for autotoxicity and reasons, thereof (b) Reasons of low yields of certain succeeding crops in rotation and (c) Determine mode of release of allelochemicals.

Lecture II. Dr. G. Dharamaraj from India spoke on sunflower allelopathy and its effects on other crops. Phenolics contents in plants increased with increase in nutrient stress, low levels of N, K, Mg, S, B, leads to increased levels of chlorogenic acid, which ultimately reduced the number of seeds in sunflower head. Allelopathic effects of sunflower leads to suppression of dicot weeds. Nitrate reductase activity, photosynthetic rate, Indoleacetic acid activity, water content of weeds were studied. Finally, he suggested the use of allelopathic crops as green manure for suppression of weeds. Allelopathic effect of sunflower leaf litter is due to presence of phenolic acids viz., *P*-Coumaric acid, *P*-hydroxybenozic acid, vanillic, chlorogenic acids and were detected in 2 N-NaOH hydrolysis. Crops susceptible to sunflower allelopathy were : Blackgram, greengram, soybean; while those resistant to sunflower allelopathy were : cotton, maize, sorghum, etc. The sunflower fields should be kept fallow at least for 75-90 days for sowing of next crops, to avoid harmful effects.

Lecture III. Dr. G. Kazinczi from Hungary reported that common milkweed (*Asclepias syriaca* L.) is a serious perennial weed in Hungary. Water extracts of its fresh roots inhibited the germination of *Amaranthus retroflexus, Chenopodium album Lepidium sativum* and subgarbeet but germination of sunflower and cotton remained unaffected.

Lecture IV. Prof. M. K. Porwal from India in his presentation reported allelopathic effects of *Avenua fatua* (wild oat) and *Phalaris minor* (Little seed canary grass) in an agroecosystem of wheat. The live study of root exudates of two grassy weeds in two tier stilt design did not influence germination and tillers of wheat but reduced the shoot length, ear length and dry matter significantly. Of the two grassy weeds, *Phalaris* was more detrimental.

Lecture V. Dr. G. L. Bansal from India indicated allelopathic potential *of Eupatorium odoratum* on *Lantana camara*. Aqueous extract of twigs at post-flowering stage (10%) was most detrimental and even did not allow *Lantana* to germinate and reduced the growth of seedlings.

Lecture VI. Prof. I. T. Breazu from Romania opined that some pasture species plant and weeds reduced the germination and stunted the shoot and root growth of white clover (*Trifolium repens*) and birds foot trefoil (*Lotus corniculatus*).

RECOMMENDATIONS

Allelopathy plays a major role in cropping systems of sub-tropical and tropical countries, because the climate is favourable for growing crops round the year. In irrigated areas of these countries, multiple cropping systems are practiced in which allelopathic interactions plays a major role. However, little information is available, therefore allelopathy research should be conducted in following areas.

1. Rice monocultures, where 1 to 3 crops of rice are grown in rotation and the yields declined in subsequent crops.
2. Sugarcane Ratoons, which continue upto 5-10 years and yields declined in subsequent ratoons.
3. Rice-wheat rotations, where yield has started declining in last ten years in East Asian countries.
4. Allelopathic interactions in crop mixtures and intercropping systems.
5. To develop crop rotations to control weeds without herbicides
6. To develop high yielding crop rotations utilizing stimulatory effects for increasing crop yields.
7. Allelopathic interactions in pastures and grasslands

SYMPOSIA 3 : AGROFORESTRY/FORESTY

Date	August 20, 1998
Time	9.00 - 12.30
Chairman	Prof. M. J. Reigosa (Spain)
Co-Chairman	Prof. A. S. Gill (India)
Rapporteur	Prof. S. S. Bisla (India)

Lecture 1. Dr. Reigosa presented the work of his laboratory on allelopathy in forests of *Acacia dealbata, Acacia melanoxylon, Quercus robur, Pinus radiata* and specially *Eucalyptus globulus*. He highlighted the methodological, ecophysiological and ecological aspects of the allelopathic relationships between trees and understorey species.

Lecture 2. Dr. Gill revealed the allelopathic interactions of MPTS in various agroforestry systems. He emphasized the need for growing multipurpose tree Sp. with arable crops to increase the total biomass production per unit. Numerous studies on *Eucalyptus* have found that it releases volatile oils, acids and tannins which are responsible for allelopathic interactions.

Lecture 3. Dr. Jacob John presented the prospects of allelopathic research in multistorey cropping systems. Considering the complexity of allelopathic interactions, the thrust areas for allelopathic research are : Separating allelopathic interference from competition, screening germplasm of crops/trees/weeds for genetic manipulation, monitoring the production of allelochemicals, exploitation of allelopathy for plant protection, to determine the threshold value of allelochemicals needed to express the inhibitory/stimulating influences for each crop/weeds.

Lecture IV. Dr. C. S. Hunsal presented the scope of allelopathy research in agroforestry systems of South India. Different combination of trees with crops have been studied to see the allelopathic influence of tree on crops. The trees were *Acacia, Casurina, Dalbergia sissoo, Eucalyptus, Tectona, Glyricidia, Leuceana leucocephala*, etc. The crops were pigeonpea, wheat, berseem, chickpea, cotton, pearlmillet, etc. The effects of tress on crops were stimulatory or inhibitory. Therefore, it is essential to determine allelopathic compatibility of crops with trees to improve the productivity of agroforestry systems.

Lecture V. Prof. S. S. Bisla presented the work on allelopathic potential of neem on germination and seedling growth of rainy season crops. The study was done to determine the allelopathic influence on different crops and to identify the compatible crops for neem based agroforestry systems. He observed that dry leaf extract had more harmful effect than green leaf extract and control. In *in-vitro* studies, both the extracts had more harmful effect on root growth then other traits. Cowpea performed better in both *in-vitro* and *in-vivo* studies, the other crops showed variable response exhibiting sensitivity in one character and tolerance in others. Cowpea may prove to be a favourable combination with neem based agroforestry systems.

Lecture VI. Dr. L. Gonzalez monitored the population of 11 microbial groups for one year in forest soil under one native (*Quecus robue*) and three exotic tree Spp. (*Eucalyptus globalus, Pinus radiata* and *Acacia melanoxylon*). He observed that metabolites released by different tree species had variable effects on the microbes present in agroforestry soil. The exotic tree species had greater influence on microbes as compared to native species.

Lecture VII. Lastly, Dr. Parihar presented his paper on toxic allelochemicals in forage legumes and their ecological significance. He emphasized that a variety of chemicals from leguminous forages involved in biochemical interactions among the living organisms belong to three major group of secondary components viz., non-protein amino acids, phenolic compounds and alkaloids. He suggested that allelochemicals can be grouped into two classes i.e., digestibility reducing allelochemicals (DRA), and toxic allelochemicals. The toxic allelochemicals are largely nitrogen based and are involved in plant interactions. Despite the negative attributes, some allelochemicals may be harnessed for therapeutic and other economic purposes.

RECOMMENDATIONS :

The multidesciplinary studies may be conducted in following areas :

1. Allelopathic effects of exotic trees spp. introduced in forest plantations on understorey and associated plant sp.
2. Allelopathic effects of exotic trees on soil microbes population and activity.
3. In coastal areas of sub-tropical and tropical countries, multistorey cropping systems are practiced since ancient times. A major role of allelopathy is conceptualized in these systems but little or no work has been done. Hence research work must be started.
4. The allelopathic interactions in agroforestry systems of irrigated and unirrigated areas needs to be studied.
5. To improve the productivity, studies should be done to screen compatible crops and their varieties suitable for Agroforestry systems.

SYMPOSIA 4 : ALLELOCHEMICALS AS HERBICIDES

Date	August 19,1998
Time	9.00 am ownards
Chairman	Prof. F. A. Macias, Spain
Co-chairman	Prof. S. J. Cutler, USA
Rapporteur	Prof. M. A. B. Mallik, UsA

Lecture 1. Dr. Macias, Spain reported that his group evaluated three bioactive sesquiterpene lactones and compared their phytotoxicity with commercial herbicides of three categories pre-, post-and mixed. Minor change in the configuration of molecules can change the mode of effectiveness. Logran was used as standard. The resutls indicated that the above named allelochemicals may prove better than commercial herbicides.

Lecture II. Dr. Culter reiterated that Global annual sales of herbicides in 2010 may reach US$ 50 Billion. Examples of phytotoxins from plants, fungi and actinomycetes were discussed. To name few from plants were : Nitropropionate (NPA)-its glycosides are found in several plants and fungi. It is highly effective. Other phytoxins mentioned were rhizobiotomell, irpexil, monersin, hydontocidin, etc.

Lecture III. Dr. C. G. Krudsen, USA narrated an interesting brief history of the discovery of a phytotoxins from an ornamental plant-bottle brush. The compound is syncarpic acid. Lagter, this and other derivatives of this acid were synthesized and tested for phytotoxicity. This led to the development of 2-benzoylcyclohexae-1-3-dions, a bleaching herbicide. This and other related compounds synthesized have unfortunately high persistence in the soil.

Lecture IV. Dr. Culter described a bioassay method using oat coleoptile to study phytotoxicity of some pharmaceuticals having herbicidal effect. *Pacelonyces* Sp. isolated from pecan fruit is the source of valium a well known pharmaceutical. Similar pharmaceuticals and agrochemicals from plants or microbes were stressed.

Lecture V. Dr. M. A. B. Mallik, USA told that microorganisms particularly actinomycetes are potential source of agrochemicals. A simple method of screening microbial isolates was described. Bioactive compounds from culture broth can be sequestered by extraction with methylene chloride.

Lecture VI. Dr. M. J. Reigosa, Spain reported that several physiological processes, e.g., chlorophyll fluorescence, phytoryuthers, etc. , are affected by allelochemicals like hydroxamic acids. Other physiological stresses enhance allelochemicals release/ production from plants, stress also increases the effect of allelochemicals on recipient plant spp.

Lecture VII. Maria Olofsdotter, Phillippines told that rice research deserves more investment and attention because it is most intensively grown crop and consumes 5000 water to produce 1 kg rice. Without any weed control measures, loss of yield could be 95%. IR64 cultivar has fair resistance against few weeds such cultivars having allelopathic activity are highly useful against weeds.

RECOMMENDATIONS:

1. Herbicides cause many problems in Agroecosystems, whereas, use of allelochemicals for weed control does not cause any such problems. Therefore, more research should be conducted to identify allelochemicals with herbicidal potential.
2. Still a large number of plant spp. have not been investigated for presence of allelochemicals. This should be done to find new and potent allelochemicals as herbicides.
3. The soil microflora particularly actinomycetes and fungi are good sources of phytotoxic compounds with herbicidal potential. This area is largely unexplored, hence, more research needs to be done.
4. The phytotoxic allelochemicals content of weed smothering crops needs to be studied. It is presumed such allelochemicals might prove to be a good herbicide or provide chemical structures to develop new herbicides.

5. The allelochemicals released in soil by allelolpathic rice varieties for weed control may be identified and their herbicidal potential need to be tested.

SYMPOSIA 5 : PLANT PROTECTION

Date	August 19, 1998
Time	14.30 pm onwards
Chairman	Prof. R. C. Saxena (Kenya)
Co-Chairman	Prof. N. C. Sukul (India)
Rapporteur	Dr. V. Mojumder (India)

Lecture 1. Dr. Saxena's lecture was very informative and thought provoking. He summarised the results of his group's research in Kenya against the three major nematode spp. viz., *Meloidogyne spp., Pratytenchus goodeyi* and *Radopholus similis*. Application of 100g neem cake/banana plant at 4 monthly interval controlled these nematodes and increased yields. In tomato 3 g neem seed powder/plant reduced the infestation of root-knot nematode.

Lecture II. Dr. Mozumder reported a new approach of nematodes management through intercropping of neem seedlings with the crops. Intercropping of neem seedling with chickpea significantly reduced the population of *M. incognita, R. reniformis, Tylenchorhynchus, Helitocylenchus indicus* and *Hoplolaimus indicus* nematodes and increased the crop yields. In case of vegetable crops, bare root dip treatment of seedlings in water extracts of neem products proved very effective to control the nematodes. Another most effective treatment is seed coating of crop seeds with neem products for control of nematodes.

Lecture III. Dr. Sukul reported that ethanolic extracts from the funicles of *Acacia auriculiformis* and flowering tops of *A. cina* effectively controlled the nematodes in mulberry and tomato, respectively. The extracts *of A. auricultiformis* not only controlled nematodes but also controlled foliar diseases in mulberry.

Lecture IV. Dr. Holihosur summarised the research of his group to control castor semilooper, *Achaea janata* using extracts of various ornamental shrubs viz., *Clerodendrum inerme, Eupatorium, Vitex negundo* and *Bougainvillea glabra*. Acetone proved to be the best organic solvent as compared to ethanol and petroleum ether for extracting the most active ingredient from leaves of *C. inerme*. Its spray @ 400 ug/ml spray effectively killed *A.janata*. The extracts of all test plant spp. proved effective at one or the other stages of insect development.

RECOMMENDATIONS:

1. To overcome the numerous problems associated with use of pesticides, the other alternative approaches should be explored.
2. Many Indian plant spp. have insecticidal properties. Till now only neem has been fully explored. Other tree spp. must be fully explored in this regard.

3. To control nematodes, agro-management practices for intercropping of nematicidal plants like *Tagetes*, neem seedling etc., should be developed.
4. In nematode control, besides lab. and pot studies, field studies need greater emphasis so that technology developed is also successful at farmers fields.

SYMPOSIA 6 : ECOLOGICAL AGRICULTURE

Date	August 20, 1998
Time	9.00 am onwards
Chairman	Prof. C.J.Itnal
Co-chairman	Prof. S. S. Narwal
Rapporteur	Prof. (Mrs) P. Challa

Lecture 1. Dr. Narwal defined Ecological Agriculture and explained strategies to minimise the use of "off-farm inputs" including agrochemicals (fertilizers pesticides) on the farm. Hc reported that more research on allelopathy and related areas could definitely develop practices to make Ecological Agriculture successful. He recommended use of crop residues, mulches, allelochemicals, resistant varieties and natural herbicides to minimise use of agrochemicals.

Lecture II. Dr. Parbha Challa, India stressed the need of discovering horticultural crops with pesticidal including herbicidal properties to control pests of agricultural crops, thus minimising use of pesticides and making sustainable agriculture successful. She described her results using mango leaves for control of weeds in rose plantation. These results have enormous potential for development of sustainable agriculture.

Lecture III. Dr. M. Hallabhavi delivered very informative talk showing that in Homeopathy, active constituents (allelochemicals) of plants are used for treatment of human ailments. While, allelopathy scientists want to use these compounds to control pests of crops and to use them as growth regulators. Both the areas are similar and complementary to each other, may be some homeopathy medicines may control pests of agricultural pests, but no studies have been done so far. Studies are needed in this direction.

RECOMMENDATIONS :

The use of agrochemicals (fertilizers, pesticides) in a short span of 50 years has resulted in numerous problems, hence, not sustainable over long periods. To develop Ecological Agriculture, which is successful over long periods, multidisciplinary researches may be conducted in following areas :

1. To overcome the problems associated with agrochemicals, minimise or eliminate the use of "off farm inputs mainly agrochemicals" by live crops, crops residues; resistant varieties, mulches, allelochemicals as natural pesticides etc.
2. To develop new technology based on allelopathy principles i.e. utilizing stimulatory or inhibitory effects of plants on each other, on pests and on soil microflora etc.
3. Exploratory investigations to study the possibilities of using medicines (Homeopathy, Ayurveda) prepared from plants for pests management and as growth regulators.

2

Allelopathy in ecological agriculture

S. S. NARWAL
Department of Agronomy,
CCS Haryana Agricultural University, Hisar – 125 004, India

Abstract

Modern agriculture is exploitive of growth resources and has caused various problems such as a environmental pollution through (a) contamination of underground drinking water resources, food and fodder with pesticides and nitrates, which is harmful to farm workers and livestock, (b) poor soil health leading to low soil productivity and (c) poor quality of life. These problems may be overcome with the adoption of Ecological Agricultural practices. The definition of Ecological Agriculture used in this chapter is "Ecological Agriculture consists of those practices, which reduces the use of outside inputs on farm. Therefore, various types of allelopathic strategies may be used for (a) maintenance of soil fertility (use of crop rotations, BNF, crop mixtures, crop residues and leaf litter etc.), (b) weed management (cover crops, crop residues as mulches, intercropping, crop rotations, phytotoxic or allelopathic varieties and natural herbicides etc.), (c) insects pests management (cropping systems, resistant varieties, insecticidal allelochemicals etc.), (d) nematodes management (plant materials, oilseed cakes, nematicidal compounds etc.), (e) diseases management (cropping systems, crop residues, organic amendments etc.) and (f) allelochemicals as growth regulators. Therefore, research efforts are needed to utilise inhibitory allelopathic effects of plants for natural control of crop pests and diseases, so that use of present pesticides could be minimised for keeping the environment clean for our future generations.

Key words: Allelochemicals, cover crops, crop mixture, crop residue, crop rotation, ecological agriculture, growth regulators, intercropping, nitrogen fixation, organic amendments, pests management, soil fertility.

1. Introduction

Modern agriculture is exploitive of growth resources and has caused various problems such as environmental pollution through (a) contamination of underground drinking water resources with pesticides and nitrates, (b) contamination of food and fodder with residues of pesticides, nitrates and antibiotics, (c) both 'a' and 'b' cause harm to farm workers (108), (d) poor soil health and soil productivity and (e) poor quality of rural

S.S. Narwal et al. (eds.), Allelopathy in Ecological Agriculture and Forestry, 11–32.

life. There is growing evidence that grain yields of cereal in cereal based rotations in modern agriculture cannot be sustained at the current levels e.g. rice-wheat rotation. In the last 16 years, there is a decline of 25-30% in the grain yields of both rice and wheat in Philippines, Indonesia, India and Pakistan despite the use of recommended cultural practices (108). Therefore, the recent emphasis in agriculture has shifted from a primary goal of maximizing yields over the short term, to a sustainable productivity over long period of time. The knowledge of ecological interactions occurring within an agroecosystem and the sustainable functioning of the system as a whole has become the overall approach. Sustainability can be achieved in an agriculture, that is ecologically sound, resource conserving and not environmentally degrading. Therefore, several definitions of sustainable/ecological/ecofriendly/organic agriculture are available, but for this chapter, the ecological sustainable agriculture means that "farmer grow the crops with the resources available on the farm, reduces dependence on off-farm inputs and maintains soil productivity and clean environment over along period of time". Hence, farmer conserves the resource base to minimize artificial inputs from outside the farm and manage pests (weeds, insects, nematodes, pathogens) through internal regulating mechanisms based on ecological principles and processes (133). Thus, ecological sustainable agriculture strives for the integrated use of a wide range of pests, nutrients and soil management technologies (108). In the last few years, some good books on similar aspects have been released (43,47,90,93,108).

All plant spp. and their residues produce secondary metabolites called allelochemicals. Allelopathy means any process involving secondary metabolites produced by plants, microorganisms, viruses, fungi that influence the growth and development of agricultural and biological systems (57). Although allelochemicals are produced by all plant parts, but the leaves and roots are mainly responsible for their production and release. The ecological agriculture maintains diversity of plant spp. on the farm through various types of multiple cropping systems viz., mixed cropping, crop rotations etc., hence, allelopathy assumes great significance. Allelopathy may be used to increase crop production through avoidance of negative impacts, exploitation of stimulatory effects, management and development of allelopathic crops and varieties to suppress pests (weeds, insects, nematodes, pathogens) and use of allelochemicals as pesticides and growth regulators (38).

The productivity of monoculture declines after few years mainly due to build up of pests and soil sickness etc. These problems could be overcome through the adoption of crop rotations and intercropping systems, which exert detrimental effects on the pests through various chemical interactions (Allelopathy) and cause physical hindrances to restrict movement of pests. Likewise, the bio-diversity provided by crop mixtures has a smothering effect on pests and diseases. These indirect synergistic effects also contribute to higher productivity in crop rotations and intercropping systems. If these effects are utilized properly it is possible to reduce the use of chemical pesticides and herbicides.

This chapter reviews allelopathic effects in crop production and possible use of multiple cropping systems, plant residues, plants and varieties rich in allelochemicals to control pests and supplement nitrogen requirement through Biological Nitrogen Fixation by legumes in ecological agriculture.

2. Soil fertility

In tropical and sub-tropical countries of the World, climate is suitable for round the year cropping. Therefore, in assured irrigated high rainfall areas, farmers grow 2-4 crops in a calendar year in multiple cropping systems (MCS). The properly planned MCS (crop rotations, crop mixtures, intercropping) aims at making ecological agriculture successful, has little harmful impact on environment conditions and maintains soil productivity over a long period of time. They maintain soil fertility, keep the pests under check and reduce soil sickness problem. Both the crop rotations and intercropping systems or crop mixtures through inclusion of legumes maintains or improves soil fertility.

2.1. CROP ROTATION

A well planned crop rotation maintains and even improves the soil fertility, prevents the build up of pests and soil sickness as compared to monoculture (8). If scientifically sound crop rotations are followed, they provide sustainability to agriculture through reducing the requirement of chemical nitrogenous fertilizers and thereby decreasing environmental pollution by substituting them with biologically fixed nitrogen of legumes.

2.1.1. *Biological nitrogen fixation*

2.1.1.1. *Legumes* : In the past when chemical fertilizers were not available, BNF, fallowing and application of farmyard manure were the only practical steps to maintain soil fertility and to provide essential nutrients to the crops. Even today in dryland areas receiving less than 300 mm annual rainfall, the farmers do not apply chemical fertilizers or manures due to uncertainty of crop production. In such regions legumes are the major component of cropping systems, which through BNF maintains soil fertility and also provide protein rich diet to the people and nutritious feed to the livestock. In irrigated areas, continuous cropping of cereals necessitates the use of large quantities of nitrogenous fertilizers resulting in the development of undesirable soil properties and contamination of underground water resources. Under such conditions, the inclusion of legumes in the cropping system and the use of Blue green algae (Azolla) is imperative, as organic nitrogen from BNF may be more suitable than fertilizer nitrogen, because it is released gradually to become available at a time when conditions are favourable both for microbial activity and plant growth. In ecological farming, legumes should alternate nitrogen demanding crops-ideally it should be possible to meet the farm's nitrogen requirements from well designed crop rotations (67) as they fixed upto 450 kg N/ha/yr (103). In sequential cropping, the legumes are generally used as green manure, forage and for grains. As such, their nitrogen contribution varies and follows the order: green manure > forage legume > grain legume, provided that factors governing BNF are similar. However, the allelopathic weeds reduce the N_2 fixation in legumes, hence, weeds must be controlled for more BNF.

The beneficial effect of preceding legumes on the succeeding non-legumes is exerted through the transfer of nitrogen and other nutrients. In crop rotation, nitrogen transfer takes place through decomposition of legumes residues and mineralisation of organic nitrogen. The mineralised forms of nitrogen are then absorbed by the succeeding crop. The quantity of nitrogen transferred depends upon the total quantity fixed and the end use of legume i.e. green manure, forage or grain crop, residues removed/added to soil and other factors.

In tropical and subtropical countries, green manuring is generally known to farmers and is practised wherever facilities are available. However, green manuring is getting less and less popular with the advent of modern farming. In the irrigated belts of India continuous cropping of cereals is adopted and green manure crops seldom find a place in the cropping system. This has caused serious soil health problems. For green manuring, fast growing legume crops are grown and incorporated *in situ* before flowering for quicker decomposition and mineralisation of plant nutrients, so that these becomes easily available to the succeeding crop. Two types of practices are followed: (a) growing of legumes especially for green manuring and (b) incorporating residues of grain legumes after the harvest of seeds at maturity. This difference causes variation in the beneficial effects of green manuring.

Therefore to reduce environmental pollution from nitrogenous chemical fertilizers, the inclusion of legumes (grain legumes, forage legumes, green manure crops) in annual crop rotations is necessary. Since forge and green manure legumes provide more nitrogen to the crop than grain legumes, hence, these should be preferably included in crop rotations.

2.1.1.2. *Blue, green algae.* In rice fields, a small fern called Azolla (*Anabena azollae*) covers the water surface but does not harm the crop growth. It fixes upto 120 kg atmospheric N\ha in the leaves and the nitrogen becomes available to the rice crop after its soil incorporation and decomposition. In Philippines, 100 days old azolla produces 57 t fresh weight/ha, which yields more than 120 kg N/ha (66). In one whole year, it may fix upto 400 kg N\ha i.e. more than tropical and subtropical legumes and thus offers opportunity to substitute inorganic fertilizers.

2.2. CROP MIXTURES OR INTERCROPPING

In terms of land use, growing of crops in mixtures is more productive than growing them separately (146); hence, it is practised traditionally in parts of Asia, Africa and Latin America. Interest in cereal-legume intercropping system is also developing in temperate and warm regions of Australia and U.S.A. because of higher grain yields, greater land use efficiency per unit land area and the improvement of soil fertility through BNF and nitrogen excretion from the legume component (146). In South American countries like Mexico etc. the interplanting of maize, bean and squash in the same planting hole is very ancient practice (48). In such soils of poor fertility, the cultivation of cereals + legumes together improves both total yields and reduces the nitrogen requirement of cereal component. In maize + cowpea mixture, 30% nitrogen taken up by the maize is obtained from the legume (2). Besides, the legume biomass may be used as mulch/green manure. Many legume spp. such as velvet bean (*Mucuna pruriens*), *Sesbania* spp. and *Tephrosia* spp. are used in many countries to conserve soil, improve soil fertility through BNF and as green manure crops. Hence, it also offers scope for developing energy efficient ecological agriculture (97).

2.2.1. *Nitrogen transfer*

It has been shown that N fixation is the source of transferred N. In intercropping systems, the transfer of nitrogen from legume to non-legume component occurs through root exudates (current transfer), via VAM (vesicular-arbuscular mycorrhizae) connections or through decomposition and mineralisation of N from fallen leaves and dead roots and nodules (residual transfer). The root exudates of legumes contain various nitrogenous components which are directly absorbed from the rhizosphere by intermingled roots of non-legume component. The residual transfer also occurs in intercropping systems, provided that non legume component grows for more than 80 days after the harvest of legume, so that nitrogen is mineralised from legume residues e.g. cowpea/greengram/clusterbean/blackgram +sugarcane intercropping etc.

The direct/current transfer of nitrogen from forage legumes to companion grasses occurs in mixed pasture swards (95). Eaglesham *et al.*(37) confirmed the current transfer of nitrogen from cowpea to maize in cowpea + maize intercropping system using N_{15}. Using replacement series designs, Patra *et al.* (99) have reported substantial transfer of nitrogen from legume component to the associated cereal in wheat + chickpea and maize + cowpea intercropping systems both in green house and field studies.

Estimates of residual transfer of nitrogen from legumes + sugarcane intercropping system are not available. However, results of studies on residual effects of legumes + cereal intercropping on succeeding crops are available. Nair *et al.* (87) found a mean wheat yield increase of about 30% after a maize + soybean intercropping and after maize + cowpea the yield increase was 34% when compared to wheat after sole maize. Singh *et al.*(129) estimated the nitrogen benefits to wheat of various preceding legume intercrops. When comparing wheat after sole sorghum with wheat after intercrops, he obtained nitrogen fertilizer equivalents of 3, 31, 46, 40 and 54 with soybean, greengram, cowpea (grain), groundnut cowpea (fodder), respectively.

Besides VAM mediated transfer of phosphorus from a non-legume to the legume crop has also been documented.

2.3. BIOMASS

Soil fertility management in ecological agriculture gives much reliance on the use of biomass (crop residues and other organic wastes) to maintain the status of organic matter in the soil and to meet the nutrients requirement of the crops. The crop residues release allelochemicals through volatiles, leaching and during microbial decomposition. The production of allelochemicals in soil affects germination, growth and yield of crops depending on plant residue type, amount, depth of placement and length of decomposing period. The allelochemicals may either be inhibitory or stimulatory to the succeeding crops (116,141).

2.3.1. *Crops residues*

Plant residues when recycled, improve the physical, chemical and biological properties of the soil. A surface mulch of plant residues ameliorates the microclimate, reduces runoff and soil erosion and makes tillage easier. In Asia, the availability of plant residues has, of late, increased substantially due to adoption of multiple cropping, reduced tillage, stubble mulch agriculture and combine harvesting of grain crops.

Few studies on stimulatory effect of legumes crop residues have been reported. For example, chopped alfalfa added to soil stimulated the growth of tomato, cucumber, lettuce and several other plants (117). The stimulatory allelochemical was identified as triacontanol. Gill *et al.* (45) have also reported stimulatory effect of mungbean, sesame and soybean residues on wheat, chickpea and lentil at lower concentration. Improvement in corn yields following legume crops has also been reported (105,135).

2.3.2. *Tree litter*

Trees form a major component of integrated ecological farming. They perform productive and protective functions in the agroecosystem. Most of the mature trees produce substantial quantity of litter, whose proper management is essential in organic agriculture. Trees due to their deep root system have the capability to improve physio-chemical condition of soil and extract nutrients from deep layers and return them to the surface through litter fall. In general, most of the litter falls underneath tree canopy.

Like crop residues, the litter of some tree species also stimulates the growth of associated crops. For example, corn, bean, black raspberry and quince (*Cydonia oblonga*) grew better within the root zone of the walnut trees than outside (28); *P. roxburghii* field soil increased the growth of blackgram (131) and dried residues of *Glyricidia maculata* significantly promoted the growth of tomato seedlings (20). Likewise, a marked increase in the productivity of pigeonpea, sesame, castor and sorghum under leucaena tree (128a) and beneficial effects of eucalyptus on sorghum (56) were also reported.

3. Weed management

There are about 250 major weed species in agriculture and many of them have allelopathic properties, which reduce crop growth and yield (102). Continuous use of herbicides for weed control has created many problems including their persistence in soil, contamination of environment, crop injury, increase in herbicide-resistant weed population, etc. Hence, non-chemical methods of weed control are preferred in ecological agriculture. Among different non-chemical methods of weed control, allelopathic suppression of weeds through the use of allelopathic plants in crop rotations and of phytotoxic mulches in soil fertility management is very effective (89).

3.1. COVER CROPS AND RESIDUE MULCHES

For weed management in ecological agriculture, the use of phytotoxic mulches and cover crops is very effective. Allelochemicals significantly contribute to weed suppression when planted no-till into residues of cover crops or previous crop residues. Cover crops of wheat, barley, oats, rye, grain sorghum and Sudan grass have been used effectively to suppress broad leaf weeds (73,125,126).

The suppressive effect of allelochemicals of rye mulch on weeds in the field is outstanding. Chou and Patrick (22) identified nine acids from ether extract of decaying rye residues in soil. Phenylacetic, 4-phenylbutyric, vanillic, ferulic, *p*-coumaric, *p*-hydroxybenzoic, *o*-coumaric and salicylic acids, all inhibited the growth of bioassay plants. Shilling *et al.* (125-127) found that β-phenyllacetic acid (PLA) and β-hydroxybutyric acid (HBA) from rye residues provided 20 to 60% inhibition of common lambsquarters and red root pigweed in no-till planted soybean, sunflower and tobacco. Barnes et al. (12) isolated two hydroxamic acids, 2-4 dihydroxy-1,4 (2H)-benzoxazine-3-one (DIBOA) and 2(3H)-benzoxazolinone (BOA) with phytotoxicity on a large number of weed test plants. Further, they also reported that a mulch of 40 days old spring planted rye reduced 69% weed biomass. Muraleedhaeran *et al.* (86) isolated a microbially transformed allelochemical, 2-2—epidioxy-1, 1-azobenzene (2, "2-oxo-1-1" azobenzene) (AZOB) from a soil supplemented with 2 ($3H^+$) – benzoxazolinone (BOA). AZOB was more toxic to curly cress and barnyard grass than DIBOA or BOA. Although there were no detectable amounts of the biotransformation product in soil under rye residues, the implications of such phytotoxic biomagnification of allelochemicals may explain allelopathic weed suppression under field conditions.

Discoveries concerning microbial transformation of certain allelochemicals from wheat and rye residues may be significant in increasing phytotoxicity of such materials to weeds. Liebel and Worsham (73) reported that ferulic acid released during the decomposition of wheat and rye residues in the presence of prickly sida seed carpels was decorboxylated by a bacterium living on the seeds to a styrene derivative, 2-methoxy-4-ethenylphenol. The styrene was more phytotoxic to prickly sida than ferulic acid and controls this weed in natural conditions under wheat and rye mulch.

In field experiments, residues of sorghum, sunflower, rapeseed, wheat and pea at five tonnes ha^{-1} selectively toxified broad-leaved and grass weeds. The response of wild oats was of particular interest. Field pea and wheat residues significantly stimulated wild oat germination and growth. The germination and growth of other grass weeds was, however, significantly inhibited by all types of residues (110). It is possible that the stimulatory compound(s) produced from wheat residues could be employed to counter the discontinuity of germination in wild oat, facilitating a more complete kill by subsequent adoption of an appropriate cultural practice.

Putnam and DeFrank (112) tested residues of several fall and spring planted crops for weed control in Michigan, USA. The plants were desiccated by the herbicides or by freezing. Wheat and rye residues reduced weed growth by upto 88%. Mulches of sorghum or Sudan grass applied to apple and cherry orchards in early spring reduced weed biomass by 90% and 85%, respectively. In a three years series of field experiments, sorghum residues reduced population of common purslane by 70% and smooth crabgrass by 98%.

Sunflower has a pronounced allelopathic effect on germination and growth of many other plants. Consequently, it strongly influences the patterning of the surrounding vegetation (116). It was observed that decomposition of sunflower residues significantly reduced the total number of weeds especially the dicotyledonous ones. In a five year field study with oats and sunflower grown in rotation, the weed density was significantly less than in control plots with oats only (70). In sunflower-wheat rotation field trials, sunflower decreased the density and dry weight of wild oat and *Cirsium arvense* in the following wheat crop (18). Similarly, sunflower reduced the population of associated weed *Trianthema portulacastrum* by 75% at flowering and 96% at maturity stage and of *Parthenium hysterophorus* by 56 and 84%, respectively. The indication was that allelopathic material was released by the roots of sunflower. The maximum inhibition was in BSH-1 variety followed by MSFH-1, Co-2 and EC-68415 in descending order (32). Likewise, preceding crop of sunflower reduced the population of broad leaf weeds such as *Cleome viscosa* and *Corchorus trilocularis* and sedges like *Cyperus iria* in succeeding crop (109).

Narwal *et al.* (92) observed suppression by pearlmillet of the weed density and growth in the succeeding sorghum crop. *Brassica campestris* also reduced the weed density in the same field in the following year owing to inhibitory effect of its residues and volatile excretions from the leaves on the germination and growth of other plants. This crop is used for weed control by the Tarahumara Indians in North Mexico (19).

Residual effects on weeds have also been reported for *Tagetes patula* L, beans, corn, cassava (6); sunflower, sweet potato, sorghum and soybean (39), forage sorghum, Sudangrass hybrid (42); sorghum, barley, oats, wheat and rye (111); crimson clover and hairyvetch (144), fescue (104), alfalfa (1) and cucumber (76).

3.2. INTERCROPPING

The mixing or intercropping of plant species with different growth habits and morphology e.g. melons + plantains provides effective weed control (94). Likewise, undersowing of wide row crops like maize with clover or other species with spreading nature control weeds (142). Bantilan *et al.* (11) reported that in maize crop, mungbean provides more weed suppression than peanut and ascribed it to more rapid early growth and uniform canopy structure in mungbean. Among mungbean cultivars, the more prostrate ones were more suppressive.

The farmers in south eastern Mexico interplant squash in maize/cowpea fields for effective weed control (71). The squash plant suppresses weeds through shade effect and selective allelochemical inhibition. Here the peasants also grow *Stizolobium pruriens* legume with maize to control weeds (49). Grechkanov and Rodionov (53) reported benefits from mixing 1-2 kg seed ha^{-1} of wild heliotrope (*Heliotropeum europeum* L.) with several legumes. This plant not only reduced weeds by 30 to 70% but also controlled other pests.

Expression of allelopathy in fields through plant to plant interactions has also been observed. Joshi and Mahadevappa (62) and Mahadevappa and Kulkarni (78) found that *Cassia sericea*, a leguminous plant, effectively controlled the parthenium weed in the fields through allelopathic activity. Bansal (10) showed that buttercup's (*Ranunculus* sp.) weed species which cause severe infestation and suppression of wheat in mid-hill conditions of Himachal Pradesh, India could be effectively controlled by planting linseed with wheat.

In a three-year field study, barley, rye and *Vicia faba* were planted in monoculture after the harvest of summer crop. These crops grew during the winter season and were ploughed in the soil in the month of March or early April. Thereafter, summer vegetables were planted by the end of May. Barley and *V. faba* and rye + *V. faba* offered almost complete weed control and the latter was most effective. It was attributed to the release of allelochemicals from root exudates during crop growth and from decomposing crop residues (46).

3.3. CROP ROTATIONS

Some weeds species are specific to some crops. Such species are : wild oat and downy brome in wheat; barnyardgrass, foxtail, sandbur and fall panicum in corn/sorghum; cocklebur and valvetleaf in soybean; field bindweed, pricky sida, spurred amoda, sicklepod and silvershade in cotton (140). In these cases, crop rotation is the most effective and economical control method. Only those crops which can compete effectively against weeds are included in such rotations. Fields with summer annual weed problems are rotated with winter grain crops; likewise, fields with troublesome winter annual weeds are rotated with spring or summer crops (145). In common poppy infested fields, rotation with wheat was less effective, while rotation with oat effectively controlled its population possibly due to allelopathic effect of oat. The poppy seeds germinated in oat field but failed to reach maturity (67).

3.4. PHYTOTOXIC VARIETIES

There is variation in the content of allelochemicals in various crops. In addition, the crop varieties also differ in the exudation or excretion of allelochemicals which may affect the degree of weed control. Fay and Duke (39) screened 3000 accessions of oat germplasm for their ability to exude scopoletin, a compound with growth inhibiting properties. Twenty five accessions exuded more scopoletin from their roots than a standard oat cultivar 'Garry'. Four accessions exuded up to three times as much scopoletin as 'Garry' oats. When one of these was grown in sand culture for 16 days with a wild mustard, growth of the mustard was significantly less than when the weed was grown with 'Garry' oats. Moreover, plants grown in close association with toxic accessions exhibited severe chlorosis, stunting and twisting indicative of chemical effects rather than competition.

Dilday (34) reported that 347 accessions of rice out of 16134 from 99 countries showed allelopathic activity against five aquatic weeds viz. ducksalad, signal grass, redstem, flatsedge and barnyard grass. Some of the accessions repelled weeds and maintained weed free area upto a radius of 12-25 cm from their base. Since rice is planted in rows at spacings of 10-25 cm, allelopathic activity would overlap the space between the rice plants or rows leading to control of these problem weeds in rice. Dilday *et al.* (35) further reported that out of the tested accessions, 347, 161 and 6 accessions demonstrated allelopathic activity to ducksalad, purple ammannia and broadleaf signal grass, respectively. Some accessions from India and Bangladesh also exhibited allelopathic activity to barnyard grass and *Cyperus iria*.

Narwal *et al.* (92) screened 13 genotypes of pearlmillet and found that HHB-67 and 8800 4A x 833-2 had greatest suppression effect on weeds particularly *T. portulacastrum*, the major weed of irrigated crops in Haryana, India. Sarmah *et al.* (122) determined the suppression effect of 11, 10 and 8 accessions of *Brassica juncea*, *B. napus* and *B. carinata*, respectively, on winter weeds of north-west India under field conditions. They observed that RH-8689, RH-8605 and RH-8693 of *B. juncea*, HNS-8902 and HNS-11-1 of *B. napus* and BCCN-5 genotype of *B. carinata* had more smothering effect on weeds compared to their other genotypes.

Superior genotypes for weed control have also been reported in cucumber (113), oats (41,70), sunflower (70) and soybean (80). It appears possible, therefore, to breed allelopathic genes into standard cultivars to aid in weed control.

3.5. NATURAL HERBICIDES

Growing awareness about the environmental and public health problems linked with the excessive use of plant protection chemicals in agriculture has stimulated interest in the search for new selective, easily degradable and environmentally safe herbicides. The ability of some natural plant compounds to effectively inhibit the development of other plants has suggested that they may be used as herbicides. Besides, recent advances in the microbial and plant biochemistry have stimulated scientific interest in the possible role of secondary plant metabolites and microbial toxins as herbicides.

Among the plant products as herbicides, juglone, isolated from walnut tree has been found effective against redroot pigweed, velvetleaf and barnyard grass (124,132,143). Caffeine derived from coffee showed considerable selectivity in inhibiting germination of *Amaranthus spinosus* L. at a concentration that has no effect on blackgram (118,119). Strigol, a sesquiterpenoid derivative from cotton roots is a potent germination stimulant of witchweed (*Striga asiatica* L. Kuntz), an obligate parasite of maize, sorghum (24) and *Orobanche minor* (132).

Dhurrin (sorghum); gallic acid (spurge); Phlorizin (apple root); trimethylxanthene (coffee) and cinch (eucalyptus) are some other important plant products having promising herbicidal activity. The commercialization and marketing of "Herbiacae" the herbicide from microbial natural product bialaphos in Japan (54) has opened up a new era in weed management. Other microbial phytotoxins found to suppress weed growth include anisomycin, tentoxin, biopoloroxin, herbimycin etc.

3.6. TREE FARMING

Very little information is available on this aspect, but it offers scope owing to availability of large quantity of tree litter during leaf fall. It may play a major role in agroforestry systems. It has been observed that in poplar (*Populus deltoides*) based agrisilviculture system the field remains almost free from weeds during the winter season (Fig. 1). It may be due to physical barrier on account of leaf fall at the time of germination of winter weed seeds and the release of catechol and benzoic acid inhibitors during the fast decomposition of the leaf litter. About 30% less weed population was recorded in wheat grown in the alleys of *Dalbergia sissoo* as compared to control plots (88).

Allelochemicals from eucalyptus could be successfully exploited for the control of noxious weeds. Oils from *Eucalyptus citriodora* and *E. globulus* completely reduced the germination of *Parthenium hysterophorus* seeds and affected the growth of mature weed plants. Similarly, the oils and other chemicals of eucalyptus checked the rooting potential of *Lantana camara* weed (65).

4. Nematode Management

The root-knot nematode (*Meloidogyne*) is ranked as number one among the ten most important phytoparasitic nematode genera with wide geographical distribution, phytophagous food habit and infecting over 2200 plant species (123). Owing to the high cost, uncertain availability, problems of application and phytotoxicity of plant protection chemicals as well as the environmental and health hazards associated with the use, other approaches to nematode management, including allelopathy, appear to be potential alternatives for ecological agriculture.

4.1. PLANT MATERIAL AS NEMATOCIDES

Chopped mature dried residues of lespedeza, alfalfa, oats and flax when incorporated @ 25 tonnes ha^{-1} into fields infested with *Meloidogyne incongnita*, significantly reduced the incidence of rootknot in tomato (60). The ploughing of rye crop in the infested soil effectively checked the *Pratylenchus penetrans* and the effect was identical to D.D. fumigation (85). The soil amended with cotton waste, lucerne pellets and lucerne hay showed reduced incidence of *Tylenchulus semipenetrans* (79). Rice straw @ 22.5 or 44.75 tonnes ha^{-1} reduced the population of *Belonolaimus longicaudatus* and other plant parasitic nematodes (137). Johnson (59) observed 75-90% reduction in root-knot incidence in potted tomato grown in soil mulched with flax, lucerne or orchard grass residues. Prasad *et al.* (107) found that wheat straw and neem cake + NPK gave maximum reduction in the plant parasitic nematodes associated with wheat and mungbean. Mishra and Prasad (83) reported good reduction in *M. incognita* incidence in tomato through application of wheat straw and paddy husk.

The oil extracts from the seeds of *Argemone mexicana* weed, when applied @ 0.2% as soil drench or foliar spray, reduced root-knot nematode infestation of okra and increased plant growth. The oil showed systematic effect and the foliar spray proved more effective than soil drench (26). *Azolla pinnata*, a biofertilizer, suppressed the infestation of *M. incognita* in okra (136). Species of polygonum weed have also been found effective in controlling nematodes (120).

Sukul (134) has recommended the planting of some trees like *Anthocephalus cadamba* Mig., *Azadirachta indica* A. Juss., Eucalyptus, *Tectona grandis* L. and *Pongamia glabra* Vent. along roads, river banks and forests for the collection and field application of their leaves to control nematodes. Mishra and Mojumdar (82) have reported that neem (*Azadirachta indica*) seed kernel is more toxic followed by its seed and seed coat. They found that addition of decomposed *A. indica* seed, seed kernel and seed coat drastically reduced the root-knot nematode population in soil and increased the yield of mungbean. Mulching with green leaves of *Pongamia* and *A. indica* reduced root-knot nematode infestation of mulberry plants in field experiments (52). The leaves of leucaena and *A. indica* after a fixed period of degradation also control root-knot nematodes (58,98).

4.2. OIL SEED CAKES AND BY-PRODUCTS

It is a common practice among vegetable and fruit growers to use oil seed cakes as a source of plant nutrients and to control nematodes. The application of neem, castor, mustard and rocket salad oilseed cakes suppressed the population of root-knot nematode and the reniform nematode (*Rotylenchulus reniformis*) in okra and improved the plant growth and water absorption capacity of its roots (7). Mainpueira, a sub product in the production of cassava flour, proved effective against *M. incognita* under green house conditions as well as in the field (43,106). The citrus fruits canning factory waste used as soil amendment gave good control of tomato nematodes (9). The use of cassava root peeling and locust bean husk as soil amendments reduced root-knot of sugarcane and increased its growth (121).

4.3. NEMATICIDAL ALLELOPATHIC COMPOUNDS

Bhatti and Nandal (14) have reviewed the information on nematicidal substances (allelochemicals). Only a few allelochemicals have been isolated, elucidated and characterized against plant nematodes. Terthienyl isolated from marigold exhibited strong nematicidal properties under laboratory conditions. It, however, failed to suppress nematode population under field conditions even at 200 ppm (139). Cucurbitacin that accumulate in bitter cucumber genotypes repelled more juveniles of *M. incognita* from infecting them than did the non-bitter genotype and the cucurbitacins were implicated with such repellent actions (55).

The compounds isolated from tobacco (nicotine), jackbean (phytolectins) and *Ocimum sanctum* (eugenol) have been found to reduce *M. incognita* infestation of host plants including tomato and okra (3,4,26,29). Juices from numerous plants or their parts and extracts with organic solvents or root exudates contain nematicidal compounds (30,51,116).

5. Insect Pests Management

Most pests spp. are naturally regulated by various ecological processes viz., competition for food or by predation and parasitism by natural enemies. Their population is stable and the damage caused is relatively insignificant in most cases. Conversely modern high input farms are planted with uniform varieties, well watered and fertilized i.e. providing ideal conditions for pests attack, for which farmers use pesticides. Pesticides may be dangerous to human and livestock health and damage the natural resources, cause pest resurgence by killing the natural enemies of target pests, can produce new pests, which were not pests in the past e.g. whitefly in cotton. Pests become resistant to pesticides so necessitating their further applications and lastly they do not provide lasting control, hence, have to be repeatedly applied. To overcome such problems, farmers are advised to use wide range of technologies based on ecological processes of predation, competition and parasitism to control pests more effectively than pesticides alone. Besides the use of Integrated Pest Management practices, not only reduces the pests population to satisfactory level, but also are sustainable and non-polluting. In ecological agriculture, insect control with non-chemicals can be achieved through the use of cropping systems, insect pest resistant varieties and antibiotic allelochemicals of plant origin.

5.1 CROPPING SYSTEMS

5.1.1. *Crop rotation*

In crop rotations, maximum use of those crops which reduce pests infestation should be encouraged. The approach is to rotate non-host crops with susceptible crops in sequence. Non-host crops reduce the pests population to very low level and then susceptible crops may be grown. The non-host crops provides a break, disrupting the relationship between a pest or pathogen and its host.

In crop rotation, the best results are usually achieved when botanically unrelated crops are rotated. For example, in upland rice-corn-cowpea-fallow rotation, corn and cowpea are unrelated, have different growth habits and attract different pests. Soil insects such as wire worm (*Elateridae*) and white grubs usually have a wide range of hosts, which restricts the choice of crops in rotation on the same land. Rotation of about 4 years duration, under host free conditions is usually necessary to reduce the pest population to a tolerance level (44).

5.1.2. *Crop mixture/intercropping*

When different crops are grown together as in mixed cropping or intercropping, the pests recognise a suitable crop either by sight or smell. In such a situation, it is possible to confuse the pests by adopting suitable structuring of the crop components.

The crop mixtures/intercropping are also very useful to reduce the pests infestation through (a) release of repellent allelochemicals as volatiles, (b) different plant spp. confuse the insect pests and (c) the various plant spp. provide a physical barrier to the movement of insect pests etc. A large number of small farmers in tropical and sub-tropical countries rely on this system. In Latin America, 60% maize is grown with beans and generally rice, cotton, beans and cassava are grown in mixtures. Generally, more diverse the agroecosystem, the less will be abundance of herbivore pests. Mixtures of cabbage + tomato, reduces the colonization of diamond backmoth, while maize + beans/squash mixtures have the same effect on chrysomelid beetles. Besides, the odours of some plants can also disrupt the searching behaviour of pests. Grass borders repel leaf hoppers from beans and the chemical stimuli from onions prevent carrot fly from finding carrots. Alternatively, one crop in mixture may act as a trap or decoy crop the "fly paper effect". Strips of alfalfa interspersed in cotton fields in California attract and trap *Lygus bugs*. The loss in alfalfa yield, offsets the cost of alternative control methods for cotton. Similarly, crucifers interplanted with beans, grass, clover or spinach are damaged less by cabbage maggot and cabbage aphids. There is less egg laying on the crucifers and the pests are subjected to increased predation (108).

5.2. RESISTANT VARIETIES

The presence of antibiotic metabolites in some plants makes them comparatively more resistant to the common insect pests. The resistant varieties identified in rice, wheat, maize, sorghum, cotton and alfalfa are being used for breeding commercial varieties. Pathak and Dale (100) have isolated some antibiotic metabolites such as saponins, phenolic acids, 6-MBOA, DIMBOA, gossypol, quercetin and 2-tridecanone. Rao (115) while screening 17 chilli genotypes, found that the pod borer resistant variety 'Loc' had the highest content of total phenols (0.61%) while the susceptible genotypes 'Tetraold' and 'Lanka-1' had the least amount (0.31%).

5.3. INSECTICIDAL ALLELOCHEMICALS

Many locally available plants have pesticidal properties and are used to repel, deter or poison pests due to presence of pesticidal allelochemicals. Many of these kills only pests and not the predators, degrade rapidly so do not contaminate environment. Increasingly scientists are determining the mechanisms of such practices. The role of plant allelochemicals in plant-insect interactions has received considerable attention in recent years. The allelochemicals are used in pest control as repellents, antifeedants, growth disrupters and toxicants. Allelochemicals commonly found in plants are toxic amino acids, protease inhibitors, alkaloids, cyanogenic glycosides, phenols, tannins, lignins, flvonoids, toxic lipids, glucosinolates, terpenoids, saponins and phytohaemagglutinins.

Among tree species, neem (*A. indica*) has received substantial attention during the pest decade particularly for insect pest control. It has been an age old practice in rural India to mix dried neem leaves with stored grains or to place them among warm clothes to repel-insects. Butterwoth and Morgan (16) isolated a substance 'azadirachtin' from neem seed that inhibited feeding in desert locust. Since then many azadirachtin based insecticides have been formulated and found effective against insects. The azadirachtin based formulations viz. Azadirachtin, Neemark, Achook, Margoside, Nimbicidine, Repelin, Parasmani, Jawan, Sukrima, Neem oil, Neem gold, Nocilneem, Neemata-2100 and Neemrich-1 and II have been found effective against cotton bollworm (33,96); castor semilooper (63); white fly of cotton (128) and rice hispa (*Dicladispa armigera* (31).

6. Disease Management

6.1 CROPPING SYSTEMS

Continuous cultivation of same or related crops leads to perpetuation and build up of soil pathogens which gradually increase the disease intensity. Crop rotation is one of the natural methods of disease prevention especially when botanically unrelated crops are included as they are affected by different pathogens. Crop rotation helps to control many soil borne diseases such as mosaic, wilt of legume crops (pigeonpea, pea, chickpea) and linseed, red rot and wilt in sugarcane, ergot and smut in pearlmillet, leaf smut and bunt in rice, bunt and molya disease in wheat and barley and root rot in vegetable crops. Chohan (21) observed that crop rotation was particularly effective in lowering the population of soil borne pathogens.

The mixtures of different crop species provide buffer against losses from diseases by delaying the onset of disease, reducing spore dissemination and/or modifying microenvironmental conditions. In soil borne pathogens some plant conditions may enhance soil fungistatis and antibiosis through indirect effects on soil organic matter content. The presence of immune or resistant plants in mixed cropping systems impedes the spread of pathogens and increases the separation between susceptible plants. Larios and Moreno (68) documented evidence of disease buffering in various intercropping systems.

6.2. CROP RESIDUES

Many plants produce chemicals either prior to or after infection by certain pathogens which render the plants resistant to diseases (116). The first scientific report on the benefits of plant and other organic materials in disease control appeared 50 years ago. They included reports on control of potato scab, wheat take all and root rot of cotton with various organics (25). In Texas, grain sorghum is a popular alternate crop with cotton. Sorghum provides 20-25 tonnes ha^{-1} of residue. This residue on incorporation in soil, controls root-rot in the subsequent crop of cotton. Green manuring with clover also controls this fungus (77). Soybean residue incorporated in soil before planting of potatoes, controlled the potato scab due to antibiotic production by *Bacillus subtilis*, a bacterium antagonistic to *Streptomyces scabies* which cause potato scab. Volatile compounds released during decomposition of crop residues also deserve attention because of their influence on plant pathogens (74). Various aldehydes from decomposing alfalfa may stimulate the germination of microsclerotia *Verticillium dahliae* and *Sclerotium rolfsii*, followed by lysis causing a net reduction in the population of these two fungi. Crucifer residues emit sulphur containing volatiles during decomposition that are inhibitory to *Aphanomyces euteiches* and may reduce the root rot of peas caused by this fungus (72).

Although residues incorporated in adequate amounts and at the right time are generally suppressive to root diseases, they may also increase some diseases. Plant residues may be colonied by a pathogen which uses them as energy source. Residues may also produce certain phytotoxic decomposition products that may predispose roots to infections. Such phytotoxins have been reported to increase the susceptibility of certain tobacco cultivars to black root-rot (101); cotton plants to root rot caused *by Theilaviopsis basicola* (75), bean to root-rot (138); sugarcane to pythium root-rot (114) or cause injury to plant roots and thereby open the way for secondary root decay (17,23).

6.3. ORGANIC AMENDMENTS ETC.

Among the common organic amendments, neem seed and cake have been found effective against many plant diseases like rice bacterial blight (40); rhizome rot of ginger (36) and other diseases caused by *Rhizoctonia solani*, *Macrophomina phaseolina*, *Fusarium solani* and *Phytophthora capsici* (130). Addition of cotton cake

@ 15 g kg^{-1} soil reduced the incidence of seedling blight of eucalyptus from 80 to 27% owing to release of inhibitor chemicals (64).

Some neem derivatives like Neemta-2100, Neemark, Nimbicidine, Sukrina, Neemoil and Jawan have been found effective against yellow mosaic virus of horsegram. Powdery mildew fungi has also shown sensitivity to Neemta-2100, Neemark and Nimbicidine (61).

7. Allelochemicals as Growth Regulators

Growth regulators are exogenous non-nutrient substances that manipulate growth, development and composition of plants and functions by interaction with the endogenous phytohormones groups. Their action include growth retardation, flower induction, hastening of maturity or senescence and stimulating biomass production etc. Allelochemicals provide a promising source for new growth regulating compounds viz., agrostemin, triacontanol and brassinolide, which have received maximum attention.

Bioprodukt (15) summarized Yugoslavian work showing that 100 g agrostemin per hectare through seed treatment or foliar spray hastened germination and increased yields of wheat, maize, sunflower and sugarbeet by 10, 15, 15 and 10%, respectively. It also enhanced the oil content of sunflower by 4%. It has proved beneficial to vegetables, flowers, fruits, pastures and forest species.

Triacontanol, a 30-carbon primary alcohol, was isolated as a growth promoting compound from alfalfa. Its foliar applications increased the yields in cucumber, carrot, rice, corn, soybean and others. Inconsistent results, perhaps due to accumulation problems and to method, rate and time of application, reduced its efficacy (69). Extensive work has been done on evaluation of brassinolide, a steroid isolated from rape (*Brassica napus* L.) pollen as a yield stimulant. Brassinolide and several analogues have been synthesized (81) but they are too expensive for use in field crops.

All the bioregulators have shown increase in yields of major crops. However, inconsistency in performance between the locations, genotypes and spraying dates, besides difficulties in formulations, has hindered their commercial use.

8. Conclusions

The crop, weed and tree residues constitute the major source of organic matter in the soil. They also release numerous organic and inorganic compounds in the soil. These compounds (allelochemicals) are generally inhibitory to the growth of other crops/trees depending on the residue amount, length of decomposition period and type of residue. However, some plant residues especially legume residues like alfalfa, mungbean, soybean and leucaena stimulate the growth of crops. In addition to this, plant residues through the release of allelochemicals and biotransformation products decrease the incidence of pests (weeds, nematodes, insect pests and diseases). Intercropping and use of phytotoxic crop varieties and natural herbicides are effective methods of non-chemical weed control in ecological agriculture. Plant diseases including those caused by nematodes and infestation by insect pests in agriculture could be successfully controlled through the use of organic amendments, allelopathic compounds and crop varieties rich in allelochemicals content. In view of the practical significance of allelopathy in ecological agriculture, research efforts are needed to make use of the inhibitory allelopathic effects of plants for natural control of crop pests and diseases.

References

1. Abdul-Rahman, A.A. and Habib, S.A. (1989). Allelopathic effects of alfalfa (*Medicago sativa*) on bladygrass (*Imperata cylindrica*). *Journal of Chemical Ecology* **15**: 2289-2301.
2. Aggarwal, P.K. and Garrity (1987). Intercropping of legumes to contribute nitrogen in low-input upland rice-based cropping systems. Proceedings International Symposium *Nutrient Management for Crop Production in Tropical Farming Systems.* Makang, Indonesia.
3. Al-Sayed, A.A. and Montasser, S.A. (1986). The role of ascorbic and glutamic acids in controlling the root-knot nematode, *Meloidogyne javanica. Egyptian Journal of Phytopathology* **18** : 143-138.
4. Al-Sayed, A.A. and Thomson, I.J. (1988). *Meloidogyne incognita* and tomato response to thiamine, ascorbic acid, L-arginine and L-glutamic acid. *Journal of Nematology*. **20**: 451-456.
5. Altieri, M.A. (1987). *Agroecology : The Scientific Basis of Alternative Agriculture*. London : Intermediate Technology, pp.227.
6. Altieri, M.A. and Doll, J.D. (1978). The potential of allelopathy as a tool for weed management in crops. *PANS* **24** : 495-502.
7. Anver, S. and Alam, M.M. (1994). Effect of oilseed cakes on *Meloidogyne incognita* and *Rotylenchulus reniformis* in okra. In *Allelopathy in Sustainable Agriculture, Forestry and Environment* (Eds., S.S.Narwal; P.Tauro; G.S. Dhaliwal and Jai Prakash) Proceedings of International Symposium. pp. 109, Hisar, India: Indian Society of Allelopathy, CCS Haryana Agricultural University,
8.Arnon, I. (1972). *Crop Production in Dry regions.* Vol. 1. *Background and Principles*. London: Leonard Hill. pp. 458-478.
9. Babatola, J.O. (1989). Effect of some organic manures on nematodes in tomato cultivation : *Pakistan Journal of Nematology* **7** : 39-46,
10. Bansal, G.L. (1994). Allelopathy of Buttercups and their alleviation. *See Reference No.* 7, pp. 16.
11.Bantilan, R.T., Palada, M.C. and Harwood, R.K. (1974). Integrated weed management. I. Key factors affecting crop weed balance. *Philippines Weed Science Bulletin* **1(2)** : 14-36.
12. Barnes, J.P., Putnam, A.R. and Burke, B.A. (1986). Allelopathic activity of rye (*Secale cereale* L.) In *The Science of Allelopathy* (Eds., A.R. Putnam and C.S. Tang). pp. 217-286. New York : Wiley Interscience,
13. Bhatt, B.P. and Todaria, N.P. (1992). Studies on allelopathic exclusion of understorey by some agroforestry tree crops of Garhwal Himalya. In : Proceedings First National Symposium, *Allelopathy in Agroecosystems*. (Eds., P. Tauro and S.S.Narwal). pp. 129, Hisar, India : Indian Society of Allelopathy, CCS Haryana Agricultural University,.
14. Bhatti, D.S. and Nandal, S.N. (1994). Nematode management through phytotherapeutic substances. In. *Nematode Pest Management in Crops*. (Eds., D.S.Bhatti and R.K.Walia). pp. 148-161. New Delhi : CBS Publishers.
15. Biprodukt (1984). *Agrostemin-Gift of Nature*. Beograd, Yogoslvia: Novi Dani,
16. Butterworth, J.H. and Morgan, E.D. (1968). Isolation of a substance that suppresses feeding in locusts. *Chemical Communications*. **23** : 4.
17. Carley, H.E. and Watson, R.D. (1967). Plant phytotoxins as possible predisposing agents to root rots. *Phytopathology*. **57** : 401-404.
18. Cernusko, K. and Borkey,V. (1992). *Rostlinna Vyroba* **38** : 608-09.
19. Chacon, J.C. and Gliessman, S.R. (1982). Use of the non-weed concept in traditional tropical agroecosystems of southern-eastern Mexico. *Agroecosystems* **8** : 1-11.
20. Chandersena, J.P.N.R., Hemalal, K.D.P. and Tillekeratne, I.M.V. (1989). Alleloapthic effects of *Glyricidia maculata* H.B.K. on selected crop and weed species. In Proceedings 12th *Asian Pacific Weed Science Conference* pp. 452-431, Taipei, Taiwan: Asian - Pacific Weed Science.
21. Chohan, J.S. (1968). Disease problems in new cropping patterns and rotations in the erstwhile Punjab state. In *Proceeding of Symposium on Cropping Systems in India*. pp. 562-564.New Delhi: ICAR,
22. Chou, C.H. and Patrick, Z.A. (1976). Autointoxication mechanisms of *Oryza sativa* L. Phytotoxic effects of decomposing rice residues in soil. *Journal of Chemical Ecology* **2**: 353-367.
23. Cochrane, V.W. (1948). The role of plant residues in the ecology of root rot. *Phytopathology* **38** : 185-196
24. Cock, C.E., Whichard, L.P., Turner, B.,Well, M.E. and Egley, G.H. (1966). Germination of witchweed (*Striga lutea* Lour) : Isolation and properties of potent stimulant. *Science* **154**: 1189-1190.

25. Cook, R.J. and Watson, R.D. (1969). Nature of the influence of crop residues on fungus induced root diseases. *Washington Agricultural Experiment Station Bullettin* **716** pp 32.
26. Das, S. and Sukul, N.C. (1988). Nematicidal effect of the oil from the seeds of *Argemone mexicana. Environmental Ecology* **6**: 194-197.
27. Davis, E. and Rich, J.R. (1987). Nictonine content of tobacco roots and toxicity to *Meloidogyne incognita. Journal of Nematology* 19 : 23-29.
28. Davis, E.F. (1928). The toxic principle of *Juglans nigra* as identified with synthetic juglanns and its toxic effects on tomato and alfalfa plants. *Journal of Botany* **15** : 620632.
29. Davis, E.L., Kaplan, D.T., Dickson, D.W. and Mitchell, D.J. (1989). Reproduction of lectin treated *Meloidogyne spp.* in two treated soybean cultivars. *Reviews of Nematology* **12**: 257-263.
30. Devakaumar, C. (1994). Nematode principles of plant origin. *See Reference No.* 14, pp. 165-189.
31. Dhaliwal, G.S., Singh, J. and Dilawari, V.K. (1994). Management of rice hispa (*Dicaldispa armigera* (OLIV) with botanical insecticides. *See Reference No.* 7, pp. 97.
32. Dharamraj, and Sheriff, M. (1994). Allelopathic activity of sunflower (*Helianthus annus* L.). *See Reference No.* 7, pp.38.
33. Dhawan, A.K. and Simwat, G.S. (1994). Field evaluation of botanical insecticides for boll worms control in cotton. *See Reference No.* 7, pp.101.
34. Dilday, R.H. (1992). *Victims noone mowons. Agricultural Research* **40** *: 10-11*
35. Dilday, R.H., Frans, R.E., Semiday, N., Smith, R.J. and Oliver, L.R. (1992). Weed control with crop allelopathy. *Arkansas Farm Research* **41** : 14-15.
36. Dohroo, N.P. (1994). Effect of organic amendments of soil on Rhizome rot, root-knot nematode and rhizosphere mycoflora of ginger (*Zingiber officinale*). *See Reference No.*7, pp.110.
37. Eaglesham, A.R.T., Ayanaba, A. and Rao, V.R. (1981). Improving the nitrogen nutrition of maize by intercropping with cowpea. *Soil Biology and Biochemistry* **13**: 169-171.
38. Einhellig, F.A. (1985) Effects of allelopathic chemicals on crop productivity . In *Bioregulators for Pest Control* (Eds., P.A. Hedin.). ACS Symp. Series No. **276** : 110-130 Washington, D.C.: American Chemical Society.
39. Einhellig, F.A.and Leather, G.R. (1988). Potentials for exploiting allelopathy to enhance crop production. *Journal of Chemical Ecology* **14** : 1829-1844.
40. Eswaramurthy, S., Algianagalingam, M.N., Marippan, V. and Muthusamy, M. (1994). Efficacy of plant products for control of Rice bacterial blight. *See Reference No.* 7, pp. 112.
41. Fay, P.K. and Duke, W.B. (1977). An assessment of allelopathic potential of *Avena* germplasm. *Weed Science* **25** : 224-228.
42. Forney, R.D., Roy, C.L. and Wolf, D.D. (1985). Weed suppression in no-till alfalfa (*Medicago sativa*) by prior cropping of summer-annual forage grasses. *Weed Science* **33**: 460-467.
43. Franco, A. and Ponte, J.J.D.A. (1988). Additional information on the use of *Manipueira* as nematicide: dosage and interference with soil fertility. *Nematologica Brasil* **12** : 35-45.
44. Gibson, K.E., Kane, M.C. and Jones, E.W.(1958). Effect of some crop rotations on wire worm population in irrigated land. *USDA Technical Bulletin* **1** : 219 pp.
45. Gill, D.S., Bhatia, R.K. and Sandhu, K.S. (1994). Allelopathic effect of *Kharif* crop residues on growth of wheat, gram and lentil. *See Reference No.* 7, pp. 43.
46. Gliessman, S.R. (1989). Allelopathy and agricultural sustainability. In *Phytochemical Ecology : Alleochemicals, Mycotoxins, Insect Pheromones and allmones.* (Eds., C.H. Chou and G.R. Walter). Monograph No. **9** : 69-80. Taipei, Taiwan : Institute of Botany, Academica Sinica.
47. Gliessman, S.R. (1990a). *Agroecology : The Basic Research for Sustainable Agriculture*. Berlin: Spinger-Verlag. pp. 380.
48. Gliessman, S.R. (1990b). Understanding the basis of sustainability for agriculture in tropics : Experience in Latin America. In *Sustainable Agriculture Systems* (Eds., Edwards *et al*). Ankeny, Iowa, USA : Soil and water Conservation Society
49. Gliessman, S.R. and Garcia, E.R. (1979). In *Tropical legumes: Resources For the Future*. Publication No. **27**: 292-293. Washington, D.C: National Academy of Science.
50. Gliessman, S.R., Garcia, E.R. and Amador, M.A. (1981). *Agroecosystem* **7** :173-185.
51. Gommers, E.J. (1973). M.Sc. Dissertation. Netherlands : Agricultural University Wageningen. pp. 73.
52. Govindaialu, N.,Suryanarayana, D.D. and Sharma, G. (1989). Effect of mulching of green leaves for the control of root-knot nematode in mulberry. *Indian Journal of Nematology* **19** : 25-28.
53. Grechkanev, O.M. and Rodionov, V.I. (1971). Interactions of the components of nectar fodder mixture with wild heliotrope. In *Physiological-Biochemical Basis of Plant Interactions in* Phytogenosis. (Ed., A.M. Grodzinsky) **2**: 88-100. Kiev: Naukova Dumka,

54. Hatzios, K.K. (1987). Biotechnology applications in weed management : Now and in the future. *Advances in Agronomy* **41**: 325-375.
55. Haynes, R.L. and Jores, C.M. (1976). Effects of the Bilocus in cucumber on reproduction, attraction and response of the plant to infection by the southern root-knot nematode. *Journal of American Horticultural Science* **101**: 422-424.
56. Igboanugo, A.B.I. (1988). Effect of eucalyptus on yield of *Vigna unguiculata* L., *Zea mays* L. and *Sorghum bicolor*. *Agriculture, Ecosystems and Environment* **24** : 453-58.
57. International Allelopathy Society (1996). Department of Biochemistry, Oklahoma State University, Stillwater, USA.
58. Jain, R.K. and Bhatti, D.S. (1988). Effect of degradation of neem leaves on incidence of root-knot nematode in tomato. *International Nematology Network Newsletter* **5** : 7-9.
59. Johnson, L.F. (1972). Effects of organic mulches upon the incidence of root-knot nematode in potted tomato plants. *Phytopathology* **62** : 767.
60. Johnson, L. F., Chambers, A.Y. and Reed, H.E. (1967). Reduction of root-knot nematodes of tomatoes with , crop residues amendment in field experiments. *Plant Diseases Reporter* **51** : 219-222.
61. Joshi, M.S., Devakar, M.P., Mandokhot, A.M. and Sardeshpande, J.S. (1994). *In vitro* fungistatic effect of neem derivatives against *Oidium mangiferae* and *Leveillula taurica*. *See Reference No.* 7, pp. 133.
62. Joshi, S. and Mahdevappa, M. (1986). *Cassia sericea* S.W. to fight *Parthenium hysterophorus* L. *Current Science* **55**: 261-262.
63. Ivani, D.B. Bharpoda, T.M. and Patel, J.R. (1994). Bioefficacy of azadirachtin (neem) based insecticides against castor semilooper (*Achaea janata* L.) *Source Ref No.* 7, pp. 129.
64. Kaushik, J.C., Gupta, V.K. and Kaushik, N. (1994). Relative efficacy of oil seed cakes, crops straw and saw dust against seedling blight of eucalypts (*E. tereticornis* S.M.). *See Reference No.* 7, pp. 113.
65. Kolhe, R.K. (1994). Allelopathic implications of Eucalyptus in Agro-ecosystems In *Allelopathy in Agriculture and Forestry*, (Eds., S.S. Narwal and P. Tauro), pp. 65-91 Jodhpur, India : Scientific Publishers.
66. Kohli, S.S. and Mitra, B.N. (1987). Effects of azolla as an organic source of nitrogen in rice-wheat cropping systems. *Journal of Agronomy and Crop Science* **159** : 212-215.
67. Lampkin, N. (1992). Rotation design for organic systems. In *Organic Farming*. pp. 125-160. Ipswick, U.K., Farming Press.
68. Larios, J.E. and Moreno, R.A. (1977). Epidemiologia de algunas enfermedades foliares de layuocajen different sistems decultiro. II. Roya Y murete descondite. *Turriabla* **27** : 151-156.
69. Laughlin, R.G., Munyon, R.L.,Ries, S.K. and Wert, V.F. (1983). Growth enhancement of plants by fematomole doses of colloidally dispersed triacontanol. *Science* **219** : 1219-1220.
70. Leather, G.R. (1983). Sunflower (*Helianthus annus*) are allelopathic to weeds. *Weed Science* **31** : 37-42.
71. Letourneau, D., Falies, G.D. and Heggeness, H.G. (1956). The effect of aqueous extracts of plant tissue on germination of seeds and growth of seedlings. *Weeds* **4** : 363-368.
72. Lewis, J.A. and Papavizas, G.C. (1975). In *Biology and Control of Soil-borne Plant Pathogens* (Ed., G.N. Burehl), pp.216. St.Paul, Minnesota : American Phytopathology Society.
73. Liebl, R.A. and Worsham, A.D. (1983). Inhibition of pitted morning glory (*Ipomoea lacunosa* L.) and certain weed species by phytotoxic components of wheat straw. *Journal of Chemical Ecology* **9** : 1027-1043.
74. Linderman, R.G. and Gilbert, R.G. (1975). Influence of volatiles of plant origin on soil borne plant pathogens. In *Biology and Control of Soil-borne Plant Pathogens* (Ed., G.W. Bruehl), pp.84-89. St. Paul, Minnesota : American Phytopathological. Society.
75. Linderman, R.G. and Toussoun, T.A. (1968). Predisposition to *Thielaviopsis* root rot of cotton by phytotoxins from decomposing barley residues. *Phytopathology* **58** : 1571-1574.
76. Lockerman, R.H. and Putnam, A.R. (1979). Evaluation of allelopathic cucumbers (*Curcumis sativus*) as an aid to weed control. *Weed Science* **27** : 54-57.
77. Lyle, R.W., Dunlop, A.A., Hill, H.O. and Hargrove, B.H. (1948). Control of cotton root rot by sweet clover in rotation. *Texas Agriculutral Experiment Station Bulletin* **699**: 621.
78. Mahadevappa, M. and Kulkarni, R.S. (1995). Managmeent of pathenium through allelopathic interaction. In *Allelopathy in Pest Management for Sustainable Agriculture* (Eds.,S.S.Narwal and P.Tauro) pp. 67-76. Jodhpur, India : Scientific Publishers.
79. Mankau, R. and Minteri, R.J. (1962). *Plant Diseases Reporter* **46** : 376-378.
80. Massantini, F.,Caporali, F. and Zellini, G. (1977). Evidence for allelopathic control of weeds in lines of soybean. In *EWRS Symposium on Different Methods of Weed Control and Their Integration* **1** : 23-30.
81. Maugh, T.H. (1981). New chemicals promise larger crops. *Science* **212**: 33-34.

82. Mishra, S.D. and Mojumder, V. (1994). Toxic behaviour of neem products on soil and plant nematodes. *Source Ref. No.* 7, pp. 130.
83. Mishra, S.D. and Prasad, S.K. (1978). Effect of soil amendments on nematodes and crop yields. II. Oilsed cakes, organic matter and inorganic fertilizers at different levels of *Meloidogyne incognita. Indian Journal of Entomology* **40**: 42-53.
84. Mishustin, B.N. and Naumov, A.N. (1955). Secretion of toxic substances of alfalfa and their effect on cotton and soil microflora. Akad. Nauk USSR *Izvestiya Ser. Biologia* **6**: 3-9.
85. Mountain, W.B. and Ellliott, J.M. (1962). Effect of summer fallowing on the root lesions nematode *(Pratylenchus penetrans* Cobb) and yield of flue-cured tobacco in Ontario. *Canadian Journal of Plant Science* **42**: 642-645.
86. Murleedharan,G.N.,Whitenack,C.J. and Putnam, A.R. (1989). 2, 2-0 x 0-1, 1-azobenzene : a microbially transformed allelochemical from 2 (3H)- benzoxazolinone. *Journal of Chemical Ecology* **15** : 2415-2429.
87. Nair, K.P.P., Patel, U.K. and Singh, R.P. (1979). Evaluation of legume intercropping in conservation of fertilizer N in maize culture. *Journal of Agricultural Science* (Cambridge) **93** : 189-194.
88. Nandal, D.P.S. and Bisla, S.S. (1994). Effect of poplar (*Populus deltoides* Marsh) and *Shisham* (*Dalbergia sissoo* Roxb.) plantation on wheat and associated weeds. *Van Anusandhan* **11** : 33-35.
89. Narwal, S.S. (1994). Interactions between plant communities. In *Allelopathy in Crop Production.* (Author, S.S. Narwal). pp. 19-75. Jodhpur, India : Scientific Publishers,
90. Narwal, S.S. (1998). Allelopathy in ecological agriculture. *Abstracts,* III *International Congress on Allelopathy in Ecological Agriculture and Forestry* (Eds., S.S.Narwal, R.E. Hogland, R.H. Dilday, M.J. Reigosa). pp. 142. Hisar, India : Indian Society of Allelopathy.
91. Narwal, S.S. and Sarmah, M.K. (1992). Suppression effect of *E. tereticornis* on the field crops. *See Reference No.* 14, pp. 111-113.
92. Narwal, S.S., Sarmah, M.K., Dahiya, D.S. and Kapoor, R.L. (1992). Smothering effect of pearl millet genotypes on weed species. *See Reference No.* 14, pp 49-50.
93. National Research Council (1989). *Alternative Agriculture.* Washington, DC, USA : National Academy Press. pp. 448.
94. Obiefuna, J.C. (1989). Biological Weed control in plantains within Equsimelon. *Biological Agriculture and Horticulture* **6** : 221-228.
95. Ofori, F. and Stern, W.R. (1987). Cereal-legume intercropping systems. *Advances in Agronomy* **41** : 41-90.
96. Panchabhavi, K.S.,Lingappa, S. and Sudhindra, M. (1994). The efficacy of neem *(Azadirachta indica* A. Juss) seed extract and *Helicoverpa* nuclearpolyhydrosis virus in the integrated pest management of cotton. *See Reference No.* 7, pp. 172.
97. Papendick, I., Sanches, P.A. and Triplett, G.B. (1976). *Multiple cropping.* Publication No. 27: 378 pp. Madison, Wisconsin : American Society of Agronomy.
98. Pruthi, I.J., Jain, R.K. and Gupta, D.C. (1987). Effect of different periods of degradation of *subabool* leaves alone and in combination with nematicide on root-knot nematodes incidence in okra. *Indian Journal of Nematology* **17**:30-32.
99. Patra, D.P., Sachdev, M.S. and Subbiash, B.V. (1986). Nitrogen uptake and efficiency in wheat-gram and maize-cowpea intercropping systems. *Fertilizer News* **31 (10)** : 21-27.
100. Pathak, M.D and Dale, D. (1983).In *Chemistry and World Food Supplies : The New Frontiers* (Ed., L.W. Schienit) pp. 129-141, New York.
101. Patric, Z.A. and Koch, L.W. (1963). The adverse influence of phytotoxic substances from decomposing plant residues on the resistance of tobacco to black root rot. *Canadian Journal of Botany* **41** : 754-758.
102. Patterson, D.T. (1986). Allelopathy. In *Research Methods in Weed Science*. (Ed., N.D. Camper). pp. 111-134. Champaign, Illinois : Southern Weed Science Society.
103. Peoples, M.B. and Herridge, D.F. (1990). Nitrogen fixation by legumes in tropical and subtropical agriculture. *Advances in Agronomy* **44**: 156-225.
104. Peters, E.J. (1968). Toxicity of tall fescue to rape and birdsfoot trefoil seeds and seedlings. *Crop Science* **8** : 650-653.
105. Phetchawee, S., Vibulsukh, N., Theppoolpon, M. and Masarugson W. (1985). Long term effect of mulching with fertilizer under corn-legumes intercropping on crop yield and improvement of soil chemical and physical properties. In *Thailand National Corn and Sorghum Program 1985 Annual Report.* Department of Agriculture, Kasetsart University and CIMMYT pp. 204-212.
106. Ponte, J.J., Franko, A. and Ponte, A.E.L. (1987). Study of the use of *Manipueira* as a nematicide under field conditions. *Nematologia Brasil* **11**: 42-47.
107. Prasad, S.K., Mishra, S.D. and Gaur, A.C. (1972). Efect of soil amendments on nematodes associated with mung and maize. *Indian Journal of Entomology* **34** : 407-311.

108. Pretty. J.N. (1995). *Regenerating Agriculture*. London : Earthscan Publications. pp. 320.
109. Prusty, J.C., Mohanty, S.K. and Behera, B. (1994). Allelopathic impact of sunflower on the growth of succeeding crops and associated weeds. *See Reference No.* 7, pp. 39.
110. Purvis, C.E., Jessop, R.S. and Lovett, J.V. (1985). Selective regulation of germination and growth of annual weeds by crop residues. *Weed Research* **25**: 415-421.
111. Putnam, A.R. (1988) Allelopathy : Problems and opportunities in weed management. In *Weed Management in Agroecosystems : Ecological Approaches*.(Eds., M.A.Altieri and M.Liebman) pp. 77-78. Boca Raton, Florida, : CRC Press..
112. Putnam, A.R. and DeFrank, J. (1983). Use of phytotoxic plant residues for selective weed control. *Crop Protection* **2**: 173-181.
113. Putnam, A.R. and Duke, W.B. (1974). Biological suppression of weeds : evidence for allelopathy in accessions of cucumber. *Science* **185** : 370-372.
114. Rands, R.D. and Doppe,E. (1938). Influence of certain harmful soil constituents on severity of Pythium root-rot of sugarcane. *Journal of Agricultural Research* **56**: 63-67.
115. Rao, S.V. (1994). Pod borer resistance in relation to certain host plant chemical constituents in chilli. *See Reference No.* 7, pp.103.
116. Rice E.L. (1984). *Allelopathy*. 2nd Ed. New York : Academic Press.
117. Ries, S.K., Wert., V.Sweeley, C.C. and Leavitt, R.A. (1977). Triacontanol, a new naturally occurring plant growth regulator. *Science* **175**: 1339-1341.
118. Rizvi, S.J.H., Mukerjee, D. and Mathur, S.N. (1980). A possible new source of natural herbicides. *Journal of Experimental Biology* **18** : 77-78.
119. Rizvi, S.H.J., Mukerjee, D. and Mathur, S.N. (1981). Selective phytotoxicity of 1, 3, 7-trimethyl-axanthine between *Phaseolus mungo* and some weeds. *Agricultural Biological Chemistry* **45** : 1255-1256.
120. Ruelo, J.S. (1983). Integrated control of *Meloidogyne incognita* in tomato using organic amendments, marigolds and nematicide. *Plant Diseases Reporter* **67**: 671-673.
121. Salawu, E.O. (1988). Effects of cassava root peeling and locust bean husk amendments in controlling *Meloidogyne incognita* of sugarcane (*Saccharum officinarium* L.) *Plant Protection in Tropics* **5** : 11-15.
122. Sarmah, M.K., Narwal, S.S. and Yadav, J.S. (1992). Smothering effect of *Brasica species* on weeds. *See Reference No.* 7, pp. 51-55.
123. Sasser, J.N. (1989). *Plant Parasitic Nematodes : The Farmers Hiddern Enemy*. A cooperative publication of Dept. Plant Pathology and the Consortium for International Crop Protection pp. 115.
124. Shettel, N.L. and Balke, N.E. (1983). Plant growth response to several allelopathic chemicals. *Weed Science* **31** : 293-298.
125. Shilling, D.G., Jones, L.A., Worsham, A.D., Parker, C.E. and Wilson, R.F. (1986). Isolation and identification of some phytotoxic compounds from aqueous extracts of rye (*Secale cereale* L.). *Journal of Agriculutral and Food Chemistry* **34**: 633-638.
126. Shilling, D.G., Liebl, R.A. and Worsham, A.D. (1985). Rye (*Secale cereale* L.) and Wheat (*Triticum aestivum* L.) mulch: The suppression of certain broad-leaved weeds and the isolation and identification of phytotoxins. In *Chemistry of Allelopathy* (Ed., A.C. Thompson), ACS Symposium Series No. **268**: 243-471. Washington, D.C.: American Chemical Society.
127. Shilling, D.G., Worsham, A.D. and Danehower, D.A. (1986). Influence of mulch, tillage and diphenemid on weed control, yield and quality in no-till flue cured tobacco (*Nicotiana tobaccum*). *Weed Science* **34** : 738-744.
128. Simawat, G.S. and Dhawan, A.K. (1994). Field evaluation of neem based formulations against whitefly (*Bemisia tabaci*) on cotton in Punjab. *See Reference No.* 7, pp. 128.
128a. Singh, G.B. (1983). Role of agroforestry in improving the environment. *Indian Farming* **35** : 15-19.
129. Singh, K., Malik, A.S. and Narwal, S.S. (1983). Effect of preceding crops and P manuring on N rquirement of succeeding mustard crop. *Journal of Research, Haryana Agricultural University* **13** : 587-592.
130. Singh, R.S. (1994). Impact of neem and its products in plant disease management. *Source Reference No.* 7, pp. 116-117.
131. Singh, S.K. and Verma, K.R. (1988). Allelopathic effects of leachates and extracts of *Pinus roxburghi* on legumes in Kumaon Himalayas. *Indian Journal of Agricultural Science* **58** : 412-413.
132. Spelce, D.L. and Muselman, L.J. (1981). *Orobanche minor* germination with strigol and G.R. Compounds. *Zeitschrift Pflanzenphysiologie* **104**: 281-283.
133. Stinner, B.R and House, G.J. (1987). Role of ecology in lower input, sustainable agriculture : An Introduction. *American Journal of Alternative Agriculture* **2** : 146-147.

134. Sukul, N.C. (1994). Control of plant parasitic nematodes by plant substances. In *Allelopathy in Agriculture and Forestry*, (Eds., S.S. Narwal and P. Tauro). pp. 183-211. Jodhpur, India : Scientific Publishers,
135. Sukhthumrong.A., Chotechaungmanirat, S., Chancharoensuk, J. and Veerasan, V. (1985). Effect of green manure-chemical fertilizer combination on soil fertility and yield of corn. *International Seminar on Yield Maximization of Food Grains Through Soil and Fertility Management*. Bangkok, Thailand.
136. Thakar, N.A., Patel, H.R. and Patel, C.C. (1987). Azolla in management of root-knot disease in okra. *Indian Journal of Nematology* **17** : 136-137.
137. Tomerlin, A.H. and Smart, G.C.Jr. (1969). *Journal of Nematology* **1**: 29-30.
138. Toussoun, T.A. and Patrick, Z.A. (1963). Effect of phytotoxic substances from decomposing plant residues on root-rot of bean. *Phytopathology* **53**: 265-270.
139. Uhlenbroek, J.H. and Bijloo, J.D. (1960). Investigations on nematicides III. Polythienyls and unrelated compounds. *Recueil Trav. Chim. des Pays-Bas* **79** : 1181-1196.
140. Unger, P.W. and McCalla, T.M. (1980). Conservation tillage systems. *Advances in Agronomy* **33**: 38-39.
141. Waller, G.R. (1987). *Allelochemicals, Role in Agriculture and Forestry.* ACS Symposium Series No. **330**. Washington, D.C : American Chemical Society.
142. Werner, A. (1988). Biological control of weeds in maize. *In Global Perspective in Agroecology and Sustainable Agricultural Systems* (Eds., P. Allen and D. Vondusia). Santacruz : University of California.
143. Weston, L.A., Burke, B.A. and Putnam, A.R. (1987). Isolation, characterization and activity of phytotoxic compounds from quackgrass *(Agropyron repens* (L.) Beauv). *Journal of Chemical Ecology* **13**: 304-431.
144. White, R.L., Worsham, A.D. and Blum, U. (1989). Allelopathic potential of legume debris and aqueous extracts. *Weed Science* **37** : 674-679.
145. Wiese, A.F. and Staniforth, D.W. (1973). In *Conservation Tillage : Proceeding of National Conference*, pp. 108-114. Ankony, Iowa: Soil conservation Society of America.
146. Willey, R.W. (1979). Intercropping - Its importance and research needs. *Field crops Abstracts* **32** : 1-10, 73-85.

3

Allelopathic strategies for eco-friendly crop protection

S. FACKNATH* and B. LALLJEE
Faculty of Agriculture, University of Mauritius, Reduit, Mauritius

Abstract

There are various strategies of allelopathic interactions, which help in reducing population of plant pests, the most popular being the 'Botanical Pesticides'. Such plant allelochemicals possessing pesticidal properties constitute a transition bridge between synthetic pesticides on one hand and totally natural pest control on the other. This aspect of allelopathy can be exploited in different ways :
(i) A large number of tropical plants, having pest control properties have been studied and in a few cases the biologically active groups or principles have been commercialised. Neem (*Azadirachta indica*) is by far the best known and documented, with over 200 commercial products in the world market.
(ii) The leaves, possessing pesticidal and/or pest repelling activity can be added to soil to control harmful soil insects, nematodes and other pests. Similarly fresh or dried leaves can be added to grain and other harvested produce for safe and effective post-harvest protection.
(iii) Appropriately designed mixed cropping systems, can help in decreasing pest damage through allelopathic effects. The allelopathic interactions between the main crop and an intercrop and their respective pest complexes keep certain pest species away from the main crop, either by acting as repellents or as attractants towards the intercrop (which is then said to act as a 'trap crop').
(iv) Another aspect of allelopathy involves insect-host plant relationships through semiochemical-mediated interactions. Allomones and kairomones play an important role in host recognition, host selection and host colonisation by insects, and have implications in insect pest management as well as in crop production.

Key words : Allelochemical, bacterieides, botanicals, crop protection, intercropping, mixed cropping, neem, nematicides, pests, semiochemicals.

1. Introduction

Selection pressure exerted by pests (insects, nematodes, pathogens and weeds), has led to the evolution of protective chemical compounds in plants. Such natural protective allelochemicals, termed Botanical Pesticides, act on the pest in many ways: preventing

---

* Correspondence author

S.S. Narwal et al. (eds.), Allelopathy in Ecological Agriculture and Forestry, 33–46.

them from infesting the plant, killing them, or rendering them less damage to the plant. Man has exploited several of these allelochemicals for pest control, notable among them are: nicotine from *Nicotiana* spp. (tobacco), rotenone from *Derris indica* and *Lonchocarpus* sp., ryanodine from *Ryania speciosa*, azadirachtin from *Azadirachta indica* (neem), pyrethrum from *Chrysanthemum cineraerifolium* and quassin from *Quassia* sp.

In 1939, the discovery of the excellent insecticidal and acaricidal properties of DDT and other organochlorine compounds ushered an era of synthetic pesticides, which has continued to this day. Although the synthetic compounds were better and more effective, but they are hazardous to humans, animal health and to the environment (11). This increasing awareness, in recent years, has led to increased research on alternate modes of pest management. Among these, allelopathic strategies (Botanicals, intercropping and mixed cropping, etc) seems to have considerable potential, hence, research in these areas is being conducted in many parts of the world. There are various mechanisms of allelopathic interactions that help in reducing pest populations. Along the same lines, following a period of increasing specialisation and monoculture, the growing realisation of the environmentally-friendly qualities of traditional agriculture has resulted in the increasing use of certain agronomic principles such as mixed cropping. The use of such eco-friendly practices as the application of botanical pesticides, intercrops and trap crops form an important part of the increasingly advocated concepts of Organic Farming and Sustainable Agriculture, which does not allow the use of synthetic agrochemicals.

Pesticidal allelochemicals are as effective as the synthetics, are non-polluting, renewable, less harmful to humans, animals and to the environment and easily available and cheap in tropical and subtropical countries. In many cases the farmer can make his own pesticides (27). Therefore, pesticidal allelochemicals constitute an important transition bridge between the synthetic pesticides on the one hand and totally natural pest control on the other.

2. Botanical pesticides

The pesticidal allelochemicals are extracted from the plant and used as conventional sprays or seed treatment. Numerous plant species of the tropical and subtropical regions have been investigated for their pesticidal potential against many pest species and their bioactive compounds have been identified and characterised. Although the first report of successful use of neem leaves in stored grain for reducing damage by pests was in 1937 (61), but the scientific study of pesticidal allelochemicals started only in the early sixties, with the discovery of the antifeedant effect of *Azadirachta indica* (neem) on locusts (59). Subsequently, however, the world-wide interest generated by the results of later studies lead to 6 International Conferences on the neem tree. Today considerable amount of research is being done world-wide on various aspects of neem and neem allelochemicals. Of the more than 2000 plant species tested world-wide, so far neem has proved the most promising. Apart from its medicinal, antiviral, antipyretic, diuretic, contraceptive qualities and its use for timber, shade, oil and fertiliser, its insecticidal, bactericidal, nematicidal properties has elevated this Meliaceae species to the status of a 'wonder-tree'. Various simple and sophisticated extracts from its different parts (roots, leaves, seeds, bark, seed coat, kernels, flowers and fruits) have been tested for their pesticidal potential against all major insect pests of tropical and temperate agriculture. Tables 1 to 4 list some of the important organisms of agricultural significance towards which neem and other plant extracts exhibit bioactivity.

2.1. ALLELOCHEMICALS AS INSECTICIDES

The bioactivity of botanicals against insect pests ranges from repellent to pesticidal and includes sterilant, ovicidal (mortality of eggs), oviposition deterrence (prevention or reduction in number of eggs laid) and regulation of growth and development. However, the most common and strongest effect of neem and some other plant allelochemicals (Table 1), on many pest species is that of antifeedance (prevention or reduction of feeding by the pest). The pest actually starves to death on a neem-treated host plant.

Table 1. Plant species possessing pesticidal allelochemicals

Plant species	Pest species	Reference
Ageratum conyzoides	*Dysdercus* spp	23
Tagetes erecta	*Meloidogyne incognita*, Tobaco Mosaic Virus (TMV)	22
Ocimum spp.	*Myzus persicae*	21
Lantana camara	*Liriomyza* spp., TMV of chilli	21, 58
Ajuga spp.	*Pieris rapae*	77, 89
Annona spp.	*Spodoptera litura*	37
Eupatorium ayapana	*Crocidolomia binotalis*	21
Ricinus communis	*Callosobruchus* spp.	35
Tephrosia purpurea	*Aphis gossypii; Dactynotus carthamii*	57
Artemisia spp.	*Schistocerca gregaria, Crocidolomia binotalis*	21
Pongamia spp.	*Spodoptera litura; Rhizoctonia solani; Pyricularia oryzae*	12, 51
Chenopodium sp	*P.xylostella; Myzus persicae*	21
Madhuca spp.	*Pyricularia oryzae*	63
Argemone mexicana	*C.binotalis; D. flavidus; TMV of chilli*	21, 58
Encelia sp.	*Heliothis zea; Plusia gamma; Peridroma saucia*	29
Cymbopogon citratus	*C. binotalis; D. flavidus*	21
Siegesbeckia orientalis	*C. binotalis; D. flavidus*	21
Cassia occidentalis	*C. binotalis; D. flavidus*	21
Piper spp.	*Sitophilus* spp.;*Tribolium* sp	21, 85
Vetivera zizanoides	*Liriomyza* spp.	21

All parts of the neem tree yield allelochemicals, but the maximum activity is in seed kernel extracts (Table 2). Among the neem allelochemicals, Azadirachtin and salannin are two alkaloids with very strong antifeedant property. Other antifeedants from plants are: tomatine from tomato, quinine from *Cinchona*, capsicin from chillies, solanine from potato, encecalin from sunflower, tephrosin and hildecarpin from *Tephrosia* and warburginal from *Warburgia*. Certain allelochemicals disturb insect growth and development by disrupting hormonal equilibrium, in particular the juvenile and moulting (ecdysone) hormones. Juvenile hormone mimics from plants include farnesol and juvabione.

Table 2. Bioactivity of neem against insect pests

Insect pest species	Activity	Reference
Plutella xylsotella	AF, IN, GR, RP, OD	18,19,20, 13
Spodoptera litura	AF, GR	38
Crocidolomia binotalis	IN, GR, AF	24,18
Schistocerca gregaria	AF, IN	91
Nephotettix virescens	IN, PDT	50
Peridroma saucia	AF, GR	40, 39
Dysdercus flavidus	IN, GR, OD	17
Myzus persicae	IN, GR	46, 90
Helicoverpa armigera	AF, GR, ST	54, 33
Pectinophora gossypiella	IN	16
Spilosoma obliqua	IN, AF	76
Nilaparvata lugens	AF, IN, GR, OD	74
Trogoderma granarium	GR, IN	78
Earias spp.	IN	16
Mamestra brassicae	IN, GR, AF	81
Trichoplusia ni	IN, GR, AF	81
Tribolium confusum	RP, IN	2
Sitophilus zeamais	RP, IN	2, 60
Callosobruchus chinensis	GR, IN, OD, RP, OV	35, 60
Dacus dorsalis	OD	84
Liriomyza trifolii	IN	25
Tetranychus urticae	RP, OD, GR, ACA	75
T. cinnabarinus	RP, OD, GR, ACA	49

AF: antifeedant; GR: growth regulating; IN: insecticidal; OD: oviposition deterrent; RP: repellent; OV:ovicidal; ST: sterilant; ACA: acaricidal

Similarly, allelochemicals mimicking the insect ecdysteroids (β-ecdysone) have also been isolated. In the seventies, allelochemicals with anti-juvenile hormone (the precocenes) activity were isolated from *Ageratum conyzoides* and anti-moulting hormone effects from neem (Azadirachtin and salannin) and *Ajuga procumbens* (ajugalactone) compounds.

2.1.1. *Plant Species Possessing Pesticidal Allelochemicals*

The allelopathic effects of various plant species of the Meliaceae, Rutaceae, Asteraceae, Annonaceae, Malvaceae, Labiatae and Cannellaceae on different insect pests have been reviewed (31). Apart from *Azadirachta indica*, the other Meliaceae plant spp. exhibiting strong allelopathic activity is the chinaberry tree (*Melia azaderach*). Its numerous allelochemicals have been isolated and identified, however, their application for pest control is limited owing to high mammalian toxicity. Other Meliaceae include *Cedrela odorata* (spanish cedar) and *Trichilia roka,* whose limonoids are antifeedants to *Spodoptera eridania,* and sendanin which is a growth inhibitor for *Heliothis virescens. Toona ciliata* yield toonacilin and 6-acetoxytoonacilin which are antifeedant and insecticidal to *Hipsiphyla grandella* .

In Rutaceae, *Citrus paradisi* limonoids are antifeedants to *Spodoptera litura*, while citrus oils deter several stored product insect species. The alkaloid zanthophylline deters feeding by *Melanoplus sanguinipes, Schizaphis graminum* and *Hypera postica.*

The Asteraceae, *Inula helenium* yields alantolactone, which reduces feeding and survival of *Tribolium confusum*, while isoalantolactone is a phagodeterrent for *Sitophilus granarium* and *Trogoderma granarium*.

Of the Canellaceae, *Warburgia* sp. and *Polygonum* sp. contain allelochemicals, namely warburganals, polygodial and mazigadial, which are strong phagodeterrents to *Helicoverpa armigera* and *Heliothis virescens*.

2.1.2. *Mechanism of Action of Pesticidal Allelochemicals*

Effective allelochemical concentrations causing antifeedance in insects, vary from 0.001% to 4% of neem seed kernel extract (NSKE). While some insect species, e.g. *Schistocerca gregaria* nymphs and adults absolutely refuse to feed on any host plant sprayed with even trace amounts of Azadirachtin. Studies on the antifeedant mode of action of the Azadirachtins have demonstrated that these allelochemicals disturb neural regulation in insects, in particular the serotoninergic system of the stomatogastric nervous system, thereby affecting gut peristalsis and feeding. The other major influence of the Azadirachtins on insect groups is on growth and development [Insect Growth Regulating (IGR) property]. All the insect species tested so far are affected to some extent by this juvenilising property, ranging from undersized, deformed juveniles or adults, prolongation of larval periods, intermediate forms to death in the juvenile stages or during moulting. Wing curling and/or distortion, twisted abdomen, dark pigmented patches on body, abnormal sclerotisation and formation of haemolymph bubbles are other typical symptoms of disruption by allelochemicals in the processes of insect development.

Another very common effect is the inability of the insect to shed its cuticle during the process of ecdysis (moulting), leading to death in that condition. Although the exact mechanism has not yet been elucidated, it is known that Azadirachtin disturbs the levels of ecdysone and juvenile hormones responsible for insect development (68). At the molecular level, the compound dramatically reduces the number of individual polypeptides of the insect endocrine organs (67).

While antifeedance and growth regulation are the two major effects of many plant allelochemicals on insects, other reported effects include (a) reduced oogenesis (egg production), (b) reduced egg viability, (c) lowered egg hatch, (d) sperm immotility and (e) failure to transfer sperm to female. In the female, synthesis of the protein vitellogenin is adversely affected and its uptake by the developing oocytes is considerably reduced, thereby preventing egg maturation. The trophocytes are also affected, leading to severely impaired embryogenesis. In the male insect, neem allelochemicals arrest seminal secretion from the accessory reproductive glands, decrease protein and lipid content of spermatophore structures, and arrest spermatogenesis (54). Another biological effect of allelochemicals is oviposition deterrence, which can be an important preventive measure in a pest management programme. Residues of Azadirachtin and other allelochemicals from neem extracts, even at fairly low concentrations, deter females from laying eggs on treated surfaces. Significantly fewer eggs are laid on treated surfaces than on untreated ones by several insect groups, among them Heteroptera, Lepidoptera, Diptera and Coleoptera.

Plant allelochemicals also exhibit direct toxicity to several insect pest groups, among them, termites, aphids, plant bugs, leaf miners, caterpillars, causing death in a dose-dependent manner. In general, the larval stages are most susceptible and the pupal stage the least.

Neem has a potentiating effect with synthetic insecticides, in some cases synergising malathion, quinalphos and monocrotophos (55), pyrethrum (87) and methamidophos, cypermethrin and deltamethrin (19).

Apart from insects, neem allelochemicals exhibit biological activity against nematodes, snails, fungi, bacteria and even vertebrate pests such as rodents and birds.

2.1.3. *Allelochemicals and Beneficial Organisms*

Neem allelochemicals have been found to be safe to the biological control agents (predators, parasitoids and pathogens) as well as to beneficials such as earthworms and honeybees (Table 3), insect parasitoids, eggs of coccinellids and chrysopids (predators) and juvenile and adult spiders. Azadirachtin and 2 neem formulations, Azatin and Neem EC did not adversely affect the egg parasitoid *Trichogramma minutum.* Facknath (20) reported the safety of neem extracts to the larval endoparasitoid, *Cotesia plutellae.* Facknath and Qayyum (unpubl.) observed that Azadirachtin, neem leaf extracts and neem seed powder had no negative side effects on *Aphidius colemani* parasitising *Myzus persicae.*

Table 3. Bioactivity of neem on beneficial organisms

Biological species	Activity	Reference
Cotesia plutellae	Non-toxic	19
Trichogramma australicum	Insecticidal on direct contact; non-toxic when host eggs treated	7
Bracon hebetor	Toxic in lab experiments, non-toxic in field trials	7
Tetrastichus israeli	Toxic in lab experiments, non-toxic in field trials	7
Telenomus remus	Non-toxic	34
Chiracanthium mildei	Non-toxic	48
Phytoseiulus persimilis	Non-toxic	48
Eisenia foetida	Increased body weight	69

Isman (30) reported that neem allelochemicals do not deter foraging honeybees, *Apis mellifera* and are safe to the beneficial mites, *Phytoseiulus persimilis* and *Amblyseius cucumeris*. The safety of neem allelochemicals to common biological control agents, coupled with their compatibility with synthetic insecticides, insect growth regulators (IGR), chitin synthesis inhibitors (CSI) and cultural practices, makes neem very suitable components of Integrated Pest Management (IPM) programmes (20, 79, 83).

2.2. ALLELOCHEMICALS AS NEMATICIDES

Although various extracts from different parts of the neem tree provide nematode control (Table 4), but neemcake mixed with soil has given the best results, either alone or in combination with other cultural practices, viz. deep ploughing, hot water treatment, soil solarization, or mixed cropping. The slow decomposition of the oil cake and the gradual release of allelochemicals make it the most efficient compared to spraying of allelochemical extracts. Oilseed cakes control plant-parasitic nematodes in vegetable fields (3) even under highly variable conditions of soil pH and organic content and temperature (5). The nematode control action of neem is both preventive as well as curative. The water soluble nematicidal and nematotoxic allelochemicals such as nimbidine, thionimone (36), furanotriterpenoids (15) leach down into the soil pores beyond the rhizosphere region of the plant, where the plant-parasitic nematodes are present and either kill them or limit their movement (5). The curative property of neem lies in certain physiological changes in treated host plant root cells which absorb phenols and amino acids released from the decomposing neemcake. Other beneficial effects of neemcake amended soils are (a) enhanced soil microbial activity, (b) increased populations of nematode natural enemies, and (c) release of plant nutrients which enhances plant vigour. Seed treatment (seed coating/soaking) with crude neem products or aqueous extracts also effectively controls the nematodes, *Meloidogyne incognita* and *Heterodera* sp. (52,53).

2.3. ALLELOCHEMICALS AS FUNGICIDES, BACTERICIDES AND ANTI-VIRAL AGENTS

Freshly chopped green neem leaves, dried neem leaf powder, leaf composts of neem and neem extracts control various important fungal diseases, viz., *Pythium aphanidermatum, Fusarium oxysporum, Sclerotinia sclerotium* (Table 4) in pot and field experiments (56). Post-harvest fungal decay of fruits and vegetables is reduced by treatment with neem leaf powder and aqueous neem leaf extracts. Fruits and vegetables so treated remain unaffected by fungi up to 5-7 days after treatment (42). The neem allelochemicals suppress sclerotia formation, inhibit mycelial and radical growth, sporulation and spore germination and arrest germ tube elongation and hence, result in low fungal populations of many species of plant pathogenic fungi and death in several cases (fungitoxicity). Extracts of cressleaf groundsel (*Senecio glabellus*) exhibits allelopathic activity on *Pythium* and rice seedling disease. Neem oilseed cake and meal controls the populations of pathogenic fungi, but stimulates the saprophytic and antagonistic fungal species, such as *Aspergillus niger, A. flavus, Trichoderma viridae, Penicillium degetatum* (86). Allelochemicals and volatile products released during the decomposition of the oilseed cake play an important role in these beneficial effects of neem amended soil.

Table 4. Bioactivity of neem against nematodes and fungi

Nematode species	Reference	Fungal species	Reference
Meloidogyne spp.	22, 70, 26	*Metarrhizium anisopliae*	1
Pratylenchus penetrans	70	*Aspergillus spp.*	62
P. thornei	26	*Fusarium spp.*	86
Tylenchulus spp.	4	*Macrophomina phaseolina*	86
Rotylenchus reniformis	43	*Rhizoctonia solani*	86
Helicotylenchus spp.	4	*Sclerotium rolfsii*	82
Heterodera cajani	26		
Tylenchorhyncus brassicae	86		
Hemicriconemoides mangiferae	86		
Xiphenema spp.	88		
Aphelenchus spp.	88		

Interestingly, these allelopathic effects of neem products on pathogenic nematodes, fungi and viruses persist even during the subsequent crop season after application, thus protecting the subsequent crop as well.

Neem oil also helps to control virus transmission by insect vectors, such as aphids and whiteflies (71). Azadirachtin, when applied as a pre-inoculation treatment, effectively controls powdery mildew (*Erysiphe pisi*) and induces resistance to this disease in pea (*Pisum sativum*) plants.

2.4. NATURE OF PESTICIDAL ALLELOCHEMICALS

In recent years, with modern analytical techniques, a large number of natural products have been isolated, identified, characterised and tested for bioactivity. This has rapidly advanced our knowledge of crop protection allelochemicals. The major plant allelochemical groups exhibiting pesticidal properties have been studied in many cases.

A. *Alkaloids* : Among the largest groups of pesticidal secondary metabolites in plants are the alkaloids, with over 6,500 compounds identified so far. The Asteraceae, Fabaceae, Liliaceae, Papaveraceae, Rubiaceae and Solanaceae are among the plant families with species rich in varying types of alkaloids with pesticidal properties.

B. *Phenolics :* The different types of phenolics with pest control potential include the glycosides, coumarins, flavonoids, quinones and tannins. Most of these compounds are water soluble, and have highly varied functions in different plant species. Many phenols act as antixenotic (repelling or deterring the pest from feeding and/or ovipositing on/in the plant) allomones.

C. *Terpenoids* : With more than 15,000 compounds identified so far, the terpenoid group constitutes the largest and most diverse class of organic compounds found in plants (8) and many of them exhibit allelochemical effects. Terpenoids are lipid soluble and may be either volatile (the essential oils) or non-volatile. They may be monoterpenes, sesquiterpenes, diterpenes or triterpenes (including the phytoecdysteroids and cardenolides).

Studies on neem tree chemistry have identified a vast array of structurally complex and diverse allelochemical compounds responsible for bioactivity against various pests. Most of the biologically active components of neem fall in the following chemical groups : protomeliacins, limonoids, azadirone and its derivatives, gedunin and its derivatives, vilasinin type compounds and the most potent and chemically most complex group – the 3 C-secomeliacins, namely nimbin, salannin and Azadirachtin (14). Twenty two members of the nimbin and salannin groups have been isolated and tested, but so far, the most active are the tetronortriterpenoids, namely the Azadirachtins. The first Azadirachtin (named as Azadirachtin A) was isolated from neem seeds by Butterworth and Morgan (10), and its chemical structure elucidated by Kraus et al (41) and Broughton et al (9). Apart from Azadirachtin A, 8 other isomers of Azadirachtin, labelled as Azadirachtin B, Azadirachtin D, Azadirachtin E, Azadirachtin F, Azadirachtin G, Azadirachtin H, Azadirachtin I, and Azadirachtin K have been isolated and tested for biological activity against insect and other pests (65, 66, 92). Azadirachtin A and B are the major components, while the other isomers are present in minute quantities only (67). Some other allelochemicals in neem include flavonoids, flavonolglycosides, dihydrochalcone, condensed tannins, coumarin (14), with significant antifeedant properties viz., meliantriol (44), salannin (28) and vepaol (73). Azadirachtin and other neem allelochemicals have been extracted even from the callus and cell cultures of neem leaves .

In several cases, visual and ultra-violet light has an important impact on allelochemicals (photoinduction, photoactivation, phototoxicity). For instance, the toxicity of furanocoumarins to *Heliothis* is directly proportional to the amount of UV light present (6). Some greenhouse vegetables have less flavonoid concentrations, which renders them susceptible to aphids, thrips and whiteflies.

3. Intercropping and mixed cropping

Certain plants grown with the main crop release allelochemicals, usually as leaf volatiles or root exudates, which have significant effects on the neighbouring crop. The effects can be stimulatory or inhibitory, either directly on the plant or indirectly through their influence on pests and/or beneficial organisms. This is one of the important components of multiple and mixed cropping. Inhibitory effects (allomones) can be exploited for pest (20, 32, 47, 72, 75) disease and weed (7, 11, 62, 65) suppression, while, stimulatory effects can be used to attract the natural enemies (predators and parasitoids) of pests attacking the crop. Facknath (19) demonstrated the repellent property of tomato, coriander and garlic allelochemicals on *Plutella xylostella* adults, when any of these plants was grown as intercrops between adjacent cabbage rows. Siddiqui and Alam (80) studied the nematicidal effect of root exudates of seedlings of neem and *Melia* sp. While considerable research has been done on the allelopathic effects of plant extracts and plant parts on crop pests, there is very little scientific literature on the pest control properties of allelochemicals released from companion plants or intercrops.

Insects make use of allelochemicals from plants to locate, recognise and select a host plant. In some species, such kairomones may be a single compound, while in others the insects respond only to the appropriate combinations of chemicals. Intercropping systems, which include different plants, result in mixing of host and non-host plant odours and affect insect response to the host plant, particularly where one odour masks another. For instance, tomato allelochemicals mask the kairomones of cabbage used by *Phyllotreta crucifera,* in a cabbage-tomato cropping system.

4. Insect-host plant relationships

The preceding sections indicate how allomones from plants can be directly utilised for crop pest control in different ways. Various kairomones, although having negative results for the source plant but positive for the receiving organisms such as insects do have potential in crop protection through selective breeding and/or genetically engineered resistant crop cultivars. Discovery of the glucosinolate-derived isothiocyanates released by cruciferous plants which are highly attractive to *Plutella xylostella,* has led to attempts to breed, or genetically engineer, cabbage cultivars free of these allelochemicals and therefore, resistant to *Plutella.* Other Brassicas release allylisothiocyanate which is used as a kairomone by the flea beetle, *Phyllotreta* and the cabbage root fly *Delia radicum* to locate their host plant. Similarly, the onion fly *Delia antiqua,* uses dipropyl disulfide released by onions to find its host, while the carrot fly *Psila rosae*, is attracted to trans-asarone released by its carrot host. Oviposition in this pest is induced by falcarindiol present in carrot leaves. Propylthiosulfinate is the kairomone produced by leek plants and used by the leek moth, *Acrolepiopsis assectella*, to recognise its host. Glucobrassicin present on the host (*Brassica*) leaf surface elicits oviposition in *Pieris* female moths.

Selective breeding and/or genetic engineering to develop resistant/tolerant crop cultivars which do not produce, or produce less, of the kairomone(s) concerned, is one potential avenue for the environment-friendly, sustainable pest control of the future. Allelochemicals released by plants affect not only the pest organisms but also the natural enemies of the pest at the tritrophic level (6). Read *et al* (64) demonstrated that the aphid *Brevicoryne brassicae*, uses the allelochemical, sinigrin, to find its cruciferous host, while the parasitoid *Diaeretiella rapae*, uses another plant allelochemical allylisothiocyanate, to locate the plant and then the aphids thereon. Lewis *et al* (45) showed that the allelochemical tricosane from maize plants is incorporated into *Heliothis zea* eggs and is then used as a cue by the egg parasitoid, *Trichogramma evanescens*, to locate its *Heliothis* host.

The type, composition and concentration of plant allelochemicals can alter the behaviour, biology and fitness of the natural enemies, leading to a greater or lesser extent of parasitism or predation. Thus allelochemicals can be used to manipulate entomophagous insects for maximum effectiveness of biological control programmes.

5. Conclusions and future areas of research

Allelochemicals with pest control properties can be used in Organic Farming, or as part of Sustainable Agriculture programmes, either in the form of Botanical Pesticides, as mulches, as intercrops/multiple crops or as leads in the breeding/selection/engineering of pest-resistant crop varieties. Variously labelled as 'eco-friendly', 'environmentally-safe', 'green', 'soft', this approach to crop protection has immense potential, in combination with biological control, in an IPM programme.

Several IPM packages being developed in some countries for the control of important agricultural pests include the use of botanical (in some cases, biocidal or microbial) pesticides.

While a large number of commercial formulations of neem exist in the market, there is a need for research on development of pesticidal formulations from the other bioactive plant species. The extraction techniques and bioassay methods need to be standardised among researchers and developers, while the present stringent pesticide registration requirements must be reconsidered for botanicals pesticides.

Farmers acceptance being very important in the successful application of any agricultural technology, it is imperative to involve farmers at an early stage in the planning, investigation and finalisation of a recommendation for crop protection. Educating farmers and increasing their awareness and understanding of eco-friendly agricultural practices must form an integral part of agricultural programmes. Attempts in this direction are ongoing in Mauritius and are beginning to show results. Sensitising the public of the health and environmental hazards of synthetic pesticides and encouraging public awareness and acceptance of organically grown produce is another very important aspect in ensuring a shift towards more environment-friendly and sustainable agriculture.

References

1. Aguda, R.M., Ranbach, M.C., Shepard, B.M. (1986). Effect of neem oil on germination and sporulation of the entomogenous fungus, *Metarrhizium anisopliae International Rice Research Newsletter* **11** : 34-35.
2. Akou-edi, E. (1984). Effects of neem seed powder and oil on *Tribolium confusum* and *Sitophilus zeamais*. In *Natural Pesticides From the Neem Tree and Other Tropical* Plants (Eds., H. Schmutterer and K.R.S. Ascher) pp. 445-451. Proceedings, II International Neem Conference. May 1983. Eschborn Germany : GTZ.
3. Alam, M.M. (1990). Control of plant-parasitic nematodes with oilseed cakes on some vegetables in field. *Pakistan Journal of Nematology* **9** : 21-30.
4. Alam, M.M., Mansood, A. and Hussain, S.I. (1975). Effects of margosa and marigold root-exudates on mortality and larval hatch of certain nematodes. *Journal of Experimental Biology* **13** : 412-414.
5. Alam. M.M. (1993). Bioactivity against phytonematodes. In *Neem Research and Development* (Eds., N.S. Randhawa and B.S. Parmar) 123-143pp. New Delhi : Society of Pesticide Science.
6. Ananthakrishnan, T.N. (1992). *Dimensions of Insect-Plant Interactions.* New Delhi : oxford and IBH Publishing Co. 184 pp.
7. Babu, K.S., Murthy, M.S.N. and Babu, T.R. (1996). Effect of botanicals on certain parasitoids. In *Neem and Environment* (Eds., R.P.Singh, M.S. Chari, A.K.Raheja and W. Kraus). pp.1117-1126. Proceedings, World Neem Conference, Bangalore, India, February 1993. New Delhi : Oxford and IBH Publishing Co.
8. Bernays, E.A. and Chapman, R.F. (1994). *Host-Plant Selection by Phytophagous Insects.* London : Chapman and Hall. 312 pp.
9. Broughton, H.B., Ley, S.V., Slawin, A.M.Z., Williams, D.J., Morgan, E.D. (1986). X-ray crystallographic structure determination of detigloyldihydroazadirachtin and reassignment of the structure of the limonoid insect antifeedant, azadirachtin. *Journal of Chemical Society Communication* : 46-47.
10. Butterworth, J.H. and Morgan, E.D. (1968). Isolation of a substance that suppresses feeding in locusts. *Journal Chemical Society Communication* : 23-24.
11. Carson, R. (1962). *Silent Spring.* Boston : Houghton-Mifflin. 360 pp.
12. Chakraborty, M.K., Prabhu, S.R., and Joshi, B.G. (1976). Isolating toxicants from pongamia oil to evaluate their insecticidal properties on tobacco caterpillar, *Spodoptera litura. Tobacco Research* **2** : 38-44.
13. Chari, M.S. and Ramaprasad, G. (1993). *Botanical Pesticides in Integrated Pest Management.* Proceedings National Symposium, 1990. Rajahmundry, India : Indian Society of Tobacco Science. 477 pp.
14. Devakumar, C. and Sukhdev (1993). Chemistry. *See Reference No.* 5, pp 63-96.
15. Devakumar, C., Goswami, B.K. and Mukherjee, S.K. (1985). Nematicidal properties from neem (*Azadirachta indica* A. Juss) Part I : Screening of neem seed kernel fractions against *Meloidogyne incognita. Indian Journal of Nematology* **15** : 121-124.
16. Dhawan, A.K. and Simwat, G.S. (1992). Field evaluation of Neemrich 20EC for management of insect pests on cotton during reproductive phase. In *Proceedings 1st National Symposium Allelopathy in Agroecosystems - 1992* (Eds., P. Tauro and S.S. Narwal). Hisar, India : Indian Society of Allelopathy, Haryana Agricultural University, pp. 152-153.
17. Facknath, S. (1991). Evaluation of a commercial nem formulation (Margosan-O) for the control of *Dysdercus flavidus. Revue Agricole et Sucriere de l'Ile Maurice* **70** : 34-38.
18. Facknath, S. (1993). The assessment of two neem formulations for the control of some economically important insects. *Revue Agricole et Sucriere de l'Ile Maurice* **72** : 44-49.
19. Facknath, S. (1996). Efficacy of neem and *Melia* in overcoming tolerance of *Plutella xylostella* to some commonly used synthetic insecticides. *See Reference No. 7*, pp. 505-515. New Delhi : Oxford and IBH Publishing Co.
20. Facknath, S. (1998a). Application of neem and intercropping for the control of important cabbage pests. In *Azadirachta indica* A. Juss. New Delhi : Oxford and IBH Publishing Co. Pvt. Ltd. pp. 165-175.

21. Facknath, S. (1998b). Botanical Pesticides : An environmentally-friendly alternative to synthetic pesticides for sustainable crop protection. *2nd Fete de la Science, University of Mauritius, 11-13 June 1998.*
22. Facknath, S. and Jadunundun, J.K. (1990). Potential of Neem and *Tagetes* plants as nematicides against *Meloidogyne incognita* infesting tomato. *Revue Agricole et Sucriere de l'Ile Maurice* **69** : 75-85.
23. Fagoonee, I and Umrit, G. (1980) : Biology of *Dysdercus flavidus* and its control by *Ageratum conyzoides. Revue Agricole et Sucriere de l'Ile Maurice.* **59** : 122-128.
24. Fagoonee, I. (1984). Effect of azadirachtin and of a neem extract on food utilisation by *Crocidolomia binotalis. See Reference No.* 2, pp. 221-224. Rauischolzhausen, Eschborn.. Germany : GTZ. pp. 211-224.
25. Fagoonee, I. and Toory, V. (1984). Contribution to the study of the biology and ecology of the leafminer, *Lriomyza trifolii* and its control by neem. *Insect Science and Application* **5** : 23-30.
26. Gupta, P. Sebastian, S. Sobita, L.R. and Mishra, S. (1996). Use of neem cake in control of lesion nematode, cyst nematode and root knot nematode. *See Reference No.* 7, pp. 622-25. New Delhi : Oxford and IBH Publishing Co. Pvt. Ltd., India.
27. Hellpap, C. (1996). The use of neem by farmers in the Dominican Republic. *See Reference No.* 7, pp. 33-45.
28. Henderson, R., Mccrindle, R. and Overton, K.H. (1964). Salannin. *Tetrahedron Letters* **24** : 1517-1523.
29. Isman, M (1989). Toxicity and fate of Acetylchromenes in Pest Insects. In *Insecticides of Plant Origin* (Eds., J.T. Arnason, B.J.R. Philogene and P. Morand). ACS Symposium Series **387** : 44-58. Washington, D.C., USA : American Chemical Society.
30. Isman, M.B. (1998). Some target and non-target effects of neem insecticides. *Proceedings International Neem Conference,* February 1996, Queensland, Australia . In Press.
31. Jacobson, M. (1989). Botanical Pesticides : Past, Present and Future. *See Reference No.* 29, pp. 1-10.
32. Jacobson,, M. Reed, D.K., Crystal, M.M., Moreno, D.S. and Soderstrom, E.L. (1978). Chemistry and biological activity of insect feeding deterrents from certain weed and crop plants. *Entomologia Experimentalis Applicata* **24** : 448-457.
33. Jhansi, B.R. and Singh, R.P. (1996). Identification of effective and inexpensive neem (*Azadirachta indica* A. Juss) seed kernel extract for management of *Heliothis armigera* (Hubner). *See Reference No.* 7, pp. 381-402.
34. Joshi, B.G., Ramaprasad, G. and Sitaramaiah, S. (1982). Effect of a neem seed kernel suspension on *Telenomus remus,* an egg parasite of *Spodoptera litura. Phytoparasitica* **10** : 61-63.
35. Ketkar, C.M. (1987). Use of tree-derived non-edible oils as surface protectants for stored legumes against *Callosobruchus maculatus* and *C. chinensis. See Reference No.* 2, pp. 535-542.
36. Khan, M.W., Alam, M.M., Khan, A.M. and Saxena, S.K. (1974). Effect of water-soluble fractions of oil-cakes and bitter principles of neem on some fungi and nematodes. *Acta Botanica Indica* **2** : 120-128.
37. Koshiya, D.J. and Ghelani, A.B. (1993). Antifeedant activity of different plant derivatives against *Spodoptera litura* (Fab.) on groundnut. *See Reference No.* 13, pp. 270-275.
38. Koul, O. (1987). Antifeedant and growth inhibiting effects of calamus oil and neem oil on *Spodoptera litura* under laboratoy conditions. *Phytoparasitica* **15** : 169-180.
39. Koul, O. (1992). Neem allelochemicals and insect control. In *Allelopathy : Basic and Applied Aspects* (Eds. S.J.H. Rizvi and V. Rizvi). pp. 489-413. London : Chapman and Hall.
40. Koul, O. and Isman, M.B. (1991). Effect of azadirachtin on the dietary utilisation and development of the variegated cutworm, *Peridroma saucia. Journal of Insect Physiology* **37** : 591-598.
41. Kraus, W., Bokel, M., Klenk, A. and Pohnl, H.D. (1985). The structure of azadirachtin and 22,23-dihydro-23β-methoxy-azadirachtin. *Tetrahedron Letters* **26** : 6435-6438.
42. Krishna, A., Prasad, A. and Ojha, N.L. (1986). Anti-fungal evaluations of leaf extracts for the control of some cucurbitaceous fruit rot diseases. *Indian Phytopathologist* **39** : 153.
43. Lal, A., Yadav, B.S. and Nandwana, R.P. (1977). Effect of chopped leaves of various plants on the plant growth and reniform nematode, *Rotylenchulus reniformis. Indian Journal of Mycology and Plant Pathology* **7** : 180-181.
44. Lavie, D., Jain, and M.K. Shpan-Gabrielith, S.R. (1967). A locust phagorepellent from two *Melia* species. *Journal Chemical Society Communication pp.* 910-911.
45. Lewis, W.J., Jones, and R.L. Sparks, A.N. (1972). A host-seeking stimulant for the egg parastie *Trichogramma evanescens* : Its source and a demonstration of its laboratory and field activity. *Annual Report, Entomological Society of America* **65** : 1087-1089.
46. Lowery, D.T., Isman, M.B. and Brard, N.L. (1993). Laboratory and field evaluation of neem for the control of aphids (Homoptera : Aphididae). *Journal of Economic Entomology* **86** : 864-870.
47. Manoranjitham, M.P., Anadhi, A.P., Sampathraj, R. and Vanithakumari, C. (1996). Alteration in testicular histoarchitecture following neem oil administration in albino rats. *See References No.* 7, pp. 1165-1172.

48. Mansour, F., Ascher, K.R.S. and Omari, N. (1987). Effect of neem seed kernel extracts from different solvents on the predacious mite *Phytoseiulus persimilis* and the phytophagous mite *Tetranychus cinnabarinus* as well as on the predatory spider *Chiracanthium mildei*. *See References No.* 2, pp. 577-587.
49. Mansour, F.A. and Ascher, K.R.S. (1984). Effects of neem (*Azadirachta indica*) seed kernel extracts from different solvents on the carmine spider mite, *Tetranychus cinnabarinus*. *See References No.* 2, pp. 461-469.
50. Mariappan, V., Jayaraj, S. and Saxena, R.C. (1988). Effect of non-edible oils on survival of *Nephotettix virescens* (Homoptera : Cicadellidae) and on transmission of rice tungro virus. *Journal of Economic Entomology* **81** : 1369-1372.
51. Mishra, M. and Tewari, S.N. (1993). In vitro studies on the mycotoxic properties of *Acacia arabica, Adhatoda vasica* and *Pongamia glabra*. *See References No.* 13, pp. 413-416.
52. Mojumdar, V. and Raman, R. (1998). Nematicidal efficacy of 'Neema-S1' _ an experimental formulation of Neem for seed treatments against *Heterodera cajani* and *Meloidogyne incognita* in cowpea and chickpea. *See References No.* 20, pp. 217-222.
53. Mojundar, V. and Mishra, S.D. (1996). Management of *Meloidogyne incognita* infesting pulse crops through seed treatments with neem products. *See References No.* 7, pp. 699-704.
54. Murugan, K., Jahanmohini, P. and Babu, R. (1996). Effect of neem kernel extract and neem oil on nutritive and reproductive physiology of *Heliothis armigera* Hubn. *See References No.* 7, pp. 321-334.
55. Parmar, B.S. and Dutta, S. (1986). Neem oil as synergist for insecticides. *Neem Newsletter* **3** : 3-5.
56. Parveen, G. and Alam, M.M. (1993). Bioactivity against plant pathogens. *See Reference No.* 5, pp 144-153.
57. Patil, K.J., Deshkar, M.M., Rane, E. and Nimbalkar, S.A. (1993). Some indigenous plant materials against *Aphis gossypii* and *Dactynotus carthami* . *See References No.* 13, pp. 238-244.
58. Peshney, N.L. and Moghe, P.C.D (1993). Inhibition of tobacco mosaic virus of chilli by plant extracts. *See References No.* 13, pp. 460-465.
59. Pradhan, S., Jotwani, M.G. and Rai, B.K. (1962). The neem seed deterrent to locusts. *Indian Farming* **12** : 7-11.
60. Prakash, A., Pasalu, I.C. and Mathur, K.C. (1981). Plant products for management of stored grain insect pests. *Bulletin of Grain Technology* **19** : 213-221.
61. Pruthi, H.S. (1937). *Report of Imperial Entomologist*. New Delhi, India : Royal Agricultural Research Institute. pp. 123-137.
62. Purnima, S. and Saxena, S.K. (1987). Effect of neem leaf powder and extract on the development of fruit rot caused by *Aspergillus niger*. *Neem News*letter **4** : 45-47.
63. Rajeshwari, E. and Mariappan, V. (1996). Efficacy of plant derivatives on rice blast pathogen *Pyricularia oryzae* cav. blast infection and enzymes of rice plants. *See References No.* 7, pp. 753-743.
64. Read, D.P., Feeny, P.P. and Root, R.B. (1970). Habitat selection by the aphid parasite *Diaeretiella rapae* (Hymenoptera : Braconidae) and hyperparasite *Charips brassicae* (Hymenoptera : Cynipidae). *Canadian Entomologist* **102** : 1567-1578.
65. Rembold, H. (1990a). Azadirachtins : Their structure and mode of action. *See Reference No.* 29, pp. 150-163.
66. Rembold, H. (1990b). Isomeric azadirachtins and their mode of action. In : *Focus on Phytochemical Pesticides : The Neem Tree* (Ed., M. Jacobson), pp. 47-67, Boca Raton, Florida, USA : CRC Press.
67. Rembold, H. (1996). Neem and its general development for pest control. *See References No.* 7, pp. 3-10.
68. Rembold, H., Uhl, M. and Muller, T. (1987). Effect of azadirachtin A on hormone titers during the gonadotrophic cycle of *Locusta migratoria*. *See References No.* 2, pp. 289-298.
69. Rossner, J. and Zebitz, C.P.W. (1987a). Effect of soil treatment with neem products on earthworms (Lumbricidae). *See References No.* 2, pp. 627-632.
70. Rossner, J. and Zebitz, C.P.W. (1987b). Effect of neem products on nematodes and growth of tomato (*Lycopersicon esculentum*) plants. *See References No.* 2, pp. 611-621.
71. Roychoudhary, R. and Jain, R.K.(1996). Neem for the control of aphid and whitefly vectors and virus diseases of plants. *See References No.* 7, pp. 763-776.
72. Sampathraj, R., Naryana, P.B.S. and Vanithakumari, C. (1996). Effect of neem oil : Structural and functional changes in the epididymis of rats. *See References No.* 7, pp. 1173-1183.
73. Sankaram, A.V.B., Murthy, M.M., Bhaskaraiah, K., Subramanyam, M., Sultana, N., Sharma, H.C., Leuschner, K., Ramaprasad, G., Sitaramaiah, S., Rukmini, C. and Rao, P.U. (1987). Chemistry, biological activity and utilisation aspects of some promising neem extractives. *See References No.* 35, pp. 127-148.
74. Saxena, R.C., Liquido, J. and Justo, H.D. (1981). Neem seed oil, a potential antifeedant for the control of the rice brown planthopper, *Nilaparvata lugens*. In : *Natural PesticidesFrom the Neem Tree (Azadirachta indica A. Juss)* (Eds., H. Schmutterer; K.R.S. Ascher and H. Rembold). Proceedings Ist International Neem Conference. June 1980, Rottach-Egern, Germany.

75. Schauer, M. and Schmutterer, H. (1981). Effects of neem kernel extracts on the two-spotted spider mite, *Tetranychus urticae*. *See References No.* 74, pp. 259-266.
76. Sharma, I.N.S., Sinha, A.K. and Singh, S.P. (1992). Allelopathic potential of some plant substances as anti-feedant against insect pests of jute. . In : *Proc. 1st National Symposium allelopathy in Agroecosystems* (Eds., P. Tauro and S.S. Narwal). pp. 166-168. Hisar, India : Indian Society of Allelopathy, Haryana Agricultural University.
77. Shin-Foon, C. (1993). Investigations on botanical insecticides in South China – an update. *See References No.* 13, pp. 134-137.
78. Siddiq, S.A. (1981). Efficacy and persistence of powdered neem seeds for treatment of stored wheat against *Trogoderma granarium*. *See References No.* 74, pp. 251-257.
79. Siddiq, S.A. and Khalafalla, A. (1996). The performance of an Integrated Pest Management package including neem for the control of potato pests in farmers fields. *See References No.* 7, pp. 79.
80. Siddiqui, M.A. and Alam, M.M. (1989). Effect of root-exudates of neem and Persian lilac on plant parasitic nematodes. *Anzeiger fur Schalingskunde Pflanzenschutz Umweltschutz* **62** : 33-35.
81. Simmonds, M.S.J. and Blaney, W.M. (1984). Some neurophysiological effects of azadirachtin on lepidopterous larvae and their feeding response. *See References No.* 2, pp. 163-180.
82. Sinha, D.C. and Prasad, R.K. (1986). Effect of the organic amendments in soil on foot and root- rot of gram caused by *Fusarium oxysporum* f. sp. ciceri and *Sclerotium rolfsii. Indian Phytopathologist* **39** : 145.
83. Sinha, S. N. (1996). Neem in integrated management of *Heliothis armigera* Hubn. Infesting chickpea. *See References No.* 7, pp. 439-446.
84. Sombatsiri, K. and Tigvattanont, S. (1984). Effects of neem extracts on some insect pests of economic importance in Thailand. *See References No.* 2, pp. 95-100.
85. Su, H.C.F. (1977). Insecticidal properties of black pepper to rice weevil and cowpea weevils. *Journal of Economic Entomology* **70** : 18-21.
86. Tiyagi, S.A. and Alam, M.M. (1996). Efficacy of neem cake against soil nematodes and fungi on chickpea. *See References No.* 7, pp. 627-641.
87. Trivedi, T.P. (1987). Evaluation of plant origin oils as synergist for pyrethrins against insect pests of rice storage. *Neem Newsletter* **4** : 35.
88. Vyas, S.C. (1996). Effect of neem cake, neem oil, Repellin and carbofuran on control of soybean nematodes. *See References No.* 7, pp. 663-667.
89. Wessner, M., Champion, B., Girault, J.P., Kaouadji, N., Saidi, B. and Lafont, R. (1992). Ecdysteroids from *Ajuga iva. Phytochemistry* **31** : 3785-3788.
90. West, A.J. and Mordue, A.J. (1992). The influence of azadirachtin on the feeding behaviour of cereal aphids and slugs. *Entomologia Experimentalis Applicata* **62** : 75-79.
91. Wilps, H., Nasseh, O., Krall, S. and Kabo, O. (1992). Lutte contre les *Schistocerca gregaria* adultes au moyen de nouvelles preparations et methodes. *SAHEL PV INFO Bulletin d'Information en Protection des Vegetaux de l'UCTR/PV (Bamako, Mali)* **44** : 7-14.
92. Yamasaki, R.B. and Klocke, J.A. (1987). Structure-bioactivity relationships of azadirachtin, a potent insect control agent. *Journal of Agriculture and Food Chemistry* **35** : 467.

4

Allelopathic interactions in soil

B. LALLJEE* and S. FACKNATH
Faculty of Agriculture, University of Mauritius, Reduit, Mauritius

Abstract

The soil is the site of most biochemical reactions and interactions affecting the plant growth. It serves as a medium for the diffusion of substances released by plants and to keep other plants away from growing too close to them. Such substances are called allelochemicals and are released in the soil through root exudation, leaching, volatilisation, biomass decay, etc. These chemicals affects various processes such as germination of seeds, nitrogen fixation, mineralisation, etc. The bioactivities of the allelochemicals are controlled by soil properties (pH, solute concentration, cation exchange capacity, fertility) and also by cultural practices such as tillage and crop rotations. The principles of allelopathy can be effectively exploited to improve soil fertility (e.g. green manuring in soil) or improve nitrogen use efficiency by plants (e.g. incorporation of neem and other plant materials for reducing nitrification rate). Similarly, certain plant materials have beneficial effects on the soil micro- and macro-fauna, which in turn influences the fertility of the soil. This paper deals with some plants and soil organisms releasing allelochemicals in the soil and the interaction of these chemicals with the soil.

Keywords : Allelochemical, allelopathic interactions, crop production, desert, forest, natural ecosystem, soil fertility, weed management.

1. Introduction

The soil is the foundation for agricultural productivity and is also the site of many biochemical interactions affecting plants (37). It serves as good medium for the diffusion of substances released by plants, to prevent other plants from growing too close to them. Such substances are known as allelochemicals and are released in the soil through root exudation, leaching, volatilisation, biomass decay, etc. The allelochemicals released by these modes finally reach the soil, where they may be (a) absorbed by plants, (b) adsorbed by the soil, (c) transformed and/or decomposed by soil microflora, (d) decomposed by sunlight, (e) leached by rain/irrigation water from the rhizosphere. A wide array of chemicals having allelopathic potential in soil have been reported, viz., oxalic acid, cyanogenic glucosides, alkaloids, lipids, terpenoids, saponins, flavanoids, tannins and lignins. These chemicals influence the various soil/plant processes, such as germination of

* Correspondence author

S.S. Narwal et al. (eds.), Allelopathy in Ecological Agriculture and Forestry, 47–58.

seeds/spores, decomposition of biomass, mineralisation of plant nutrients and biological nitrogen fixation. The bioactivities of the allelochemicals are influenced by chemical properties of the soil (pH, solute concentration, cation exchange capacity) and also by cultural practices and farming systems, e.g. conventional and minimum tillage, monocropping and crop rotations.

The principles of allelopathy may be efficiently and effectively exploited to improve soil fertility (e.g. biological nitrogen fixation) or to improve nitrogen use efficiency by plants (e.g. use of nitrification inhibitors, such as neem and other plant material, to reduce nitrification rate). Some plants may indirectly affect soil properties by exerting beneficial or detrimental effects on soil microflora and micro and macrofauna, for instance, the earthworms. In this paper specific examples of allelopathic interactions in soil in natural and manmade ecosystems are discussed.

2. Allelopathic Interactions

The soil is a growing medium for plants, hence the allelopathic interactions occurring in soil greatly influence the germination, growth and development of plants. However, the magnitude of this influence varies with the physico-chemical and biological properties of the soil, the type of plant species and the nature of the ecosystem etc.

2.1. ALLELOPATHIC INTERACTIONS IN NATURAL ECOSYSTEMS

There is large diversity in plant and animal groups in a natural ecosystem and the species interact among and within themselves at various levels and ultimately establish equilibrium. The nutrient cycling in such systems is usually the result of several processes and allelopathic relationships which are balanced in time and space. Any disturbance to this natural cycle leads to disruption and loss in soil fertility and ultimately to loss of species diversity.

2.1.1. *Forest Ecosystems*

It has been shown that vegetation in certain ecosystems inhibits nitrification in soil by producing phenolic compounds and other chemicals that inhibit oxidation of ammonia by nitrifying organisms. Studies done in Mauritius (26) have shown that ammonification, nitrification and total mineralisation rate in a natural ecosystem and an adjoining pine forest ecosystem followed monthly and seasonal patterns and were highest during the summer months as compared to the winter months. The ammonification, nitrification and total mineralisation rate was always lower in the pine forest than in the undisturbed natural ecosystem (net mineralisation rate 1.65 times higher in undisturbed ecosystem). pH, organic matter and earthworm populations were at optimal levels in the natural ecosystem as compared to the pine forest ecosystem. Apart from environmental effects, these differences have been attributed to allelopathic interactions in the soil by the diverse vegetation present in the two systems, for instance, the liberation of terpenes during decomposition of pine needles. Poor establishment and reduced seedling growth of Scots pine (*Pinus silvestris* L.) in Northern Sweden was related to an allelopathic inhibition by the dwarf shrub *Empetrum hermaphroditum* (32). Indoor bioassays with green and brown leaves of *Empetrum* had strong negative effects on rooting ability, radical elongation and growth of Scots pine seedlings. Bioassays with soil samples showed that phytotoxic substances leached from *Empetrum* foliage accumulate in the soil. Field experiments revealed that chemical inhibition by *Empetrum*, causing high mortality and slow growth of pine seedlings, can be reduced by adding activated charcoal to the soil. Indoor and outdoor experiments demonstrated that allelopathy is an important factor in seed regeneration of Scots pine in forest floor vegetation dominated by *Empetrum*. Zackrisson and Nilson (40) indicated that low doses and short exposure times of seeds to leachates had strong negative effects on seed germination and early root development. Field experiments have shown that allelopathic effects by *Empetrum* were strong during early spring. Extracts passed through soils collected from *Empetrum* were detoxified, while those passed through sterilised soil were not. Micro-organisms could, therefore, detoxify allelochemicals under some environmental conditions.

Twenty years old experiments on *L. lecucocephala* at the University of Mauritius have proved that the yield of maize grown between rows of *Leucaena* was substantially higher than maize grown in control plots (without *Leucaena)*. Lalljee *et al* (27) demonstrated that the fertility of the soil was much better under *Leucaena* than under normal cultivation. The respiration rate was significantly increased (1.98 X 10^{-9} g CO_2 / g soil / sec.) in the treated soil as compared to 1.70 X 10^{-9} in the control. Microbial biomass C, microbial biomass N, and microbial biomass P were significantly higher in the soil under *Leucaena* (0.041% , 0.051% and 9.0 X 10^{-5} % respectively) than in the soil under normal cultivation (0.015%, 0.031% and 6.3 X 10^{-5}% respectively). These results could be partly due to the positive effects of the allelopathic chemicals released from *Leucaena* leaves and roots.

2.1.2. *Desert Ecosystems*

The bare zones observed around the shrubs *Ceratiola ericoides* in Florida indicate the possibility of allelopathy by scrubs controlling the distribution of grasses invading from adjacent sand hills. This observation led to allelopathic studies on the interactions between different plant species. Hydrocinnamic acid (HCA), an allelochemical, has been identified as a breakdown product of ceratiolin released from the shrub *Ceratiola ericoides*. Williamson *et al* (38) investigated the effects of HCA on shoot, root and biomass of the grass *Schrizachyrium scoparium* in a greenhouse study. Linear increases in the concentration of HCA from 0 to 200 ppm resulted in exponential decreases in root and shoot biomass of the grass. Concurrent investigations of reduced nutrient levels indicted greater inhibition by HCA in a nitrogen and potassium deficient treatment relative to HCA inhibition in the full nutrient treatment. The effects of HCA in a reduced phosphorus treatment and a reduced phosphorus and potassium treatment, were not significantly different from the effects of HCA in the full nutrient treatment. However, extraction of the soil at harvest indicated no build up of HCA at the end of the experiment. The experiment showed the sensitivity of *S. scoparium* to HCA in general, and increased sensitivity under low nitrogen and low potassium levels.

Hegazy *et al* (15) conducted laboratory experiments to investigate the autotoxic effects of *Anastatica hiertochuntica* and possible effects on 5 other desert plants. Seed germination, seedling growth and cell division of all species tested were inhibited by the aqueous extract of the shoots of *A. hiertochuntica*. A gradual increase in the percentage of prophase and decrease in the other mitotic stages as well as the mitotic index, were observed with increasing extract concentrations. The inhibitory substances were apparently released to the soil by repeated washing of the standing plants by rain and dew.

2.2. ALLELOPATHIC INTERACTIONS IN MANMADE ECOSYSTEMS

Natural ecosystems are changed by man for production of food, as pastures, or for leisure. It disrupts the normal nutrient cycling, as well as the defensive capacity, of such ecosystems and then requires extensive inputs in terms of pesticides and fertilisers.

2.2.1. *Crop Production*

Alfalfa (*Medicago sativa* L.) is known to be both autotoxic and allelopathic. Hedge and Miller (16) studied these properties on alfalfa itself and sorghum (*Sorghum bicolor* L.) on a silt loam soil previously cropped to alfalfa and sorghum. Bioassays of seedling exudate and shoot leachate of alfalfa and sorghum were also conducted. Plant height and fresh weight per plant of sorghum were lower in alfalfa soil than in sorghum soil. The preceding crop did not affect the germination of both alfalfa and sorghum and plant height of sorghum. Soil incorporation of fresh alfalfa roots only, or both roots and shoots, reduced alfalfa emergence, plant height and dry weight/ plant. Primary effects of water-soluble inhibitory compounds from alfalfa shoot appeared on germination and radical elongation. Alfalfa allelopathy proved more severe than autotoxicity. Ells and McSay (11) reported the toxic effects of alfalfa residues on cucumber (*Cucumis sativus* L.) seed germination and seedling growth. Ground alfalfa roots (0.5% w/w) inhibited germination when added to the growing medium.

Poor emergence of lettuce has been observed when planted immediately after harvest of a celery crop. Schilling *et al* (34) conducted a greenhouse study to evaluate the possible allelopathic effects of celery residue on the emergence and growth of lettuce. Celery root tissue was 1.8 and 1.6 times more toxic to lettuce seedling growth than celery petiole or laminar tissue, respectively. Lettuce shoot growth was significantly inhibited when grown in sand amended with celery residue, than in amended vermiculite or potting soil. Incubating celery root residue in soil for 4 weeks increased phytotoxicity at 1% (v/v) and decreased it at 40% (v/v). The addition of activated charcoal to the medium increased the growth of lettuce exposed to celery residues.

Cast *et al* (7) extracted the allelochemicals from the soil of no-tillage and conventional tillage wheat plots and evaluated their biological activity. They found that the most significant inhibition in bioassays of extracts was from soil collected immediately after harvest. No tillage soils produced significant inhibition even during the fallow period. Mass spectrometry showed that fatty acids were the most abundant compounds. However, in bioassays, the authentic samples of the five free fatty acids did not show significant biological activity against wheat.

Kalburtji and Mosjidis (22) incorporated residues of *Sericea lespedeza* in soil in laboratory and greenhouse experiments and found that the residues inhibited growth of bermudagrass and bahiagrass but not their seed germination and emergence. In the greenhouse, no differences were found among the cultivars of bermudagrass and bahiagrass in response to *S. lespedeza* residues, but in the laboratory, differences were observed among bermudagrass cultivars for tolerance to *S. lespedeza* residues. The harmful effects of *S. lespedeza* residues were small (17% and 16% reduction in dry weight of bermudagrass and bahiagrass, respectively) compared to the positive effects of N-fertilisation.

In field studies, sweet potato greatly reduced yellow nutsedge growth when both species were grown together using standard cultural practices. But the presence of yellow nutsedge did not markedly affect sweet potato growth. The most polar fractions of serially extracted sweet potato periderm tissue were highly inhibitory to yellow nutsedge growth. The above results showed that sweet potato interference under field conditions is partly due to allelopathy (14).

Blum *et al* (4) used various soil extracts to estimate the levels of allelopathic compounds in soil, especially with respect to pH, solute potential and iron content of extracts. They used a water autoclave extraction procedure to extract allelochemical compounds from a wheat no-till soil and found that although the extract did not affect the germination of clover and morning glory, and the radical and hypocotyl length of morning glory; but it reduced the radical and hypocotyl length of clover. Individual phenolic acids did not inhibit germination, but reduced the radical and hypocotyl length of both clover and morning glory. 6-MBOA (6-methoxy-2,3-benzoxazolinone) a conversion product of 2-O-glycoxyl-7-methoxy-1,4-benzoxacine-3-one, a hydroxalic acid (HA) in living wheat plants inhibited germination and radical and hypocotyl length of clover and morning glory. 6-MBOA, however, was not detected in wheat debris, stubble, or soil extracts. The levels of total phenolic acids in wheat-conventional till-soil extracts were not related to germination and hypocotyl length of two varieties of morning glory. Total phenolic acids, solute potential and acidity of wheat no-till soil extracts appeared to be additive on the action of clover radical and hypocotyl length.

2.2.2. *Rangelands and Golf Courses*

Low vigour of seedlings and stand failure is the main problem in revegetated, semi- arid and arid rangelands. Phytotoxicity, sandbur (*Cenchrus incertus*) competition, seed bed preparation (ploughing versus disking) and nitrogen fertilisation were studied as reasons for low vigour of weeping lovegrass (*Eragrostis curvula*) seedlings on sand shinnery oak (*Quercus harvardii*) range in West Texas. Oak leaf residue and sandbur-dominated grass extracts did not affect seed germination, but the residue extracts reduced root length by 92 and 21% respectively (29). The annual bluegrass (*Pra annua* L.) has long been observed as a weed on creeping bentgrass (*Agrostis stolonifera* L.). The aqueous leachates from annual bluegrass and creeping bentgrass were used to irrigate creeping bentgrass for two years. The results did not show any effect on turf colour, foliar ground cover, shoot density or disease incidence in the creeping bentgrass to indicate allelopathy (5).

2.2.3. *Weed Management*

Rye (*Secale cereale*) provided excellent weed suppression through allelopathic and competitive mechanisms. Its residues maintained on the soil surface released 2,4-dihydroxy-1,4(2H)-benzoxaxin-3-one (DIBOA) and a breakdown product 2(3H)-benzoxalinone (BOA), both of which were strongly inhibitory to germination and seedling growth of dicotyledenous annual weeds. In addition, soil fungi converted BOA to 2,2`-oxo-1,1`-azo-benzene which was ten fold more phytotoxic than BOA. A variety of natural products seem to contribute to the herbicidal property of rye residues (33).

2,2`-oxo -1,1` - azo-benzene (AZOB), a compound with strong herbicidal activity was isolated and characterised from a soil supplemented with 2,3-benzoxazolinone (BOA) (31). A similar experiment with 6-methoxy-2,3-benzoxazolinone (MBOA) yielded AZOB as well as its mono-(MAZOB) and dimethoxy-(DIMAZOB) derivatives. These compounds were produced by soil microorganisms. Although BOA and 2,4-dihydroxy-1,4 (2H)-benzoxazin-3-one (DIBOA) were leached out from rye residues, there were no detectable amounts of the biotransformation products in the soil. When BOA was mixed with soil and rye residue, AZOB was detected. Levels of free BOA in the soil were greatly reduced by incubation with rye residues. AZOB was more toxic to curly cress (*Lepidium sativum*) and barnyard grass (*Echinochloa crusgalli*) than either DIBOA or BOA.Three allelochemicals (AZOB, BOA, DIBOA) from rye or its breakdown products were evaluated for activity on gardencress, barnyard grass, cucumber and snapbean. BOA, DIBOA and AZOB were applied singly at 50, 100 and 200 ppm. AZOB at 200ppm produced 38 to 49% more inhibition than DIBOA. All combinations were slightly antagonistic to barnyard grass, while several combinations produced synergistic responses to garden cress germination and growth. Cucumbers and snapbeans exhibited both types of responses depending upon combinations and rates of allelochemicals applied. The benzoxazinones were more inhibitory to crops than to weeds (8).

A unique pattern of weed exclusion was found under the canopy of *Delonix regia,* an ornamental tree in South Taiwan (10). The number of species and the coverage of understorey species were significantly lower in the region of *D. regia* than that of grassland, indicating growth suppression of understorey species by *D. regia.* A series of aqueous extracts of leaves, flowers and twigs of *D. regia* were bioassayed against three species to determine phytotoxicity, and results showed highest inhibition in flowers. The phytotoxicity of fallen leaves and flowers of *D. regia* were not significantly affected by temperature. However, the phytotoxicity of other plant parts was decreased when the plant material was subjected to temperatures above 70°C, indicating that the allelopathic nature of *D. regia* could easily be decomposed by fire. The responsible phytotoxins were identified as 4-hydroxybenzoic, chlorogenic, 3,4-dihydroxybenzoic, gallic, 3,4-ihydroxycinna-mic,3,5-dinitrobenzoic and L-azetidine-2-carboxylicacids and 3,4 -dihydroxybenzaldehde. The findings of the bioassays and the number and amounts of responsible allelochemicals compounds in *D. regia* were well co-related, confirming that the exclusion of understoried plants under the canopy of *D.regia* trees was due to the allelopathic effects of the fallen leaves, flowers and twigs.

Hairy vetch residue (0-616 g/m^2) had no effect on corn, slightly reduced velvetleaf and greenfoxtail establishment and severely inhibited common lambsquatters establishment under full sunlight conditions. Day/night temperatures of 24/16°C or 32/16 °C had no effect and soil moistures of 50-133% field capacity had little effect on all the species and on residue rates (36).

Anaya *et al* (3) demonstrated the allelopathic potential of *Ipomea tricolor*, used in traditional agriculture for weed control. They measured the inhibitory activity of aqueous leachates and organic extracts of the plant material on seedling growth of *Amaranthus leucocarpus* and *Echinochloa crusgalli.* Bioactivity-guided fractionation of the most active extracts led to the isolation of mixture of glycosides having jalapinolic acid, which were the allelopathic principles.

Schreiber (35) investigated the interaction of tillage, crop rotation and herbicide use levels on weed seed population, weed population and crop yield in a long term study. Increasing herbicide use levels above the minimum reduced giant foxtail seeds in the 0-2.5 cm soil depth. No-tilling increased giant foxtail seed over conventional tillage. Growing corn in a soybean or soybean-wheat-corn rotation reduces giant foxtail seed bank in all 3 soil depths sampled. Crop rotation significantly reduced the giant foxtail stand with maximum reduction in the soybean-wheat-corn rotation. This may be due to the allelopathic influence of wheat straw (35).

Gerig and Blum (13) studied the effects of phenolic acids (ferulic, caffeic, *p*-coumaric, *p*-hydroxybenzoic, protocatechuic, sinapic, syringic and vanilic) at concentrations of 0 to 0.5 moles /g soil on cucumber (*Cucumis sativus)* seedlings. The action of *p*-hydroxybenzoic acid on leaf expansion rates was inhibited by the presence of ferulic, *p*-coumaric and vanilic acids. Howcvcr, no othcr antagonisms or syncrgisms among the acids were reported.

A number of plant pests are suppressed by the incorporation of cruciferous material in the soil. Although this effect is attributed to the conversion of glucosinolates into toxic products, little is known about the degradation of glucosinolate in the soil (6). Soil amended with rapeseed (*Brassica napus* L.) meal produces isothiocyanate, which repels the late instar wire worms in bioassays. Hence the rapeseed meal is non-toxic, but the isothiocyanates in the seeds repels the insect.

2.2.4. *Soil Fertility*

The biologically fixed nitrogen (BNF) in legume and non-legume plants is released into the soil as allelochemicals (nitrogenous compounds, amides, amines, and amino acids). These allelochemicals are stimulatory and are absorbed by associated plant species in cropping mixtures/intercropping or in agroforestry systems and by succeeding crops in multiple cropping systems. Therefore, these allelochemicals increase the growth and yields in such cropping systems. Incorporation of green manure (*Arachis glabrata, Phaseolus vulgaris* and *Leucaena leucocephala*) resulted in substantial improvement in soil properties, soil respiration rate, soil microbial biomass and consequently substantial increase in yield of maize. This is due to the plant nutrients made available by the green manures, as well as the release of beneficial allelochemicals (24).

The efficiency of nitrogenous fertilisers is very low (<40%). In Mauritius about 80% of nitrogenous fertiliser applied to sugarcane fields is not recovered by the sugarcane plant. There is a risk of the leachates finally reaching the ground water, streams, lakes, rivers and the sea, where they may cause eutrophication. Allelopathic nitrification retarders, for example neem cake, can slow the rate of nitrification down. Neem cake applied to potted soil, at 10 to 30 kg/ha, inhibited nitrification significantly upto 80 days following incubation (25). Fresh neem leaves, when incorporated and incubated in an oxisol, substantially reduced the number of nitrifying organisms. The numbers of these microorganisms in the treated soil were significantly lower as compared to the untreated control.

Allelopathic activities are more important in soils of low fertility. Gerig and Blum (12) investigated the inhibitory effects of nitrogen and ferulic acid levels on growth of cucumber cv. Early Green Cluster seedlings grown in growth chambers at 5, 10, 15, 20 and 25 μg N/g in a Portsmouth B-1 horizon soil treated with ferulic acid (FA) at 0 or 10 μg/g soil. Both low N (5μg N/g soil) and FA treatments reduced shoot dry weight, rates of leaf expansion and increased the root : shoot ratio. High N treatments reduced shoot dry weight. Ferulic acid inhibited cucumber seedling growth over a range of N concentrations, suggesting that the allelopathic activity of phenolic acids may be important in both nutrient limiting and non-limiting soils for some species.

Kholderbarin and Oertli (23) investigated the effects of cotyledon powder from seeds of tea and several oats varieties rich in phenolic compounds, on biological oxidation of ammonium and nitrite to nitrate. Treating culture solutions with cotyledon powder resulted in rapid disappearance of ammonium and nitrite. Such losses were believed to be due to fixation of ammonium and volatilisation of nitrite by phenolic substances and reactions with other organic compounds present in cotyledon tissues.

McCarty *et al* (30) evaluated the effects of three phenolic acids, namely ferulic acid, caffeic acid and *p*-coumaric acid, on nitrite production by three genera of terrestrial autotrophic nitrifying organisms (*Nitrosospira*, *Nitrosomonas* and *Nitrosolobus* spp.) grown on a defined medium containing NH_4^+. They found that nitrite production by *Nitrosospira* was not inhibited by ferulic acid, caffeic acid or *p*-coumaric acid at concentrations of 10^{-6} or 10^{-5} M and was slightly inhibited at a concentration of 10^{-4}M. Ferulic acid caused inhibition at a still lower concentration of 10^{-3}M. They concluded that these allelochemicals did not significantly affect the soil microbes.

3. Allelochemicals in Soil

Soil is the store house of allelochemicals, their transformation and degradation products. This is became, the allelochemicals released through root exudates and biomass decomposition (largest source of allelochemicals in agriculture and foresty) are directly added to the soil. While those released as leachates during precipitation (rainfall, snow fog, mist, dew) or as volatiles also comes directly or indirectly to the soil. In the soil, these allelochemicals under go different types of transformations (physical, chemical, microbial) and biodegradation etc. Some allelochemicals are lost through runoff or percolation losses to deep layers of soil with rain or irrigation water. However, very little information is available on these aspects. From the soil pool, allelochemicals are aborbed by plants and thus exert allelopathic effects in receiver plants.

Inoue *et al* (21) reported the presence of allelochemicals in root exudates of *Polygonum sachalinense* and found that they caused significant inhibition in growth of lettuce seedlings. Bioassay with 80% acetone extract of its rhizomes and roots showed that the inhibition was due to the presence of emodin and physcion (Anthraquinone compounds). Emodin also occurred in effective concentrations in the soil underneath this plant. The anthraquinone allelochemicals were also present in the soil. Yun and Kil (39) conducted field and laboratory studies to examine the differential phytotoxicity of residues of wormwood (*Artemisia princeps* var. *Orientalis)*, using its own plants as test species. Seedling elongation and dry weights of receptor plants were inversely proportional to concentration and incubation time of dry leaves of *A. princeps* in vermiculite. The elongation and dry weight of seedlings grown in soil from underneath wormwood plants were severely inhibited, suggesting that certain growth inhibitors were released from its plants and were present in the soil. Sasa (*Sasa cernua*) is a very serious weed. Li *et al* (28) studied its allelopathy using lettuce, wheat, timothy and green amaranth as test species. Cultured in the rhizosphere soil of *Sasa cernua*, the seedlings were inhibited by 42-80% as compared to controls cultured in normal soil and in vermiculite. The phenolic fraction from the rhizosphere soil of *S. cernua* caused significant inhibition in seed germination and seedling growth of lettuce, timothy, green amaranth and barnyard grass. HPLC and NMR identified *p*-coumaric, ferulic, vanilic and *p*-hydroxybenzoic acids and *p*-hydroxybenzaldehyde as the main allelochemicals at 5640, 1060, 860, 810 and 630 λg/100g soil, respectively, in the rhizosphere. Inderjit and Dakshini (18) isolated formononettin 7-O-glocoside (ononin), an isoflavanoid from soils infested with the perennial weed *Pluchea lanceolata*. Aqueous solution of this compound significantly inhibited the root and shoot growth of mustard and the level of inhibition was similar to that of hesperidin and taxifolin –3-arabinoside. Water soluble compounds synthesised by *P. lanceolata* and released into the soil significantly reduced seed germination, number of nodes, internode length, shoot and root length, nodule number and weight, chlorophyll a and chlorophyll b in asparagus bean plants. The concentrations of Mg^{++}, Zn^{++}, $PO4^{3-}$ were higher and K^{+} was lower, in shoots of the test plants grown with treated soil as compared to the untreated control. In roots the accumulation of Mg^{++} and NO_3^- was higher (19).

Under field conditions, *Euphorbia prostata* plants interfered with *Cynodon dactylon*. Soil collected from underneath *E. prostata* stands was very inhibitory to seed germination and seedling growth of *C. dactylon*. Aqueous extracts, decaying residues and root exudates of *E. prostata* were also found to be inhibitory to the test species (2). *Ailanthus altissima* contains one or more phytotoxic compounds in roots and leaves. Its powdered root bark and leaflets strongly inhibited the growth of garden cress (*Lepidium sativum*) when mixed with soil in petridishes. Pieces of root bark mixed with soil at 2,1 and 0.5 g /pot reduced cress biomass in the greenhouse, whereas, methanol-extracted root bark did not. The inhibitory effect of *Ailanthus* tissues in soil was short-lived. Fresh *Ailanthus* root segments in or on the soil surface reduced the growth of nearby cress seedlings. The results suggested that allelopathy caused by toxic exudation may contribute to the aggressiveness and persistence of *Ailanthus* in certain habitats (17). The effect of leachates of leaves and root/rhizomes of cogon grass on seed germination and seedling characteristics of radish, mustard, fenugreek and tomato were examined to asses the qualitative and quantitative differences in phytochemical components. The leachates and the soils from three sampling sites away from cogon grass were analysed by HPLC. There were no differences in nutrient availability, but qualitative and quantitative differences were found in the phenolic fractions (20).

Aerts *et al* (1) studied the inhibition of seed germination by quinoline alkaloids present in *Cinchona*. The alkaloids, when applied at concentrations higher than 0.3mM, strongly inhibited the germination of *Ocimum*, *Spermacoce* and *Catharanthus* and *Cinchona* itself. To test the allelopathic significance, the soil in which 2-year-old *Cinchona* plants were grown was tested for its quinoline alkaloid content. The plant roots contained high concentrations of these alkaloids (about 10mM), but very low concentrations (0.02mM) were found in the soil, indicating that the allelopathic interaction may not play a role under field conditions.

Chase *et al* (9) conducted transformation studies of BOA and AZOB in sterile and non-sterile soils and demonstrated that *Acinetobacter calcoaceticus*, a gram negative bacterium isolated from field soil, was responsible for the biotransformation of AZOB. Production of AZOB increased linearly with concentration of BOA in sterile soil, but showed a quadratic trend in non-sterile soil.

4. Conclusions and future areas of research

The principles of allelopathy have been applied in traditional agriculture since time immemorial, however, the formal study of allelopathy has gained interest in recent years. The increasing concerns about environmental pollution and contamination from pesticides have prompted the search for safer alternatives to present-day, high-input, unsustainable agricultural practices. Alternative natural means exist, such as allelopathy and organic farming, for safer crop protection, improved soil fertility and increased crop production, for instance plants with pesticidal properties, plants to improve soil processes such as nitrification, plant extracts or leachates to increase efficiency of nitrogenous fertilisers by inhibiting nitrification.

The challenge facing us in the coming millennium is the development of eco-friendly agricultural practices and sustainable means of increasing crop production to feed today's global population, while preserving the environment and especially the soil, in a condition that will continue to feed future generations.

Information on allelopathy for soil management is relatively scanty. Research on allelopathic interactions in soil requires the collaboration among soil scientists, agronomists, entomologists, plant pathologists and other plant scientists and organic chemists etc. Beneficial allelopathic relationships in soil, such as weed control, nitrification inhibition, nitrogen fixation must be exploited and integrated in a holistic agricultural/food production systems. For modern agriculture to be sustainable, natural methods of fertilisation, weed control, crop improvement must be given due importance. Although the effects of these beneficial interactions are slow and not very dramatic, the long-term effects can have a major influence on the sustainability of the system.

References

1. Aerts, R.J., Snoeijer, W., Meijden, E. and van der, Verpoorte, R. (1991). Allelopathic inhibition of seed germination by *Cinchona* alkaloids. *Phytochemistry* **30** : 2947-2951.
2. Alsaadawi, I.S., Sakeri, F.A.K. and Al-Dulaimy, S.M. (1990). Allelopathic inhibition of *Cynodon dactylon* L Pers and other plant species by *Euphorbia prostata* L. *Journal of Chemical Ecology* **16** : 2747-2754.
3. Anaya, A.L., Calera, M.R., Mata, R., and Pereda-Miranda, R. (1990). Allelopathic potential of compounds isolated from *Ipomea tricolor* cav. (Convulvulaceae). *Journal of Chemical Ecology* **16** : 2145-2152.
4. Blum. U., Gerig, T.M., Worksham, A.B. Holopper, L.D. and King, L.D. (1992). Allelopathic activity in wheat conventional and wheat no-till soils : development of soil extract bioassays. *Journal of Chemical Ecology* **18** : 2191-2221.
5. Brede, A.D. (1991). Field apparatus for testing allelopathy of annual bluegrass on creeping bent grass. *Crop Science* **31** : 1372-1374.
6. Brown, P.D., Morra, M.J., Mccaffrey, J.P., Auld, B.L. and Williams, L. (1991). Allelochemicals produced during glucosinolate degradation in soil. *Journal of Chemical Ecology* **17** : 2021-2034.
7. Cast, K.G., Mcpherson, J.K., Pollard, A.J., Krenzer, E.G. and Waller, G.R. (1990). Allelochemicals in soil from no-tillage versus conventional tillage wheat (*Tritum sativum*) fields. *Journal of Chemical Ecology* **16** : 2227-2289.

8. Chase, W.R., Nair, M.G. and Putnam, A.R. (1991). 2,2`-oxo-1,1`-azobenzene : selective toxicity of rye (*Secale cereale*) allelochemical to weed and crop species. *Journal of Chemical Ecology* **17** : 9-19.
9. Chase, W.R., Nair, M.G., Putnam, A.R. and Mishra, S.K. (1991). 2,2$^{/}$ -Oxo 1,1$^{/}$ - azobenzene : microbial transformation of rye *Secale cereale* allelochemical in field soils by *Acinetobacter calcoaceticus*. *Journal of Chemical Ecology* **17** : 1575-1584.
10. Chou, C.H. and Liu, L.L. (1992). Allelopathic substances and interactions of *Delonix regia*. *Journal of Chemical Ecology* **18** : 2285-2303.
11. Ells, J.E. and Mcsay, A.E. (1991). Allelopathic effects of plant residues on emergence and growth of cucumber seedlings. *Applied Horticultural Science* **26** : 368-370.
12. Gerig, T.M. and Blum, U. (1990). Effects of soil nitrogen and ferulic acid and inhibition of cucumber seedlings. *Journal of Chemical Ecology* **16** : 1371-1383.
13. Gerig, T.M. And Blum, U. (1991). Effects of 4 phenolic acids on leaf area expansion of cucumber seedlings grown in a Portsmouth B soil. *Journal of Chemical Ecology* **17** : 29-40.
14. Harrison, H.F. and Peterson, J.K. (1991). Evidence that sweet potato (*Ipomea batatas*) is allelopathic to yellow nutsedge (*Cyperus* exulentus). *Weed Science* **39** : 308-312.
15. Hegazy, A.K., Mansour, K.S. and Abdel-Hady, N.F. (1990). Allelopathic and autotoxic éffects of *Anastatica hierochuntica*. *Journal of Chemical Ecology* **16** : 2183-2193.
16. Hegde, R.S. and Miller, D.A. (1990). Allelopathy and autotoxicity in alfalfa : characterisation and effect of preceding crops and residue incorporation. *Crop Science* **30** : 1255-1259.
17. Heisey, R.M. (1990). Evidence for allelopathy by tree-of-haven (*Ailanthus altissima*). *Journal of Chemical Ecology* **16** : 2039-2055.
18. Inderjit and Dakshini, K.M.M. (1992a). Formononetin –7-O-glucoside (ononin), an additional growth inhibitor in soils associated with the weed *Pluchea lanceolata*. *Journal of Chemical Ecology* **18** : 713-718.
19. Inderjit and Dakshini, K.M.M. (1992b). Interference potential of *Pluchea lanceolata* growth and physiological responses of asparagus bean (*Vigna unguiculata)*. *American Journal of Botany* **79** : 979-981.
20. Inderjit and Dakshini, K.M.M. (1992c). Investigation on some aspects of chemical ecology of cogon grass (*Imperata cylindrica)*. *Journal of Chemical Ecology* **17** : 343-352.
21. Inoue, M., Nishimura, H., Li, H.H. and Mizutami, J. (1992). Allelochemicals from *Polygonum sachalinense*. *Journal of Chemical Ecology* **18** : 1833-1840.
22. Kalburtji, K.L. and Mosjidis, J.A. (1992). Effects of *Serica lespedeza* residues on warm season grasses. *Journal of Range Management* **45** : 441-442.
23. Kholdebarin, B. and Oertli, J.J. (1992). Allelopathic effects of plant seeds on nitrification effects and ammonium oxidisers. *Soil Biology and Biochemistry*, **24** : 59-64.
24. Lalljee, B. and D'Costa, V.P. (1997).Effect of green manures on soil fertility. *Sols Africain/African Soils* **28.** *Proc. 3rd African Soil Sc. Soc. Conf.*, Ibadan, Nigeria, 21-26 Aug., 1995 : 441-449.
25. Lalljee, B. (1998). Effect of neemcake on nitrification inhibition in an oxisol of Mauritius. In : *Azadiracta indica* A. Juss. (Eds., R.P. Singh and R.C. Saxena). New Delhi : Oxford and IBH Publishing Co. pp. 291-301.
26. Lalljee, B. and Facknath, S. (1998). Nutrient cycling in forest ecosystems of Mauritius : rate of nitrification. *Proc. 3rd Annual Meeting of Agricultural Scientists, Mauritius, Nov. 1998.* Mauritius : Food and Agricultural Research Council. In Press.
27. Lalljee, B., Facknath, S. and Osman, A. (1998). Improvement of soil properties under long term *Leucaena leucocephala*. In : *Leucaena _ Adaptation, Quality and Farming Systems* (Eds., H..M. Shelton, R.C. Gutteridge, B.F. Mulen and R.A. Bray).Australian Centre for International Agricultural Research **86** : 178-180.
28. Li, H.H., Nishimura, H., Hasegawa, K. and Mizutani, J. (1992). Allelopathy of *Sasa cernua*. *Journal of Chemical Ecology* **18** : 1785-1796.
29. Matizha, W. and Dahl, B.E. (1991). Factors affecting weeping lovegrass seedling vigour on shinnery oak range. *Journal of Range Management* **44** : 223-227.
30. Mccarty, G.W., Bremmer, J.N. and Schmidt, E.L. (1991). Effects of phenolic acids on ammonia oxidation by terrestrial autrotrophic nitrifying organisms. *FENS Microbiology Letters - Federation of European Microbiological Societies*, **85** : 345-350.
31. Nair, M.G., Whitenack, C.J. and Putnam, A.R. (1990). 2,2$^{/}$ - Oxo-1,1$^{/}$-azobenzene : a microbially transformed allelochemical from 2,3-benzoxazolinone. *Journal of Chemical Ecology* **16** : 353-364.
32. Nilsson, M.C. and Zarkrisson, O. (1992). Inhibition of scots pine seedling establishment by *Empetrum hermaphroditum*. *Journal of Chemical Ecology*. **18** : 1857-1870.

33. Putnam, A.R., Nair, M.G. and Barnes, J. P. (1990). New directions in biological control : Alternatives for suppressing agricultural pests and diseases. (Eds., R.R. Baker and P.E. Dunn). *Proceedings UCLA Symposium on Molecular and Cellular Biology,* January 1989, Fresco, Colorado. pp. 317-322.
34. Schilling, D.G., Dusky, J.A., Mossler, M.A., Bewick, T.A. and Alendria, V.A. (1992). Allelopathic potential of celery residues on lettuce. *Journal of American Society for Horticultural Science* **117** : 308-312.
35. Schreiber, M.M. (1992). Influence of tillage, crop rotation and weed management on giant foxtail (*Setaria faberi*) population dynamics and corn yield. *Weed Science* **40** : 645-653.
36. Teasdale, J.R. (1993). Interaction of light, soil moisture and temperature with weed suppression by hairy vetche residue. *Weed Science* **41** : 46-51.
37. Warkentin, B.P. (1992). Soil Science for environmental quality : How do we know what we know ? *Journal of Environmental Quality,* **21** : 163-166.
38. Williamson, G.B., Obee, E.M. and Weidenhamer, J.D. (1992). Inhibition of *Schizachirium scoparium* by the allelochemical hydrocinnamic acid. *Journal of Chemical Ecology* **18** : 2095-2105.
39. Yun, K.W. and Kil, B.S. (1992). Assessment of allelopathic potential in *Artemisia principes* var. orientalis residues. *Journal of Chemical Ecology* **18** : 1933-1940.
40. Zackrisson, O. and Nilsson, M.C. (1992). Allelopathic effects of *Empetrum hermaphroditum* on seed germination of two boreal species. *Canadian Journal of Forest Research* **22** : 1310-1319.

5

Eco-friendly technologies for management of phytoparasitic nematodes in pulses and vegetable crops

VIJAYALAKSHMI MOJUMDER
Division of Nematology
Indian Agricultural Research Institute, New Delhi -110012, India

Abstract

The traditional use of neem is to apply its parts and products like leaf, cake etc. to the soil as amendments. But this requires large quantities. The alternate application methods have been developed in recent years. These are row, spot or nursery bed treatments either as soil amendments or as soil drenching where bulk quantifies may be reduced. Bare root dip treatments in water extracts of neem products reduced the nematode populations. This method and nursery bed treatment are useful in case of transplantable vegetable crops through these methods are to be standardized before any recommendation. Seed treatments with neem products are effective against various phytonematodes and these are also cost effective. This effective, economical and easy-to-handle seed treatment technology with neem products had been recommended for pulse crops (Mojumder and Mishra, 1997). *Tagetes* spp. is another plant which can be successfully used as inter-crop or alternative crop in the nematode infested field as its root exudates are effective against nematodes and also this method of management may prove economical to farmers around cities as this is a cash crop.

Key words: Allelochemicals, botanicals, crops, ecofriendly, intercropping, nematodes management, nursery, oilseed cakes, plants, root exudates.

1. Introduction

The eco-friendly approach of pest management is the order of the day and nematode management is no exception. Phytonematodes feed mainly on roots, reducing water and nutrient uptake by the host plant and thus cause yield losses. Their management is more challenging due to their subterranean nature and microscopic size. The use of botanicals as soil amendments for nematode management started in 1937, when addition of chopped pineapple leaves to the soil reduced root knot nematode (14). The neem (*Azadirachta indica*) plant has broad-spectrum uses for pesticidal, medicinal, industrial and other purpose (50). Its nematicidal value is well established (21, 22). Other plants with nematicidal potential are *Argemone mexicana, Calotropis gigantea, Cuscuta reflexa, Datura metel, Eclipta alba, Embelia ribes, Melia azadirach, Tagetes patula* and many de-oiled seed cakes (20, 32, 43, 59).

S.S. Narwal et al. (eds.), Allelopathy in Ecological Agriculture and Forestry, 59–69.

The parts or products of some plants, tested against nematodes and the cakes used for nematode management in India have been listed in Table 1 and nematicidal potential of *C. gigantea* is shown in Table 2.

Table 1. Important botanicals having nematicidal properties

Sr.No.	Botanical name	Nematode spp.	Reference
A. Botanicals			
1.	*Argemone mexicana* Linn.	*Heterodera cajani*	43
2.	*Azadirachta indica* A.Juss.	*Meloidogyne incognita* *Rotylenchulus reniformis* *H. cajani* etc	17, 33,56,59
3.	*Calendula officinalis* Linn.	*M. incognita*	10, 11
4.	*Calotrpis gigantea* Linn.	*M. incognita*	32,56,59
5.	*Cuscuta reflexa* Roxb.	*M. incognita*	32
6.	*Datura metel* Linn.	*M. incognita*	56,59
7.	*Eclipta alba* Hassk.	*M. incognita*	56,59
8.	*Embelia ribes* Burm.	*M. incognita*	56,59
9.	*Melia azadirach* Linn.	*M. incognita, R. reniformis*	51
10.	*Tagetes spp.* Linn.	*Pratylenchus spp*	62
B. De-oiled cakes			
1.	*Arachis hypogea Linn.*	*M. incognita* and other phytonematodes	18, 33, 58
2.	*Azadirachta indica* A. Juss.	*M. incognita,* *Rotylenchulus reniformis* and other phytonematodes	2, 17, 21, 22, 54
3.	*Brassica compestris* Linn.	*M. incognita*	33, 58
4.	*Calophyllum inophyllum* Linn.	*M. incognita*	9, 10, 11
5.	*Carthmus tinctorius* Linn.	*M. incognita*	9, 10, 11
6.	*Gossypium spp.* Linn.	*M. incognita* and other phytonematodes	18
7.	*Linum usitassimum* Linn.	Phytonematodes	54
8.	*Madhuca indica* J.F. Gmel.	*M. incognita*	9, 10, 11
9.	*Pongamia glabra* Vent.	*M. incognita* and other phytonematodes	18, 56, 61
10.	*Ricinus communis* Linn.	*M. incognita* and other phytonematodes	18
11.	*Sesamum indicum* Linn.	*M. incognita* and other phytonematodes	18
12.	*Shorea robusta* Gaertn.	*M. incognita*	9, 10, 11
C. Root exudates			
1.	*Azadirachta indica* A. Juss.	*M. incognita*	2, 3, 26, 46
2.	*Brassica compestris* Linn.	*M. incognita*	55, 62
3.	*Crotolaria juncea* Linn.	-	55, 62
4.	*C. spectabilis* Roth.	-	55, 62
5.	*Sesamum indicum* Linn.	-	55, 62
6.	*Tagetes erecta* Linn.	*Pratylenchus penetrans* and other phytonematodes	55, 62
7.	*T. patula* Linn.	*Pratylenchus penetrans* and other phytonematodes	55, 62

Table 2. Nematicidal potential of *Calotropis gigantea* against *Meloidogyne incognita*

Treatment	Concentrations	% mortality of juveniles		% reduction in hatching		% reduction*		Reference
		Exposure time		Exposure time		Exposure time		
		24 h	48 h	24 h	48 h	24 h	48 h	
Fresh flowers	S**	100	100	94	100	98	100	32
	S/2	100	100	89	99	95	100	
	S/4	100	100	82	97	94	98	
	S/8	88	95	66	94	52	94	
	S/16	60	70	63	83	29	45	
Dry flowers (powder)	S***	100	100	-				56, 59
	S/2	100	100	-		-		
	S/10	9	74					
	S/20	0	68	-				

* % reduction in penetration of juveniles from treated egg mass in tomato
** S = 100 g in 25 ml water, *** 25 g in 100 ml water.

2. Economical application methods

The traditional method of application of neem and other botanicals for controlling the phytoparasitic nematodes population in soil, is to use these either green or in dry form as amendments to the soil (2, 15, 16, 54). Satisfactory control of nematodes has been also recorded with de-oiled cakes of *Azadirachta indica* (neem), *Pongamia glabra, Madhuca indica, Carthamus tinctorius, Calophyllum inophyllum, Shorea robusta, Sesamum indicum, Arachis hypogea* (groundnut), *Ricinus communis* etc. and botanicals as soil amendments (9, 10, 12,18, 54, 56, 60, 61). The effective doses have been calculated either on the basis of percentage of soil weight [1-2 % (w/w) of organic material to the soil] or N requirement of the crop (1.5-2.0 t/ha for non-legumes). These approaches were neither practical nor economical, hence, alternate methods have been developed to reduce the doses of plant materials needed, particularly, for neem products (Table 3) as discussed in following section.

2. 1 NURSERY BED TREATMENT

This may be done either as soil amendment with neem products or as soil drenching with aqueous extracts at the time of nursery bed preparation. The soil amendment @ 0.25-0.50 % (w/w) of soil with powdered neem seed, neem cake and Achook® (a neem-based pesticide) reduced the root knot nematode incidence when seedlings of brinjal from treated nursery were transplanted in the infested soil (29). The treatment of nursery bed with neem and *Lucaenea lucocephala* leaves effectively managed the *Meloidogyne javanica* in tomato (13). Seedbed treatment with neem cake powder reduces the root knot nematode incidence in tomato in pot trial (Varela, Personal Commun.).

Soil drenching with 5 % (w/v) aqueous extracts of neem seed kernel and neem seed coat reduced root knot galling in chickpea (40). Higher doses of aqueous extracts of neem seed kernel and neem seedcoat in split application were also effective against *M. incognita* in chickpea (38). Aqueous extracts of neem leaves also reduced galling of *M. javanica* in chickpea (49). The use of neem products for drenching entire field soil with irrigation water in the entire field is not economical, however, this method may be used in the nursery beds.

Table 3. Economical methods of application of neem products

Application method	Effective Dose	Crop and test nematodes	Reference
1. Nursery bed treatment			
a. Soil amendment	Neem seed powder, neem cake etc. @ 0.25 – 0.5% of soil (125 –250 g /m^2).	*Meloidogyne incognita* in brinjal (*Solanum melongena*).	29
b. Soil drenching	-	-	
2. Row / spot treatment	15 g per spot or 100 g per furrow / m^2.	*M. incognita* in tomato (*Lycopersicon esculentum*) and brinjal.	7
3. Bare root dip treatment	2-3 h in 1-5 % w/w aq. extracts of neem seed, neem cake etc.	*M. incognita* in tomato and brinjal.	27, 29, 58
4. Seed treatments (Seed sown crops) a. Seed soaking	6 h in 12.5 – 25 % w/v of aq. extracts neem seed, neem seed kernel, neem seed coat, neem cake etc.	Mungbean (*Vigna radiata*) and chickpea (*Cicer arietinum*) against *M. incognita* and *Rotylenchulus reniformis*. In pigeon pea (*Cajanus cajan*) against *Heterodera cajani*.	24, 28, 34, 37
b. Seed coating	10–20 % w/w of powdered neem formulations and 5 % v/w of liquid neem formulations	In all important pulse crops mungbean, chickpea, cowpea (*Vigna unguiculata*), pigeon pea etc. against *M. incognita*, *R. reniformis*, *H. cajani* and other phytoparasitic nematodes.	23, 25, 30, 31, 41, 42, 44, 45

2. 2. ROW OR SPOT TREATMENT

Row or spot treatments also reduces the quantum of the organic matter used as soil amendment or for soil drenching. The application of non-edible cakes reduces the nematode population and improves the plant growth. The application of neem cake @ 15 g per spot or 100 g per furrow/m^2 three weeks prior to transplanting of tomato/ brinjal reduces the gall index to 2.3 -3.0 in treated against 4.23 in untreated plants (7).

2. 3. BARE -ROOT DIP TREATMENT

Bare root dip treatment in aqueous extracts of neem products or other oil seed cakes reduces nematode population in transplanted crops like tomato and brinjal. The dipping of roots of tomato seedlings in 10 and 20 % (w/v) aqueous extracts of neem, mustard, groundnut, *Madhuca indica* and sesame cakes reduces the number of galls by

Meloidogyne incognita and its soil population (Table-4) (58). There is reduction in root

Table 4. Root dip treatments of tomato seedlings in aqueous extracts for reduction in root knot incidence

Treatments	Concentration	Percent reduction	
		No. of galls	Soil population
Arachis hypogea	S	63	31
	S/2	53	23
Azadirachta indica	S	70	69
	S/2	60	57
Brassica rapa	S	80	69
	S/2	75	58
Madhuca indica	S	64	64
	S/2	56	42
Sesamum indicum	S	66	46
	S/2	59	36

Source : 58. S=25 g in 100 ml water

knot incidence, when infested seedlings were dipped in aqueous extracts of neem cake for different periods (19). A significant control of *M. javanica* was found with neem extracts of neem cake for different periods (19). A significant control of *M. javanica* with neem cake extract has been reported in tomato and eggplant (1). Bare root dip treatments in aqueous extracts of neem leaf also reduces the root knot nematode in brinjal and root knot, reniform and stunt nematodes in other vegetable crops (12, 51). The neem seed kernel, neem cake, margocide and neem oil have been used for root dipping of various crops (4, 47). Root dip in 1-5 % (w/v) aqueous extracts of neem seed, neem cake and Achook®, reduces root knot nematode galling, nematode population in roots and soil and significantly increases the plant growth of tomato (28) and brinjal in pot experiments (29). In the root dip treatments, time of exposure as well as the concentration of the extract are very crucial. After judging the efficacy, treatment at lower concentration for longer period is desirable.

2. 4. SEED TREATMENTS

Seed treatment with crude and commercial neem formulations have proved effective against important phytonematodes in seed sown crops. For pulse crops, a 3-E Technology (Effective, Economical and Easy-to-handle) package of seed treatments with eco- friendly neem has been proposed (42). Seed treatments are done in following ways.

2.4.1 Seed soaking

This is done in aqueous extracts of crude neem products or in water soluble portions of neem based formulations. The seed is treated in bulk and must be sown immediately. The concentration of extracts and treatment (dipping) time is critical. Seed soaking for 3-6 h in aqueous extracts of neem products @ 25 % (w/v) or even lower concentrations for more time were effective against *M. incognita* in mungbean and *M. incognita* and *R.*

reniformis in chickpea and against *Heterodera cajani* in pigeon pea (24, 27,33, 34, 37, 41). Seed soaking of chickpea in extracts of fresh leaves of neem, *Datura metel, Argemone mexicana, Calotropis gigantea* etc. and in extracts of de-oiled cakes of neem, groundnut, *Pongamia glabra*, mustard etc. have also been found effective against *Meloidogyne incognita* (32). Boiled leaf extracts of neem are also used in Nigeria for crops like maize (6). Neem oil and other liquid formulations of the neem have successfully been used against M. *incognita* in tomato and eggplant (5, 47). Some experimental results are presented in Table 5 and 6.

Table 5. Seed soaking of chickpea in aqueous extracts of oil seed cakes and plant leaves against juvenile penetration of *Meloidogyne incognita*

Treatments	Concentrations (S= 25 g/ 100 ml water)	% reduction in penetration of *M. incognita* juveniles
	Oil-seed cake	78
Arachis hypogea	S	73
	S/2	73
Azadirachta indica	S	62
	S/2	67
Brassica rapa	S	62
	S/2	67
Pongamia glabra	S	67
	S/2	
	Plant leaves	
Argemone mexicana	S	88
	S/2	88
Azadirachta indica	S	88
	S/2	86
Cannabis sativa	S	100
	S/2	93
Datura metel	S	96
	S/2	93
Eclipta alba	S	88
	S/2	88

Source : 33. *S = 25g/ 100 ml water, ** S = 100g fresh material/ 25 ml water

2.4.2. Seed coating

Seed coating with crude or formulated neem products were found effective in reducing penetration, development, multiplication etc. of phytonematodes and thereby reduced nematode population and increased crop yields (Table 7). Seed coating of pulse crops was done using gum as adhesive and chalk powder as drier, while in liquid forms only drier, was used. We have determined the dose, method of coating etc. and effects on mungbean, chickpea, cowpea and pigeonpea in greenhouse and field conditions (42). The effective dose against phytonematodes was found to be 10-20 % (w/w) for

Table 6. Seed soaking of pulses in aqueous extracts of neem products for nematode management

Neem product	Dose	Per cent reduction			Reference
		No. of galls/ plant	Soil nematode population: Root knot	Soil nematode population: Reniform	
		Chickpea (*M. incognita, Rotylenchulus reniformis*)			
Neem cake	S*	74	68	60	24
	S/2	32	55	38	
Neem seed kernel	S	74	71	54	
	S/2	47	68	31	
Neem seed coat	S	32	68	31	
	S/2	21	45	9	
		Pigeonpea *(Heterodera cajani)*			
Neem seed	S*		64**		28
	S/2		48		
Neem seed kernel	S		79		
	S/2		61		
Neem seed coat	S		49		
	S/2		41		
Neem cake	S		73		
	S/2		59		
Achook®	S		46		
	S/2		42		

* S= 25g/100 ml water, ** % reduction in number of cysts per plant

powdered forms and 5 % (v/w) for liquid forms. We first reported that seed coating of mungbean with neem oil and neem cake reduced nematode penetration (57). Later, the efficacy of neem products in seed coating of pulse crops has been reported (23, 25, 30,31, 35, 36,39, 41, 44, 45). In transplanted vegetable crops (tomato, brinjal, cabbage and cauliflower) also seedcoating proved effective against *M. incognita* and *R. reniformis* (51-53). Seed coating of wheat with neem seed kernel reduces *Anguina tritici* juveniles reaching the growing point (8). In groundnut also, seed treatments with neem seed kernel reduces penetration of juveniles and gall formation of *M. arenaria* (48).

2. 5. INTERCROPPING

Intercropping of neem seedlings with chickpea under field conditions reduces penetration and galling of *M. incognita,* reduces plant parasitic nematodes in mungbean and chickpea and also improves their plant growth (26, 46). In field studies (4 m^2 plots) neem intercropping with two varieties of chickpea considerably reduces plant parasitic nematodes. A reduction in *Tylenchorhynchus brassicae* in tomato, eggplant, cabbage and cauliflower has also been reported (3).

Some other plants like *Tagetes* spp., *Sesamum indicum, Brassica compestris* etc. are also known to secrete root exudates which are toxic or repellent to plant parasitic

Table 7. Effect of seed coating with neem products on plant parasitic nematodes (PPN) and grain yield

Neem Products	Dose (% w/w)	% reduction in total PPN (%)	Increase in grain yield (%)
	Mungbean (*Vigna radiata*)		
Seed kernel	20	71	107
Seed cake	20	58	86
Seed coat	20	42	56
Achook ®	20	51	68
Neemark ®	5*	37	40
Nimbecidine ®	5*	31	38
	Chickpea (*Cicer arietinum*)		
Seed kernel	20	55	126
Seed cake	20	50	118
Seed coat	20	24	51
Achook ®	20	40	56
Neemark ®	5*	26	47
Nimbecidine ®	5*	24	45
	Cowpea (*Vigna unguiculata*)		
Seed kernel	20	65	123
Seed cake	20	58	108
Seed coat	20	42	80
Achook ®	20	38	83
Neemark ®	5*	33	66
Nimbecidine ®	5*	28	56
	Pigeonpea (*Cajanus cajan*)		
Seed kernel	20	60	53
Seed cake	20	52	46
Seed coat	20	37	21
Achook ®	20	48	28
Neemark ®	5*	35	12
Nimbecidine ®	5*	31	6

Source : 42. * % (v/w), PPN = Population.

nematodes (62). Among these plants, *Tagetes* spp. has been successfully used as intercrop or alternate crop in crop rotation for nematode management. Its root exudates contain α-terthinyl, an antagonist to nematode invasion (55). The inclusion of *Tagetes* spp. in cropping systems (intercropping/crop rotation) may be more economical, as it is a cash crop in urban areas. Two other crops viz. sesame and mustard, used for intercropping since ancient times are also known to release nematode repellent root secretions (Table 1). *Crotolaria* spp. is a trap crop which allows nematode penetration but restrict their further development. Unfortunately, use of these crops for nematode management has not been fully developed. It need to be explored on priority to reduce or minimise use of nematicides.

3. Conclusions

The use of antagonistic crops, trap crops and resistant crops for the nematode management depends upon the allelochemicals of the root exudates. Neem and other botanicals as soil amendments release certain chemicals in the soil, which may influence the nematode population. In technologies described (nursery bed treatments, bare root dip treatments, seed treatments), there may be direct as well as indirect action of chemicals against nematodes. In these cases, the neem chemicals might be absorbed by plant roots or seeds and these in turn may make the host crop resistant to the nematode parasites. Despite the interesting results with these application methods, the concept of induced resistance due to treatments with neem is still hypothetical. Although, the nematode management with these eco-friendly technologies may become effective and economical, but further research is needed to make farmer-friendly packages of eco-friendly technologies.

References

1. Abid, M. and Maqbool, M. A. (1991). Effect of bare root dip treatment in oil cake and neem leaf extract on the root knot nematode development and growth of tomato and eggplant. *Pakistan Journal of Nematology* **9**: 13- 16.
2. Alam, M. M. (1993). Bioactivity against phytonematodes. In *Neem Research andDevelopment* (Eds., N.S. Randhawa, and B. S. Parmar). pp. 123-143. New Delhi : Society of Pesticide Science.
3. Alam, M.M., Saxena, S. K. and Khan, A.M. (1977). Influence of interculture of marigold and margosa with vegetable crops on plant growth and nematode population. *Acta Botanica Indica* **5**: 33- 39
4. Banu, J.G. and Vadivelu, S. (1990). Efficacy of seed soaking with nematicides or botanicals for the control of *Meloidogyne incognita* infesting tomato. *Indian Journal of Nematology* **25**: 108-109.
5. Dash, B. K. and Padhi, N. N. (1990). Response of root knot nematode infesting tomato seed treatments by some neem products. *Orissa Journal of Agricultural Research* **2**:148-150
6. Egunjobi, O. A. and Onayemi, S. O. (1981). The efficacy of water extracts of neem *(Azadirachta indica)* leaves as a systemic nematicide. *Nigerian Journal of Plant Protection* **55**: 70-74.
7. Gill, J. S. and Jain, R. K. (1995). Nematode problems of vegetable crops in India. In *Nematode Pest Management: An Appraisal of Eco-friendly Approaches* (Eds., G. Swarup, D. R. Dasgupta, and J. S. Gill,), pp.166-178. New Delhi: Nematological Society of India.
8. Gokte, N. and Swarup, G. (1988). Effect of neem (*Azadirachta indica*) seed kernel powder treatment on wheat. *Indian Journal of Nematology* **18**: 149.
9. Goswami, B. K. and Vijayalakshmi, K. (1986). Nematicidal properties of some indigenous plant materials against root-knot nematodes, *Meloidogyne incognita* on tomato. *Indian Journal of Nematology* **16**: 65-68.
10. Goswami, B. K. and Vijayalakshmi, K. (1986). Efficacy of some indigenous plant materials and non-edible oil seed cakes against *Meloidogyne incognita* on tomato. *Indian Journal of Nematology* **16**: 280-281.
11. Goswami, B. K. and Vijayalakshmi, K. (1986). Effect of some indigenous plant materials and oil cake amended soil on the growth of tomato and root knot nematode population. *Annals of Agricultural Research* 7: 363-366.
12. Hussain, S., Kumar, R., Khan, T. and Titoo, A. (1984). Effect of root dip of seedlings with plant extracts nematicides, oilcake extracts and anthelmenthic drugs on plant growth and root knot development. *Pakistan Journal of Nematology* **2**: 79-83.
13. Jain, R.K. and Bhatti, D. S. (1988). Effect of degradation of neem leaves on incidence of root knot nematode in tomato. *International Nematology Network Newsletter* **55**: 7-9.
14. Linford, M. B. and Oliveira, J.M. (1937). Stimulated activity of natural enemies of nematodes. *Science* **85**: 123-124.
15. Mishra, S. D. and Mojumder, V. (1994). Management of plant parasitic nematodes through soil amendments of soil. In *Nematode Pest Management in Crops* (Eds., D.S. Bhatti and R.K. Walia) Pp. 132-149. Delhi: C.B.S. Publishers.
16. Mishra, S. D. and Mojumder, V. (1995). Soil amendments in nematode management. In *Nematode Pest Management: An Appraisal of Eco-friendly Approaches* (Eds., G. Swarup, D. R. Dasgupta and J. S. Gill). pp.106-114. New Delhi : Nematological Society of India.
17. Mishra, S. D. and Mojumder, V. (1996). Toxic behaviour of aqueous extracts of neem products on soil and plant nematodes. *Current Nematology* **7**: 13-15.
18. Mishra, S. D. and Prasad, S. K. (1974). Effect of soil amendments on nematodes and crop yield. I. Oil seed cakes, organic matter, inorganic fertilizers and growth regulators on nematodes associated with wheat and their residual effect on mung. *Indian Journal of Nematology* **4**: 1-19.

19. Mishra, S. D. and Prasad, S. K. (1977). Effect of neem cake water extract on root knot nematode infesting tomato. *Indian Journal of Entomology* **39**: 227-231.
20. Mojumder, V. (1994). Botanicals - An alternative for nematode management. *Abstract. National Symposium on Pests of Agricultural Importance and Their Management*, (Eds., B. K. Dwivedi,G. Pande, S.M. Srivastava and A.K. Choube), pp 35. Allahabad, India : Bioved Research Society of India.
21. Mojumder, V. (1995). Nematoda, Nematodes. In *The Neem Tree: Azadirachta indica A. Juss. and Other Meliacious Plants; Source of Unique Natural Products for Integrated Pest Management Medicine, Industry and Other Purposes*, (Eds., H. Schmutterer) pp. 129-150. Weinheim, Germany: V.C.H. Publications.
22. Mojumder, V. (1997). Nematode management. In *Neem in Sustainable Agriculture* (Eds., S. S. Narwal, P. Tauro and S. S. Bisla), pp. 239-253. Jodhpur: Scientific Publishers.
23. Mojumder, V. (1998). Effect of seed coating of mungbean with crude neem products and neem based pesticidal formulations for the management of *Meloidogyne incognita* and *Rotylenchulus reniformis. IARI Neem Newsletter* **15**: 12-13.
24. Mojumder, V. (1998). Effect of seed soaking in aqueous extracts of neem seed kernel, neem seedcoat and neem cake against root knot and reniform nematodes infesting chickpea. *IARI Neem Newsletter* **15**: 14-15.
25. Mojumder, V. (1998). Seedcoating of pigeonpea with neem formulations for the management of *Heterodera cajani* and other phytoparasitic nematodes. *Abstracts, III International Congress on Allelopathy in Ecological Agriculture and Forestry*, held at Dharwad, (Eds., S. S. Narwal, C.J. Itnal, R.E. Hoagland, R.H. Dilday and M.J. Reigosa)p. 127. Hisar, India : Indian Society of Allelopathy.
26. Mojumder, V. (1998). Effect of neem seedlings on the major phytoparasitic nematodes associated with chickpea and grain yield of the crop. *Abstracts III International Congress on Allelopathy in Ecological Agriculture and Forestry*, held at Dharwad, Aug. 18 -21, 1998, (Eds., S. S. Narwal, C.J. Itnal, R.E. Hoagland, R.H. Dilday and M.J. Reigosa) p. 128. Hisar, India: Indian Society of Allelopathy.
27. Mojumder, V. (1998). Effect of root dip treatment of tomato in aqueous extracts of neem products for the management of *Meloidogyne incognita. Abstracts, National Symposium on Rational Approaches in Nematode Management for Sustainable Agriculture*, (Eds., S.C. Dhawan, K.K.Kaushal and S. Ganguly), held at Anand, November 1998, p 4.
28. Mojumder, V. (1998) Management of *Heterodera cajani* in pigeonpea through seed soaking in aqueous extracts of neem formulations. *Abstracts, National Symposium on Rational Approaches in Nematode Management for Sustainable Agriculture*, (Eds., S.C. Dhawan, K.K.Kaushal and S. Ganguly), held at Anand, November, 1998, p. 68.
29. Mojumder, V. (1999). Efficacy of neem formulations for the management of plant parasitic nematodes associated with vegetable crops. *Abstracts, World Neem Conference, NEEM'99*, held at University of British Columbia, Vancouver Canada, May 1999.
30. Mojumder, V. (1999). Effect of seed treatments of chickpea with crude and commercial neem based pesticides on multiplication of plant parasitic nematodes and crop yield. *International Journal of Nematology* : (In press).
31. Mojumder, V. and Basu, R. (1999). Effect of seedcoating of chickpea with neem based pesticidal formulations on the management of *Meloidogyne incognita.Indian Journal of Nematology* **29**: (In press).
32. Mojumder, V. and Goswami, B. K. (1987). Effect of aqueous extracts of *Madar (Calotropis gigantea)* and *Amarbel (Cuscuta reflexa)* on larval mortality, hatching from egg-masses and subsequent penetration into tomato roots. *Annals of Agricultural Research* **8**: 285 –289.
33. Mojumder, V. and Mishra, S. D. (1991). Nematicidal efficacy of some plant products and management of *Meloidogyne incognita* in pulse crops by soaking seeds in their aqueous extracts. *Current Nematology* **2**: 27 –32.
34. Mojumder, V. and Mishra, S. D. (1992 a). Effect of seed soaking in aqueous extracts of neem seed on germination of mungbean and penetration of second-stage juveniles of *Meloidogyne incognita* . *Annals of Agricultural Research* **13**: 297-299.
35. Mojumder, V. and Mishra, S. D. (1992 b). Management of root-knot nematode *Meloidogyne incognita* infecting chickpea with neem seed coat. *Annals of Agricultural Research* **13**: 388-390.
36. Mojumder, V. and Mishra, S. D. (1993). Management of nematode pests. In: *IARI Research Bulletin No. 40: Neem in Agriculture*, (Eds., B.S. Parmar and R.P.Singh), pp. 40-48. New Delhi: Indian Agricultural Research Institute.
37. Mojumder, V. and Mishra, S. D. (1993). Management of *Meloidogyne incognita* infesting mungbean by seed soaking in aqueous extracts of neem seed kernel and neem seed coat. *Current Nematology* **4**: 25-28.
38. Mojumder, V. and Mishra, S. D. (1993). Soil application of aqueous extracts of neem products for the management of *Meloidogyne incognita* in chickpea. *Current Nematology* **4**: 105-108.
39. Mojumder, V. and Mishra, S. D. (1994 a). Comparative efficacy of neem products and Marshal as seed treatment against *Meloidogyne incognita* infecting chickpea. *Current Nematology* **5**: 145- 148.

40. Mojumder, V. and Mishra, S. D. (1994 b). Effect of aqueous extracts of neem seed kernel and seed coat used as soil drenching against *Meloidogyne incognita* chickpea. *Current Nematology* **5**: 173 –175.
41. Mojumder, V. and Mishra, S. D. (1996). Management of *Meloidogyne incognita* infesting pulse crop through seed treatments with neem products. In *Neem and Environment* (Eds., R. P. Singh, M. S. Chari, A. K. Raheja and W. Kraus) Proceedings of World Neem Conference, 1993, pp. 669 –703, New Delhi: Oxford and IBH Publishing Co.
42. Mojumder, V. and Mishra, S.D. (1997). *Neem Products for Nematode Management* in *Pulse Crops.* Technical Brochure. pp. 11. New Delhi : Division of Nematology, Indian Agricultural Research Institute.
43. Mojumder, V., Mishra, S. D., Haque, M. M. and Goswami, B. K. (1989). Nematicidal efficacy of some wild plants against pigeonpea cyst nematode, *Heterodera cajani. International Nematology Network Newsletter* **6**: 21-26.
44. Mojumder, V. and Raman, R. (1999). Nematicidal efficacy of NEEMA-S1, an experimental formulation for seed treatments against *Heterodera cajani* and *Meloidogyne incognita* in cowpea and chickpea. In *Azadirachta indica, A. Juss.* (Eds., R. P. Singh and R.C. Saxena). *Proceedings of International Neem Conference*, 1996, Australia. pp. 217-222. New Delhi : Oxford & IBH Publishing Co.
45. Mojumder, V. and Raman, R. (1999). Comparative efficacy of neem based nematicidal products for the management of nematode pests of pulse crops. In: *Azadirachta indica, A. Juss.* (Eds., R. P. Singh and R.C. Saxena). *Proceedings of International Neem Conference*, 1996, Australia. Pp. 223-232, New Delhi : Oxford & IBH Publishing Co.
46. Mojumder, V. and Mohan, S. (1996). Use of neem seedlings for the management of plant parasitic nematodes infesting mungbean. *Abstracts. National Symposium on Prospects and Problems of Technology Transfer and Rural Reality*, (Eds., B.K. Dwivedi, G. Pandey, A.K. Pandey and A.K. Choube), p. 48, Allahabad, India : Bioved Research Society.
47. Pradhan, P., Das, S. N. and Dora, D. K. (1989). Effect of some organic oils on the infectivity of root knot nematode (*Meloidogyne incognita*) affecting tomato. *Indian Journal Nematology* **19**: 162-165.
48. Prasad, D. and Sethi, C.L. (1986). An approach to manage *Meloidogyne arenaria* in groundnut by seed treatment with neem seed kernel powder. *Abstract, National Conference, Plant Parasitic Nematodes of India: Problems and Progress,* (Eds., G. Swarup and D.R. Dasgupta), p. 91. New Delhi: Nematological Society of India.
49. Ram, K. and Gupta, D. C. (1980). A note on the efficacy of fresh neem leaf extract in the control of *Meloidogyne javanica* infesting chickpea *(Cicer arietinum). Indian Journal of Nematology* **10**: 96-98.
50. Schmutterer, H. (1995). *The Neem Tree, Azadirachta indica A. Juss., Source of Unique Natural Products for Integrated Pest Management, Medicine, Industry and Other Purposes.* Weinheim, Germany: VCH Publications, pp. 716.
51. Siddiqui, M. A. and Alam, M.M. (1988a). Control of root knot and reniform nematodes by bare root dip in leaf extracts of margosa and persian lilac. *Z. Pfllanzzenk Dflanzenschut* **95**: 138-142.
52. Siddiqui, M. A. and Alam, M.M. (1988b). Effect of seed treatments with azadirachtin on root knot development and on growth of some vegetables. *Annals of Applied Biology* **112** (Suppl.) **9**: 20-21.
53. Siddiqui, M. A. and Alam, M.M. (1990). Control of root knot, reniform and stunt nematodes by nimbin seed dressing. *Nematologia Mediterranea* **18**: 19-22.
54. Singh, R. S. and Sitaramaih, K. (1970). Control of plant parasitic nematodes with organic amendments. *PANS* **16**: 287-297.
55. Swarup, G. and Gokte, N. (1986). Biological control. In: *Problems and Progress* (Eds., G. Swarup and D.R. Dasgupta), pp. 476-489. New Delhi: Indian Agricultural Research Institute.
56. Vijayalakshmi, K. (1976). *Studies on the Effect of Some agricultural Chemicals and Soil Amendments on the Plant Parasitic Nematodes*. Ph. D. Thesis. New Delhi: Indian Agricultural Research Institute.
57. Vijayalakshmi, K. and Goswami, B. K. (1986). Effect of seed treatments with neem cake and neem oil on the germination of *moong* (greengram) and its vulnerability to root knot nematodes. *International Nematology Network Newsletter* **3**: 8-9.
58. Vijayalakshmi, K and Goswami, B. K. (1987). Effect of root dip treatments of tomato seedlings in aqueous extracts of some oil seed cakes on root-knot nematode infestation. *Annals of Agricultural Research* **8**: 168-171.
59. Vijayalakshmi, K., Mishra, S. D. and Prasad, S. K. 1979). Nematicidal properties of some indigenous plant materials against second-stage juveniles of *Meloidogyne incognita* (Kofoid and White) Chitwood. *Indian Journal of Entomology* **41**: 326-331.
60. Vijayalakshmi, K. and Prasad, S. K. (1979). Effect of some plant products and alum as soil amendments on *Meloidogyne incognita* on tomato. *Vegetable Science* **6**: 139-141
61. Vijayalakshmi, K. and Prasad, S. K. (1982). Effect of some nematicides, oilseed cakes and inorganic fertilizers on nematodes and crop growth. *Annals of Agricultural Research* **3**: 133-139.
62. Webster, J. M. (1972). Biological control. In: *Economic Nematology* (Ed., J. M. Webster), pp. 469-496. Academic Press: London.

SECTION II

ALLELOCHEMICALS AS HERBICIDES

6

Plant and microbial compounds as herbicides

ROBERT E. HOAGLAND* and STEPHEN J. CUTLER[1]
*USDA-ARS, Southern Weed Science Research Unit,
Stoneville, Mississippi 38776 USA*

Abstract

There are numerous reports of secondary compounds derived from plants and microorganisms that are phytotoxic. The most commercially successful of these allelochemic natural products are bialaphos and its active ingredient phosphinothricin, which led to the development of the herbicide glufosinate. There are many other natural products with herbicidal potential. Pyridazocidin, a phytotoxin produced by a soil *Streptomyces* species, causes rapid chlorosis and necrosis of several weeds. This is the first natural product that appears to act via reversible oxidation/reduction linked to photosynthetic electron transport, analogous to synthetic compounds (e.g., bipyridiniums). The natural product 3-nitropropionate (NPA) and its glycosides are found in several plants and in several fungi (phytopathogens and non-phytopathogens). NPA has been shown to be phytotoxic. Sorgoleone, isolated from grain sorghum root exudates, is phytotoxic to grass and broadleaf species and acts via inhibition of photosynthetic electron transport (PS II). Monensin, from *Streptomyces cinnamonensis*, is a carboxylic polyether antibiotic compound used in agriculture that also exerts herbicidal activity on a variety of plants, as does a related antibiotic, nigericin. Hydantocidin and cornexistin are new herbicidal compounds isolated from *Streptomyces hygroscopicus* and *Pacilomyces variotii*, respectively. Artemisinin, a plant (*Artemesia annua* L.) sesquiterpene with antimalarial properties, is also a potent phytotoxin. Some allelochemicals have sufficient potential to be herbicides, while others with lower herbicidal efficacy may serve as chemical templates for new synthetic herbicides. The continued search for allelochemicals from plants and microbes is important in the developement of improved herbicides with novel molecular modes of action. Selected allelochemicals with phytotoxic properties that have been isolated from plants and microbes are presented and discussed in this review.

Key words: Allelochemicals, herbicides, microbial compounds, microrganisms, plants, weeds.

1. Introduction

Plants and microbes produce a multitude of secondary compounds, some of which have been implicated in allelopathy. Many of these compounds have been tested and shown to possess

---

* Corresponding author
[1] Mercer University, Southern School of Pharmacy, Atlanta, Georgia 30341 USA

S.S. Narwal et al. (eds.), Allelopathy in Ecological Agriculture and Forestry, 73–99.

phytotoxicity and herbicidal activity. Plants may be expected to be a less rich source of herbicidal compounds than of allelochemicals that possess insecticidal or fungicidal activity. This is true because in nature, selection pressure from other plants is generally less than that from insects or pathogens. Defensive mechanisms have generally evolved against organisms that possess differing biochemistry and physiology. Futhermore, although many plants are known to produce phytotoxins, these allelochemicals are generally not potent herbicides.

There is great interest to discover efficacious herbicides that have low non-target toxicity and that are more environmentally friendly than some of the synthetic herbicides that have been found as contaminants in ground water and in ecosystems. Herbicides comprise the major portion of total pesticide sales (77) and are necessary to provide high quality food and fiber to an ever increasing world population. There is also an incentive to find new phytotoxins with novel chemistries and/or compounds that possess molecular modes of action that differ from those of commercial herbicides presently used. This latter reason stems from the fact that weeds have evolved resistance to many of the herbicides now used. There are over 200 weed biotypes from 125 different species that have become resistant to synthetic herbicides (65). The pesticide industry is interested in new naturally-occurring phytotoxins that can be directly used as herbicides, be used as templates for new herbicide classes, or that can provide information useful aspects of biorational design such as new target sites of action.

There are numerous reports of secondary compounds derived from plants and microorganisms that are phytotoxic. Several selected books and review articles are available (30, 31, 39-42, 70). The most commercially successful of these allelochemic natural products is phosphinothricin (PPT) (73, 76), the discovery of which led to the development of the herbicide glufosinate (9). There is also a variety of other compounds with herbicidal potential. Some allelochemicals have sufficient potential to become herbicides, while others with lower herbicidal efficacy may serve as chemical templates for new synthetic herbicides. Herein, we present an overview of some selected natural products from plants and microbes that exhibit herbicidal activity. We also attempt to show an array of different chemistries and modes of action, in instances where this parameter is known.

2. Allelopathic compounds from plants with herbicidal activity

2.1. ARTEMISININ [I]

Artemisia is the largest genus of the Asteraceae which belongs to the tribe Anthemideae. This genus contains several bitter and toxic compounds (84, 137). Annual wormwood (*A. annua*) is a medicinal plant of Chinese origin and leaves and flowers of this species contain artemisinin (an endoperoxide sesquiterpene lactone) that is used against *Pasmodium falciparum*, the causative agent of malaria (86). Artemisinin is also known for phytotoxic properties and inhibits seed germination and growth of various seedlings (12). Other related sesquiterpene lactones from *A. annua* have plant growth regulatory and allelopathic properties (27, 43, 102).

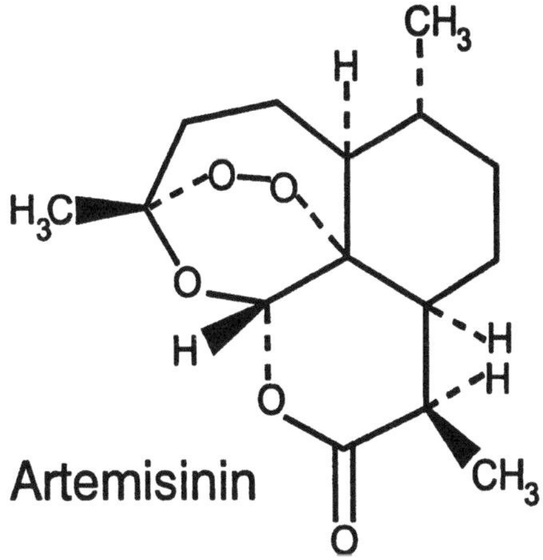

I. Artemisinin

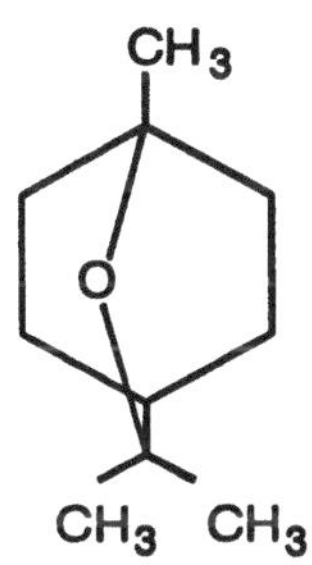

II. 1,8-Cineole

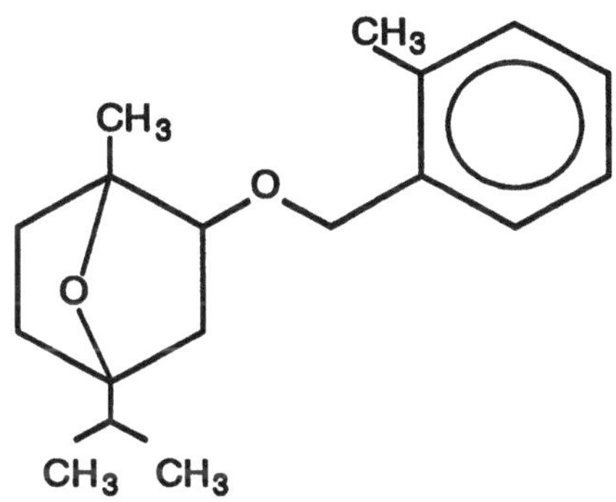

III. Cinmethylin

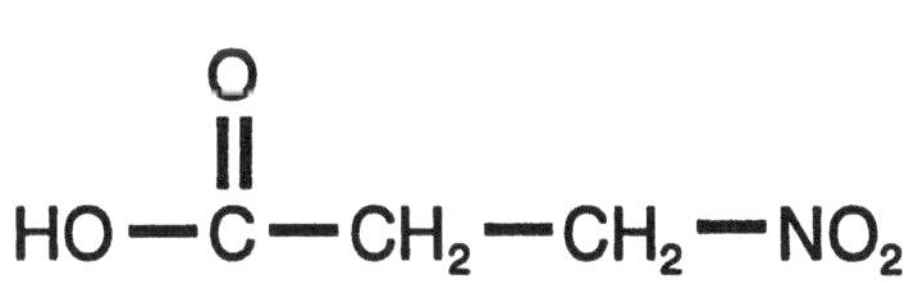

IV. 3-Nitropropionic acid

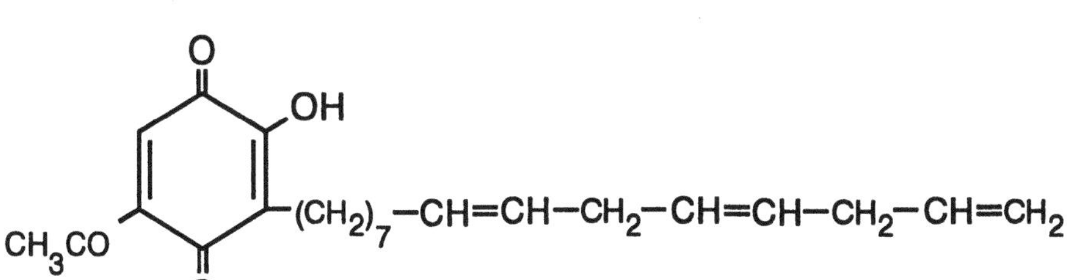

V. Sorgoleone

Arteether, obtained by etherification of dihydroartemisinin is a reduction product of artemisinin. Arteether was phytotoxic to several plant species at 1 ppm and at lower concentrations exhibited growth promoting effects in several plants (12). Radish (*Raphanus sativus*) and annual wormwood were highly tolerant to the phytotoxic effects of artemisinin and related compounds (12). Artemisinin exhibited autotoxic effects on annual wormwood at 33 ppm (43). Tissue localization of artemisinin in annual wormwood showed that artemisinin is present in foliar tissues localized entirely in the subcuticular space of capitate glands (38).

The mode of action of this selective phytotoxin is not fully understood, however, data suggest that it acts by a mechanism other than disruption of mitosis, alteration of amino acid biosynthesis, or inhibition of respiration (43). Although the antimalarial property of artemisinin in animal cell cultures is associated with large reductions in putrescine levels, plants treated with this compound displayed only slight reductions in polyamine levels (36). This suggested that the primary biochemical activity in plants is not blockage of the polyamine biosynthetic pathway. The effects of artemisinin on the plasma membrane of *Lemna* spp. also does not explain its effects on inhibition of germination and growth of various plant species (27). The complex structure of this natural product renders it unsuitable for use as a commercial herbicide.

2.2. CINEOLE [II]

One of the most potent phytotoxins from plants that was initially studied is 1,8-cineole. This compound is produced by many plant species and plays a role in allelopathic interactions in some plants (115). Although the compound has potent phytotoxicity, its very high volatility renders it unsuitable as a commerical herbicide. An Analogue of cineole was synthesized, i.e., cinmethylin **[III]** which went through some product development, but was ultimately not marketed as a commercial herbicide (60). The mode of action of compounds of this chemistry is not known. As in the case artemisinin which has structural characteristics similar to cinmethylin, no dramatic effect on polyamine biosynthesis was noted in plant tissues treated with cinmethylin (36).

2.3. NITROPROPIONIC ACID [IV]

The toxicity of some plant species such as *Astragalus*, *Indigofera* and *Hiptage* to animals has long been known (114). Some of the toxic principles in these species are nitro-compounds. Nitro-compounds can be very toxic, but *in vivo* toxicity in the producer can be sequestered by chemical linkages to glycosides. Hiptagenic acid was probably the first nitro-compound isolated from plants in 1920 and was later identified as β-nitropropionic acid (22). Miserotoxin [a poisonous glycoside in Wasatch milkvetch (150)] is metabolized to 3-nitropropanol by rumen of cattle and sheep (183). Miserotoxin is present in many species of *Astralagus* that occur in North and South American and Asia (180-182). Environmental factors such as temperature play a role in the amount of toxin production (129) and application of several herbicides to Wasatch milkvetch were found to lower the concentration of miserotoxin *in planta* (184). Nitro-propanol and nitro-propionic acid have been found as the free alcohol and acid, respectively (114).

β-Nitropropionic acid is also produced by several fungi, including *Aspergillus flavus, A. oryzae, A. wentii,* and *Penicillium atrovenetum* (45, 79). It was also found in a pathogen of *Zinnia* spp. and was implicated in disease symptomology (82). More recently, NPA was found to be produced by *Septoria cirsii*, a fungal pathogen of the weed Canada thistle (*Cirsium arvense* L.). In culture, this potential biocontrol pathogen of Canada thistle produced large amounts of this allelochemical. NPA was found to inhibit seed germination of Canada thistle (68). In leaf puncture tests with over 25 plants species, NPA caused lesions in several other plant species, but only 3 weed species tested: sowthistle (*Sonchus oleraceus*), common mallow (*Malva neglecta* Wallr.), and spurge (*Euphorbia* sp.). *Septoria cirsii*, a host-specific pathogen of Canada thistle, that causes leaf spot and is an effective biocontrol agent. *S. cirsii* was recently reported to produce 3-nitropropionic acid (NPA; also called β-nitropropionic acid) as a major phytotoxin. NPA inhibited germination and induced chlorosis and necrosis in this host. Furthermore, several other plant species were sensitive to NPA when it was applied to leaf puncture wounds. NPA is also reported to inhibit germination of red alder seeds (*Alnus rubra*) (45).

NPA has been examined for phytotoxicity to several other weed and crop species (71). Generally, NPA (2×10^{-4} to 5×10^{-3} M) was an ineffective germination inhibitor of 9 species of weeds and crops. Weak (−8 to 20%) inhibition was found in sicklepod (*Cassia obtusifolia* L.), cucumber (*Cucumis sativa* L.), wheat (*Triticum aestivum* L.), showy crotalaria (*Crotalaria spectabilis* Roth) and okra (*Hibiscus esculentus* L.); inhibition in light was slightly greater than in the dark. No significant germination effects were noted on hemp sesbania [*Sesbania exaltata* (Raf.) Rydb.], grain sorghum (*Sorghum bicolor* Pers.), radish (*Raphanus sativus* L.), or mung bean (*Vigna radiata* L.). Tests of NPA (10^{-4} to 8×10^{-3} M) with intact 4-d-old dark-grown seedlings indicated that foliar applications reduced shoot elongation 3 to 46% in mung bean, 13 to 40% in sicklepod, and 6 to 30% in hemp sesbania after 72 h. Root-feeding of these NPA concentrations reduced shoot elongation 4 to 94% in mung bean, 20 to 90% in hemp sesbania, and 0 to 55% in sicklepod. Root elongation was inhibited 100% when NPA (5×10^{-4} to 8×10^{-3} M) was root-fed to these three species. NPA at 10^{-4} M NPA inhibited root elongation 50 to 75%. Excised leaf segments of greenhouse-grown dandelion (*Taraxacum officinale* Weber), curly dock (*Rumex crispus* L.), and crimson clover (*Trifolium incarnatum* L.) exhibited much more chlorosis/necrosis after 24 and 48 h when exposed to NPA (10^{-3}M) in light (200 $\mu Em^{-2}s^{-1}$) than in darkness. In greening tests, NPA (10^{-3} M) completely blocked chlorophyll production when excised stems and cotyledons of 4-d-old dark-grown sicklepod and hemp sesbania were incubated in the light. Results indicate that NPA is an effective phytotoxin and that light plays a role in phytotoxic expression in some systems.

The mode of action of nitro-propionic acid is inhibition of succinate dehydrogenase, whereby the enzyme is irreversibly inhibited (7). However, the inactivation was not due to nucleophlic addition of the β-position of nitropropionate to the flavin acceptor. The actual inactivator has been shown to be the oxidation product of nitropropionate (29).

2.4. SORGOLEONE [V]

Use of sorghum as green manure or application of its residues to soil for weed control has been practiced for many years. Furthermore, various sorghum species have been shown to have allelopathic activity on both weed and crop species (44, 125). Sorgoleone was first reported by Netzley and Butler (121) to be a major constituent of root exudates of *Sorghum bicolor*. Since then, this compound has been shown to be a potent phytotoxin in a variety of species using several different bioassays.

Sorgoleone and two synthetic herbicides {metribuzin [4-amino-6-(1,1-dimethyl)-3-(methylthio)-1,2,4-triazin-5(4*H*)-one] and diuron [*Nr*-(3,4-dichlorophenyl)-*N,N*-dimethylurea]} inhibited competitive binding with atrazine [6-chloro-*N*-ethyl-*Nr*-(1-methylethyl)-1,3,5-triazine-2,4-diamine] in thylakoids of susceptible species, but no competitive binding in thylakoids of resistant species (176). The binding affinity of sorgoleone was intermediate compared to that of diuron and metribuzin (123). Sorgoleone incorporated into soil at rates above 40 ppm reduced shoot fresh weight, but did not affect root fresh weight. Mode of action studies of sorgoleone showed that it is a potent inhibitor of electron transfer between Q_a to Q_b at the reducing side of PSII (58, 124).

Sorgoleone also has residual activity in soil. Recovery of soil amended with sorgoleone was high (ca 85%) 1 h after application, using acetonitrile:water (80:20, v/v) as the extraction solvent. Although recovery declined to 45% at 24 h after treatment, recovery decreased with time after application, but the compound could still be extracted and detected 6 weeks later (176). Some sorgoleone-derived metabolites were also extracted and detected.

3. Allelopathic compounds from microorganisms with herbicidal activity

3.1. ANISOMYCIN [VI]

Anisomycin (from *Streptomyces* sp.) showed selective root growth inhibition at 12.5 ppm and shoot growth inhibition at levels above 50 ppm in monocotyledons [rice (*Oryza sativa* L.), barnyardgrass (*Echinochloa crus-galli*), crabgrass (*Digitaria adscendens*)] and dicotyledons [leucerne (*Medicago sativa*) and tomato (*Lycopersicon esculentum*)] (186). Another analogue, deacetyl anisomycin [VIII], inhibited shoot and root growth in rice and barnyardgrass more strongly than anisomycin. An analogue, *N*-acetyl anisomycin **[VII]**, showed little or no phytotoxicity at 100 ppm. The total synthesis of this antibiotic has been achieved (46). Derivations based on this structure led to diphenylmethane analogues and eventually to 4-methoxy-3,3'-dimethylbenzophenone, a selective herbicide (NK-049; Nippon Kayaku Co.) for rice (167). Anisomycin thus became the first microbial product that led directly to the development of a synthetically-based commercial herbicide.

3.2. CORNEXISTIN [IX]

Cornexistin was isolated from the culture broth of the basidiomycete *Paecilomyces variotii* SANK 21086 and the organism was isolated from deer dung (118). It is a

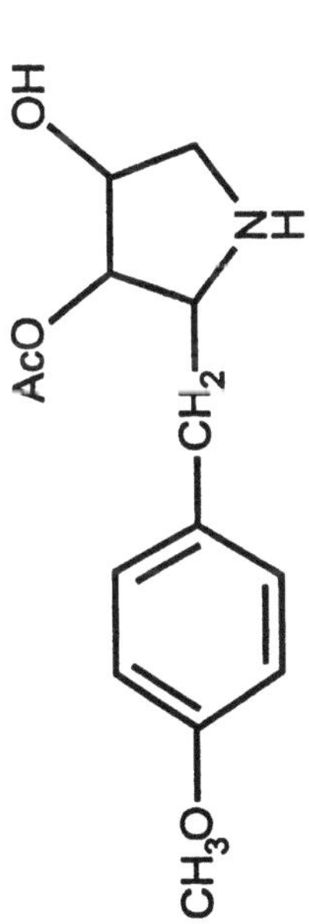

VI. Anisomycin

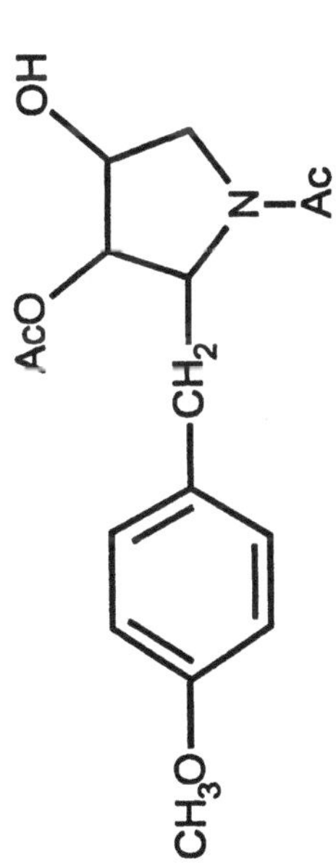

VII. N-Acetyl anisomycin

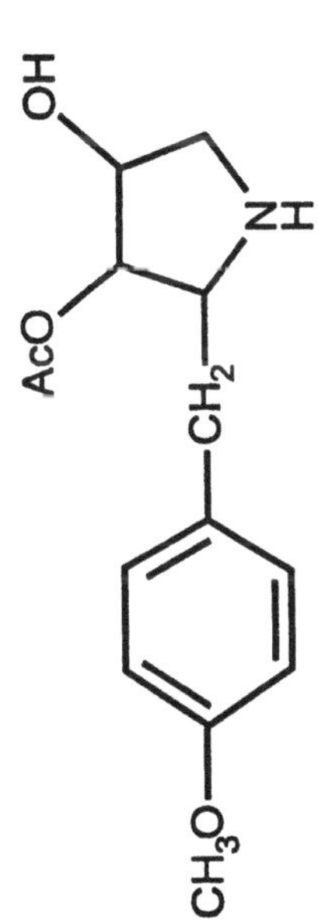

VIII. Deacetyl anisomycin

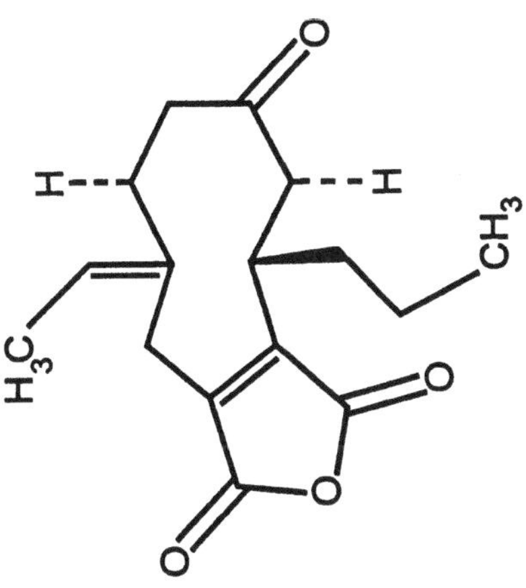

IX. Cornexistin

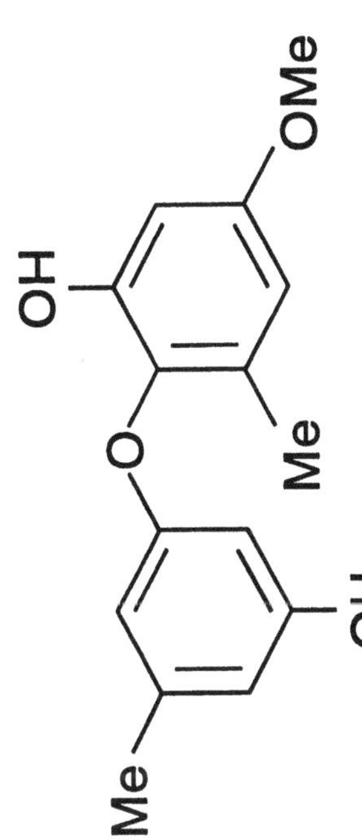

X. Cyperin

nonadride compound (nine membered carboxcyclic ring compound that has one or five-membered rings). It has excellent postemergence herbicidal activity against dicot and monocot weeds, but corn (*Zea mays* L.) is tolerant to the compound. In a duckweed (*Lemna pausicostata*) bioassay, cornexistin at 333nM caused strong growth inhibition within 72 h after exposure. The mechanism of action of cornexistin is unclear, but there is evidence that it acts on an enzyme of carbon metabolism and that it is a proherbicide whose active form inhibits an aspartate transferase isozyme (8).

3.3. CYPERIN **[X]**

Cyperin is a natural product isolated from fungal cultures of *Preussia fleischhakii* (174). *Ascochyta cypericola* (152) and *Phoma sorghina* (172). The latter two fungi are pathogens of purple nutsedge (*Cyperus rotundus)* and pokeweed (*Phytolacca americana*), respectively. This compound is a diphenyl ether and the pure compound was found to be highly phytotoxic to purple nutsedge (152) and several plant species (172). Other studies have shown that cyperin inhibited purple nutsedge root growth grown on agar, but not in soil. (64). This product also inhibited growth of mouse-ear cress (*Arabidopsis thaliana*) and marsh bentgrass (*Agrostis palustris*) grown in agar and was shown to have a mode of action different than the commercial diphenyl ether herbicides that inhibit protoporphyrinogen oxidase.

3.4. FUMONISINS **[XI]**

The fumonisin B mycotoxins are a group of structurally related metabolites produced by *Fusarium moniliforme* (Sheldon) (23), which is one of the most prevalent fungi associated with corn (*Zea mays* L.), a major food source for humans and animals (104). The major fumonisin is FB_1, which is produced under both natural conditions and in culture. Although *F. moniliforme* produces several toxins, only the fumonisins have been shown to produce symptoms of animal diseases such as pulmonary oedema and hydrothorax in swine and equine leukoencephalomalacia in horses (74). FB_1 is one of a series of long-chain polyhydroxyl alkylamines with propane tricarboxylic acid moieties (17). Other fumonisins include FA_1, FA_2, FB_2, FB_3 and FB_4. The initial interest in these compounds was their mammalian toxicity, but recently, their phytotoxicity was discovered (1-3). The phytotoxicity of FB_1 and other analogues has been examined and summarized (4, 5). The plant growth regulator activity of FB_1 has been demonstrated in the *de novo* induction of adventitious roots in excised shoots of tomatoes (11). Fumonisins are produced not only by *F. moniliforme* (5), but also by other members of the Liseola section of the *Fusarium* genus such as *F. proliferatum* and *F. nygamai* (120). FB_1 has also been isolated from *Alternaria alternata*, the cause of stem canker on certain varieties of susceptible tomatoes (26). The mode of action of the fumonisins is the inhibition of sphingolipid synthesis at the enzyme site, ceramide synthase (4).

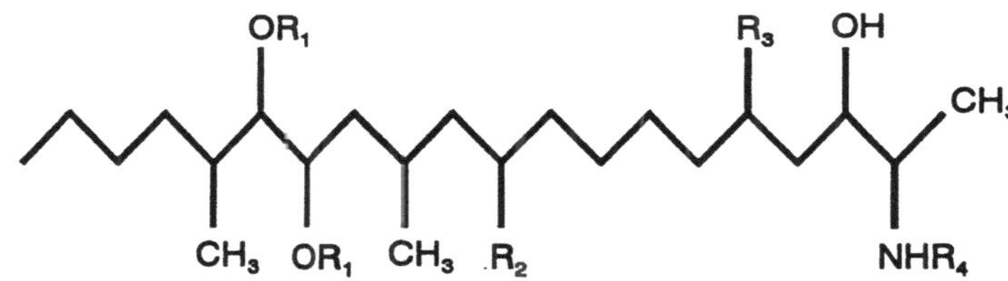

Toxin	R_1	R_2	R_3	R_4
FA_1	$CO\text{-}CH_2\text{-}CH(CO_2H)\text{-}CH_2CO_2H$	OH	OH	$COCH_3$
FA_2	$CO\text{-}CH_2\text{-}CH(CO_2H)\text{-}CH_2CO_2H$	H	OH	$COCH_3$
FB_1	$CO\text{-}CH_2\text{-}CH(CO_2H)\text{-}CH_2CO_2H$	OH	OH	H
FB_2	$CO\text{-}CH_2\text{-}CH(CO_2H)\text{-}CH_2CO_2H$	H	OH	H
FB_3	$CO\text{-}CH_2\text{-}CH(CO_2H)\text{-}CH_2CO_2H$	OH	H	H
FB_4	$CO\text{-}CH_2\text{-}CH(CO_2H)\text{-}CH_2CO_2H$	H	H	H
AP_1	H	OH	OH	H
AP_2	H	H	OH	H

XI. Fumonisins

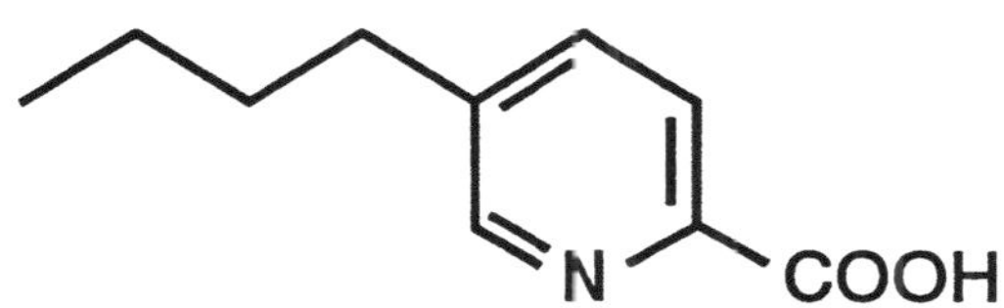

XII. Fusaric acid

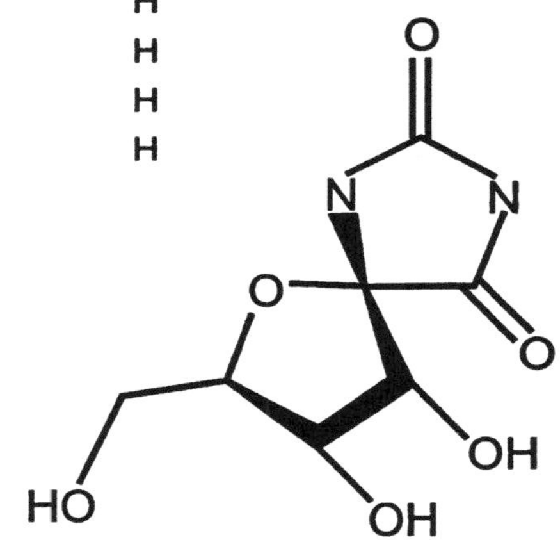

XIII. Hydantocidin

3.5. FUSARIC ACID [XII]

Fusaric acid (5-*n*-butyl picolinic acid) is produced as a cultural metabolite by several f. sp. of *Fusarium oxysporum*. This compound was first described as a wilt toxin from *F. oxysporum* f. sp. *lycopersici, F. oxysporum* f. sp. *vasinfectum* and *Gibberella fujikuroi* (54). Dehydrofusaric acid and some other related compounds are also produced with fusaric acid (85). *F. oxysporum* f. sp. *lycopersici* produces substantial quantities of fusaric acid; i.e., 200 mg/L after 20 days, or after 6 days in still or shake cultures, respectively. High production of fusaric acid has been correlated with the virulence of plant pathogenic strains of *Fusarium* species (24, 93). Fusaric acid can cause wilt disease symptoms in pepper (*Capsicum* spp.) (142) and has been used to screen for wilt resistance. The compound has antibiotic, pharmacological and insecticidal activities, with moderate toxicity to animals (21). Fusaric acid inhibits respiratory activity (10), decreases ATP levels (32) and alters membrane electrical potential and increases electrolyte leakage in several plant species (10, 32, 53). However, the specific mode of action of this compound is still unclear.

3.6. HYDANTOCIDIN [XIII]

Hydantocidin is a relatively new phytotoxin discovered in the culture broth of *Streptomyces hygroscopicus* SANK 63584 (160). The water-soluble compound was crystallized and found to be more phytotoxic against perennial than annual plants. Confirmation of the stereochemistry and the reconfirmation of the structure of hydantocidin involved a series of experiments that provided total synthesis of hydantocidin and all of its stereoisomers (108-110).

Potent herbicidal activity of this compound has been reported (117). The compound showed nonselective herbicidal activity against mono- and dicotyledonous perennial and annual plants. Deoxy-derivatives of hydantocidin have been synthesized, but had no phytotoxicity (109, 110). In comparative tests of hydantocidin, glyphosate [*N*-(phosphonomethyl)glycine], and bialaphos on 17 weed species, the herbicidal activity of hydantocidin on both annual monocots and dicots was equal to glyphosate and slightly greater than bialaphos (117). Similar effects were found when these three compounds were tested on a series of perennial weeds. Hydantocidin had no antimicrobial activity when tested against 35 species of bacteria, yeasts, and fungi, even at 1000 μg/ml using a disc-agar diffusion assay and has low toxicity to mammals and fish. Due to its low non-target toxicity and high contact herbicidal activity with systematic action to plant roots, the compound may have very high potential for development as a natural product-based herbicide.

Mode of action studies of hydantocidin indicated that it is metabolized in plants to hydantocidin-5'-phosphate, which is a potent inhibitor of adenylosuccinate synthase (48, 66, 155). Thus, this metabolite appears to be the actual phytotoxic compound. Crystal structure analysis of adenylosuccinate synthase bound to hydantocidin-5'-phosphate may lead to the design of inhibitors of greater potency (130).

3.7. IRPEXIL [XIV]

Irpexil is a herbicidal compound produced by the basidiomycete *Irpex pachyodon* (47). This compound has structural features similar to aminooxy-acetate [XV] and the herbicide benzadox [XVI]. Benzadox and analogues that are structurally closer to irpexil, have been reported to have potential as herbicides. Benzadox is a relatively strong inhibitor of alanine aminotransferase (119). The mechanism of action of irpexil is unknown, but it and benzadox translocate basipetally in plants and produce symptoms similar to those produced by the synthetic herbicide glyphosate (47). Benzadox has been shown to be metabolized in plants to produce aminooxy-acetate which is a potent inhibitor of pyridoxyl phosphate requiring enzymes,i.e., amino-transferases (18, 81). Aminooxy-acetate has been patented as a herbicide (169). Aminooxy-acetate also lowered phenylalanine ammonia-lyase activity, anthocyanin, chlorophyll and hydroxyphenolic compound levels and caused severe injury in soybean [*Glycine max* (L.) Merr.] seedlings (75).

3.8. MONENSIN [XVII]

Monensin is a carboxylic polyether antibiotic isolated from the mycelium of an actinomycete, *Streptomyces cinnamonensis*. The compound is an ionophore that binds potassium, sodium, and hydrogen ions (162). Its lipophilic character renders it soluble in lipid components of biological membranes (14, 156) and it can induce changes in mycelial lipid and sterol composition (49). Monensin causes swelling, disruption and cytological changes in dictyosomes (Golgi apparatus) of animals (163) and plants (112, 113, 136, 145). The effects of monensin on Golgi apparatus function make it a potent inhibitor of protein secretion in animals (162) and plants (69, 107, 151, 162).

Until the mid-1980's, there had been little work on the effects of monensin on plant growth. This lack of plant-based data and the fact that monensin is introduced into the environment via animal wastes from its extensive use as an antibiotic in chickens (139) and as a feed additive in cattle (59), prompted a study of its effect on plant growth (111). In that study, monensin was tested on only one plant species, ryegrass (*Lolium multiflorum* Lam), and found to inhibit germination and growth when seeds imbibed aqueous solutions of 10^{-4} M monensin. Monensin also inhibits synthesis of cellulose or matrix polysaccharides and supresses auxin-induced elongation of pea (*Pisium sativum* L.) stems (20, 92). Monensin inhibited the incorporation of β-glucans into cellulose and callose of fibers of cotton (*Gossypium hirsutum* L.) ovules *in vitro* (50). Recently, elongation of rice coleoptiles was supressed by monensin at very low concentrations (10^{-7} M) (78). This compound can also ameliorate galactose suppression of auxin-induced cell elongation in oats (*Avena sativa* L.) (187, 188).

Monensin has been examined for effects on the growth of a variety of plant species, including crops and major agricultural weeds (72). The agricultural antibiotic monensin caused herbicidal injury to 1- to 2-wk -old seedlings of seven weed and two crop species when applied at 10^{-4} M as a foliar spray in the greenhouse. The non-ionic surfactant tergitol TMN (0.01%) aided foliar absorption of monensin increased herbicidal injury. Fresh weight reductions ranged from 20% to 75% (compared to control seedlings treated with tergitol alone) 65 h after spray application of 10^{-4} M monensin with tergitol. When root-fed at 10^{-4} M, monensin (without tergitol) caused death in all nine species within 24 to 72 h after treatment. Substantial growth effects also occurred in hemp sesbania and sicklepod (33 and 15% reductions in fresh weight, respectively); jimsonweed (*Datura stramonium* L.) and johnsongrass (*Sorghum halepense* L.) had the least fresh weight reductions. Injury included chlorosis, necrosis, desiccation and leaf abscission. Cotton and okra were the most tolerant of all species tested; however, three malvaceous weeds [spurred anoda (*Anoda crisata* L., velvetleaf (*Abutilon theophrasti* Med.) and prickly sida (*Sida spinosa* L.) were severely damaged by the compound. Monensin exhibited some degree of selectivity among these crop and weed species and caused greater injury in light-treated than in dark-treated plant tissues.

3.9. MONILIFORMIN **[XVIII]**

Moniliformin was discovered as a mycotoxin from *Fusarium moniliforme* (28) and structurally characterized as the sodium or potassium salt of 1-hydroxycyclobut-1-ene-3,4-dione (149). Moniliformin is produced by at least 15 *Fusarium* species (25, 104) and is highly toxic to animals (132). Phytotoxic activity of moniliformin has been demonstrated (28, 153) and various analogues have been synthesized in attempts to develop commercial herbicides (47). The proposed mode of action of moniliformin is inhibition of the pyruvate dehydrogenase complex (52).

3.10. NIGERICIN **[XIX]**

Some *Streptomyces* spp. produce a polyether antibotic nigericin that inhibits Gram-positive bacteria and some phytopathogenic fungi (63, 147). This compound is an analogue of monensin **[XVII]** which was discussed earlier and it has been shown to be phytotoxic to several species. Nigericin had preemergence activity on garden cress and large crabgrass [*Digitaria sanguinalis* (L.) Scop.] and postemergence activity on garden cress (*Lepidium sativum* L.) and velvetleaf (67). The postemergence activity on velvetleaf was very high, i.e., at doses as low as 0.3 kg/ha, leaves showed injury within 24-48 h. Inhibition of seed germination and seedling growth with some plant mortality were the main herbicidal effects.

3.11 OXETIN **[XX]**

Oxetin was isolated from fermentation broth of the soil isolate *Streptomyces* sp. strain OM-2317 (128). Its chemical structure was determined as (2R,3S)-3-amino-2-oxetane carboxylic acid via IR, NMR, X-ray analysis and FD-MS spectral data. This compound was the first natural product to be identified that contained an oxetane ring system. It possesses very low antibiotic activity against bacteria or fungi and only inhibited *Bacillus subtilis* and *Piricularia oryzae* when incorporated into a minimal, rather than a complex media. This inhibition in minimal media could be partially reversed by individual supplementation with various amino acids, including either L-glutamine, L-glutamate, or L-alanine.

Oxetin was phytotoxic toward alfalfa (*Medicago sativa*) and turnip (*Brassica rapa*), but at 125 µg/ml, injury symptomology was not apparent until 4 or more days after treatment. Oxetin was found to be a potent inhibitor of glutamine synthetase (GS) from *B. subtilis* and spinach (*Spinacia oleracea* L.) leaves. Oxetin is structurally related to 5-oxoproline, which is formed in the GS reaction on glutamate plus ATP; and this may explain its inhibitory action on the enzyme (96).

3.12. PYRIDAZOCIDIN [XXI]

Pyridazocidin is a new compound isolated from shake-flask cultures of a *Streptomyces* sp. from soil (55). Various methods, such as solvent partitioning, size-exclusion chromatography, and HPLC, were used in the isolation process. This compound had significant phytotoxic activity on several weed species {velvetleaf (*Abutilon theophrasti* Medik.); ivyleaf morningglory [*Ipomoea hederaceae* (L.) Jacq.]; barnyardgrass [*Echinochloa crus-galli* (L.) Beauv.]; and giant foxtail (*Setaria faberi* Herrm.)} when applied postemergence. Giant foxtail was the most sensitive weed tested. An analogue of this compound i.e., pyridazomycin [XXII] has been reported to be fungicidal (61).

The herbicidal action of pryidazocidin was further quantified in dose-response tests of *Arabidopsis thaliana* (L.) Heynh on agar plates. Pyridazocidin inhibited growth in this species by about 30% and 100% at 10 and 100 µg ml^{-1}, respectively.

Mode of action studies with pyridazocidin showed that it induced the Mehler reaction (autooxidation by molecular oxygen manifested by light-dependent oxygen consumption) in isolated spinach chloroplasts (55). The effects of pyridazocidin on the Mehler reaction were similar to that caused by the herbicide diquat [6,7-dihydrodipyrido(1,2-α:2',1'-c)pyrazinediium ion], but a higher concentration was required. This mode of action is shared by other chemical families including the herbicidal bipyridiniums (154).

3.13. RHIZOBITOXINE [XXIII]

This allelochemical is produced by some bacterial strains of *Rhizobium japonicum* and causes chlorosis and other signs of injury in many plant species other than soybean which is its host. Rhizobitoxine is an analogue of cystathione and irreversibly inhibits β-cystathionase which catalyzes conversion of cystathionine to homoserine (57), the immediate precursor of methionine. This phytotoxin can reduce growth in microoganisms that can be alleviated by supplemental addition of methionine, however, there is no strong evidence that β-cystathionase is the sole site of action (56). Rhizobitoxine also inhibits ethylene biosynthesis from methionine (98). An analogue, 2-amino-4-methoxybut-3-enoic acid, produced by certain *Rhizobium* and *Streptomyces* species has weak herbicidal activity (47). This analogue also disrupts ethylene biosynthesis and amino acid metabolism.

3.14. TABTOXIN [XXIV]

Tabtoxin is a dipeptide produced by several pathovars of *Pseudomonas syringae*. Hydrolysis of tabtoxin in plants or by the host producer yields tabtoxinine-ß-lactam which

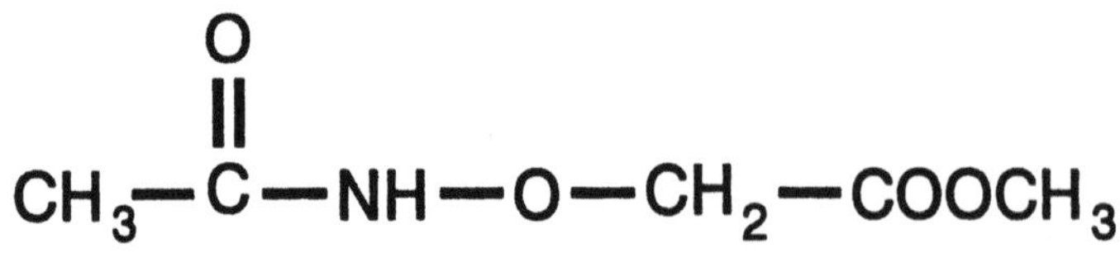

XIV. Irpexil
methyl-2-(acetamidooxy)acetate

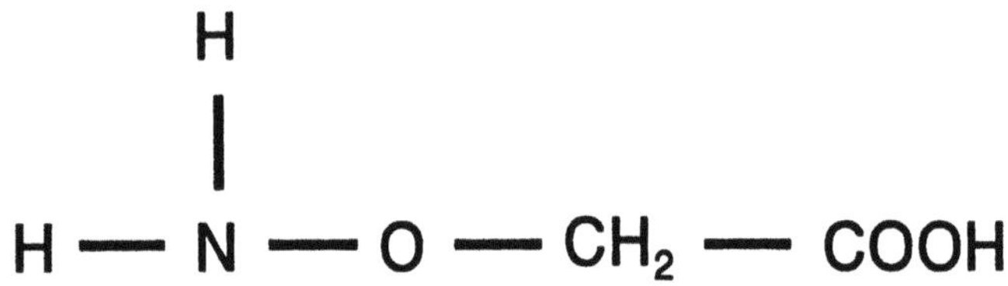

XV. Aminooxy-acetic acid

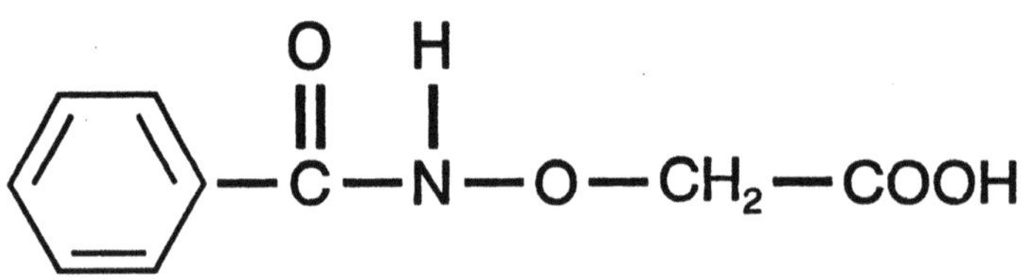

XVI. Benzadox [(benzamidooxy)acetic acid]

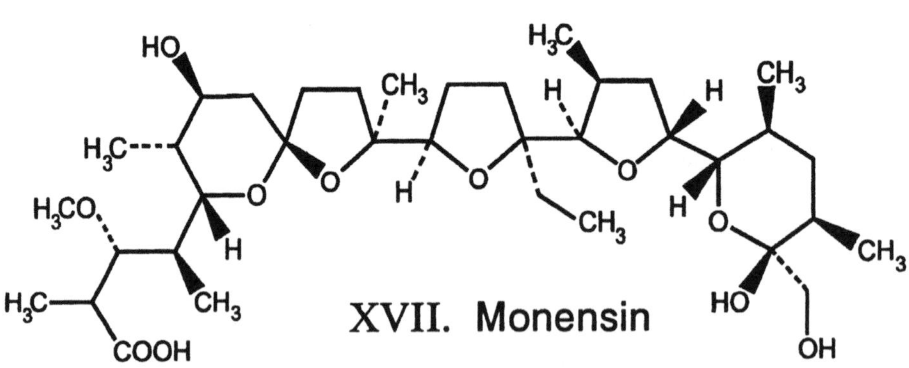

XVII. Monensin

is an active GS inhibitor and the toxin responsible for causing wildfire disease (165). Oat varieties that are resistant to the *P. syringae* pathogen were found to have GS that was less sensitive to tabtoxinine-ß-lactam (87). Tabtoxinine-ß-lactam increases nitrogen fixation of legume root nodules via selective inhibition of one GS isoenzyme (88). Tabtoxinine, a breakdown product of tabtoxinine-β-lactam, does not inhibit GS, which indicates that the β-lactam ring is required for enzyme inhibitory activity (94).

3.15. BIALAPHOS [XXV] and PPT [XXVI]

Of all the numerous phytotoxic natural products isolated from plants and microbes, bialaphos and phosphinothricin produced by *Streptomyces* [marketed as glufosinate, the synthesized ammonium salt of phosphinothricin] constitute the most successful, since they have been developed into widely-marketed major commercial herbicides (73, 76). An overview of some aspects of the history and development of these compounds as herbicides follows.

During 1971, two laboratories discovered a naturally-occurring compound called bialaphos. This tripeptide contained a unique amino acid, L-2-amino-4-[hydroxy-(methyl)-phosphinyl]butyric acid (called phosphinothricin or PPT) linked to two L-alanyl moieties. The compound, isolated from *Streptomyces viridochromogenes* (15) and from *Streptomyces hygroscopicus* cultures (88), was the first reported naturally-occurring amino acid containing a phosphinic group and occurs as L-PPT. Bialaphos was initially found to have some antifungal (*Botrytis cinerea*) and antibacterial (Gram-negative and Gram-positive) activity (15, 122), but this was attributed to PPT. Later examination of L-PPT for phytotoxicity by Hoechst AG showed that it possessed herbicidal activity, resulting in a patent (140). Synthesis of the DL-PPT ammonium salt (common name, glufosinate) resulted in the commercial herbicidal formulation of this active ingredient. The free acid and the ammonium salt were initially numbered compounds, HOE-35956 and HOE-391866, respectively. Glufosinate is marketed globally as Basta®, Buster®, Challenge®, Finale®, Harvest®, Ignite®, Liberty® and Rely®. Bialaphos has been patented as a herbicide by Meiji Seika Kaisha (161) and is marketed in Japan as Herbiace7 (105). Bialaphos also exhibits herbicidal activity when release of L-PPT occurs via hydrolytic cleavage of the alanine moieties by plant peptidases. Other herbicidal peptide analogues of bialaphos have been isolated, i.e., phosalacine (PPT-ala-leu) from *Kitasatosporia phosalscinea* KA-338 (126, 127) and trialaphos (PPT-ala-ala-ala) from *Streptomyces hygroscopicus* sp. KSA-1285 (83). Again, peptidases can also cleave the amino residues from these molecules, yielding PPT.

PPT and glufosinate are highly stable molecules with extremely high water solubility (>1,350 g/l at pH 7.0), with much lower solubility in common organic solvents. Many studies on mammals, other animals and insects have demonstrated low toxicity of these compounds (76). Generally, the data indicate no genotoxic, carcinogenic, or teratogenic potential, or other specific toxicological hazards.

Bialaphos is rapidly degraded in soil to PPT (158) and PPT is also rapidly degraded, with half-lives of 4 to 7 days in soils (51, 145). Of 300 bacterial isolates from soil, all strains degraded L-PPT to the 2-oxo analogue of PPT via transamination (13). Studies in

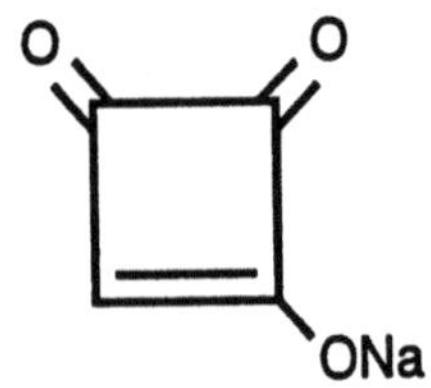

XVIII. Moniliformin

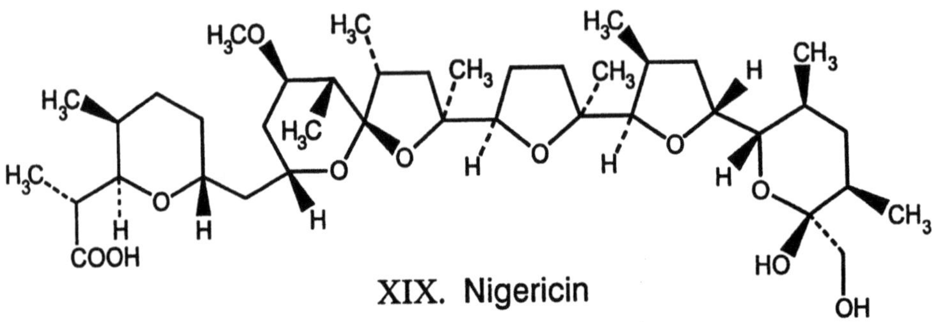

XIX. Nigericin

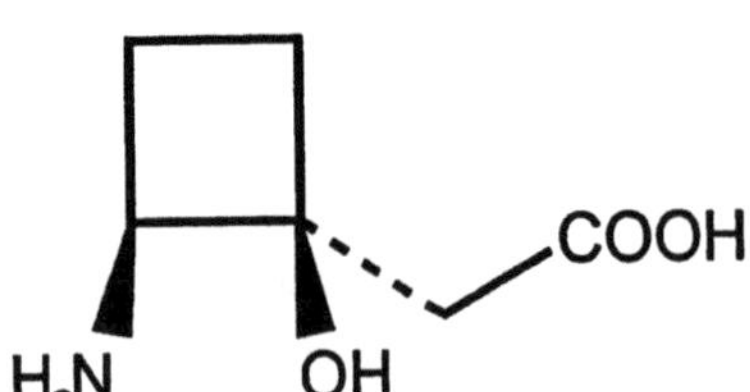

XX. Oxetin

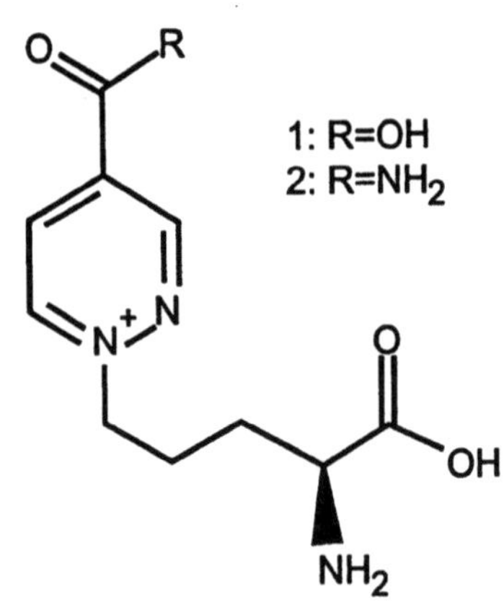

XXI. Pyridazocidin (1)
XXII. Pyridazomycin (2)

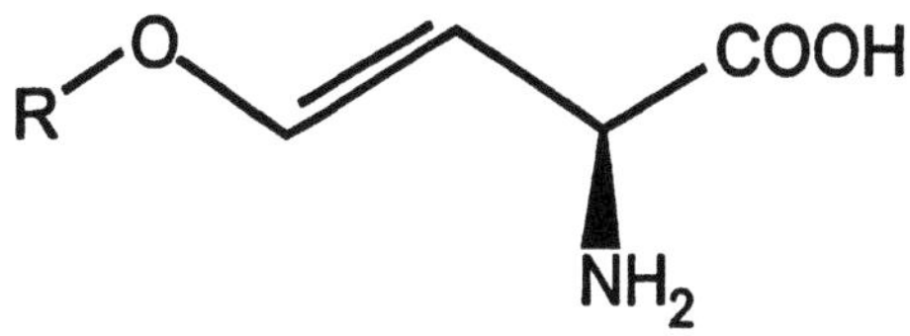

XXIII. Rhizobitoxine

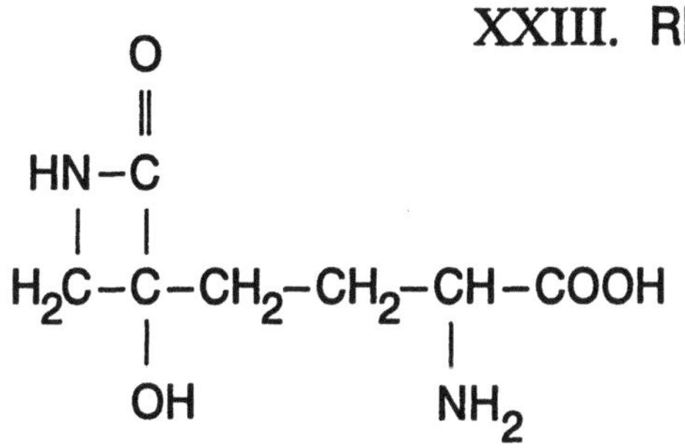

XXIV. Tabtoxinine-ß-lactam

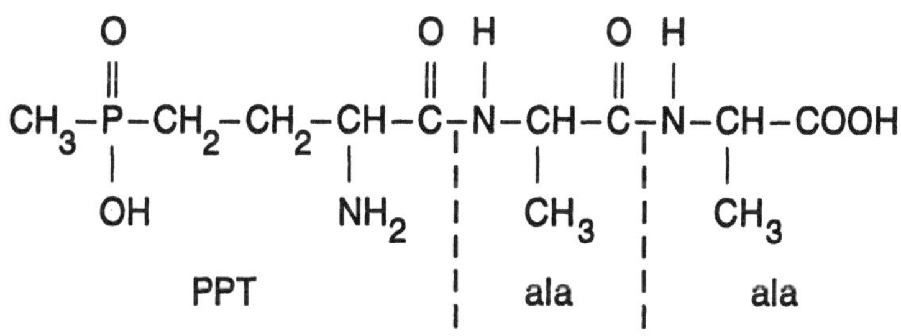

XXV. Bialaphos

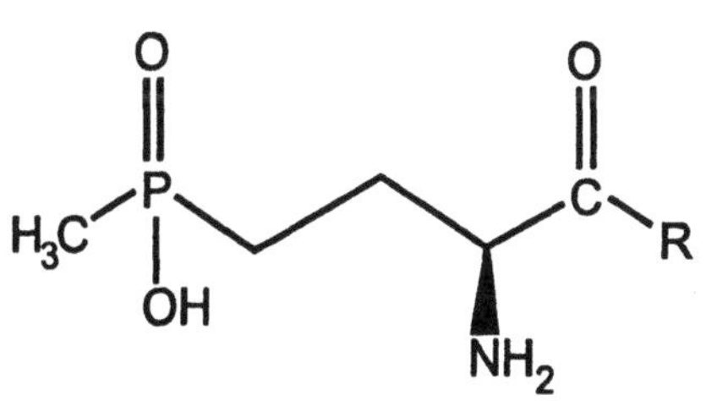

XXVI. Phosphinothricin (PPT)

soils have shown that glufosinate does not leach below a soil depth of 15 cm (16, 148).

Glufosinate (PPT) is a nonselective, postemergence herbicide used for weed control in orchards and vineyards, in chemical fallow situations, as a preharvest desiccant, as a burn-down herbicide of cover crops and/or weeds prior to no-till planting (9) and for weed control in transgenic crops resistant to the herbicide (171). Dicots are usually more sensitive than grasses and both annual and perennial weeds are controlled.

Bialaphos controls a wide range of weeds, including perennials when applied to foliage (158). It is generally used in the same situations as PPT. Bialaphos is absorbed through plant leaves and some translocation (of bialaphos or its metabolites) occurs (158). Bialaphos is metabolized in plants soon after absorption to yield PPT. PPT can also be rapidly absorbed by plant tissues, but translocation to other plant parts is minimal. In cogongrass (*Imperata cylindrica*) and sour paspalum (*Paspalum conjugatum*), half the applied ^{14}C-PPT was absorbed within 4 h, but there was little translocation (89). Three days after application of ^{14}C-glufosinate to five plant species, ^{14}C was transported and detected in untreated plant tissues at levels of 2 to 4% of the initially applied dose (62).

Many mode of action studies have been published on phosphinothricin. Although PPT was initially found to be a potent inhibitor of GS in *E. coli* (15), the kinetic properties of GS inhibition on a wide variety of plants showed that K_i values varied by a factor of five (6, 135). Furthermore, whole plant susceptibility to glufosinate ranged over two orders of magnitude (135). GS is inhibited and ammonia levels rise soon after treatment with PPT. Although the subsequent cascade of events leading to plant death has not been totally elucidated, it has been explained in several ways. One explanation is that a lack of amino donors causes disruption of photorespiration and then photosynthesis, minutes after herbicide treatment (143). An alternative proposal is a decoupling of photophosphorylation by excess ammonia (95). Substantial increases in ammonia levels occur very early after PPT application in a number of species, for example, tall morningglory (*Ipomoea purpurea*) and johnsongrass (*Sorghum halapense*) (89). Although GS inhibition and accumulation of ammonia are generally thought to be pivotal to glufosinate mode of action, as substantiated by GS gene amplification for glufosinate tolerance (33), other molecular target sites have been suggested.

PPT caused a more rapid inhibition of photosynthesis in C_3 than in C_4 plants under normal atmospheric conditions; but under non-photorespiratory conditions, photosynthesis was not inhibited (175). Also, ammonia accumulation was lower in C_4 plants. Addition of glutamine or glutamate significantly alleviated photosynthetic inhibition, even though ammonia levels were greatly increased. This suggests that ammonia accumulation may not be the primary cause of photosynthesis inhibition by glufosinate. Transamination of glyoxylate to glycine in photorespiration was inhibited due to lack of amino donors, which could cause glyoxylate accumulation, thereby inhibiting ribulose-1,5-bisphosphate carboxylase and, consequently, CO_2 fixation (175). The most complete picture of the events occurring in plants after treatment with PPT (or other GS inhibitors) has been discussed elsewhere (178).

Bialaphos does not inhibit GS, but is rapidly metabolized by peptidases in plant tissues yielding PPT (157, 179). The D-isomer of PPT also does not inhibit GS (103), is non-herbicidal (171) and is not degraded in transgenic plants (37). Other naturally-occurring

GS inhibitors with phytotoxicity have been discovered, that are structurally similar to PPT (73). L-Methionine sulfoximine (MSO) was initially synthesized (164), then found as a natural product in the bark of a tree, *Cnestis glabra* (80). Another PPT Analogue, L-(N^5-phosphono)methionine-S-sulfoximine (PMSO) is a potent GS inhibitor and a metabolite of L-(N^5-phosphono)methionine-S-sulfoximinyl-L-ala-L-ala (PMSO-ala-ala) (138) isolated from a *Streptomyces* (146). PMSO and MSO can result from the action of peptidase on PMSO-ala-ala or phosphatase action on PMSO, respectively (131, 144). MSO has been patented as a herbicide (*173*), but its inhibitory activity on GS is much less than that of PPT (97, 177). In whole plant tests, PPT was generally 5- to 10-fold more phytotoxic than MSO (96). Although numerous GS inhibitors that are Analogues of PPT have been synthesized, none have been found to be more herbicidal than PPT (100, 101, 106).

There have been numerous developments in the biotechnological aspects of PPT. The biosynthetic pathway of bialaphos has been elucidated (116). Starting with precursors containing three carbon atoms, bialaphos is produced in a complex series of over a dozen steps, as summarized elsewhere (116). One step involves an acetyl CoA-dependent reaction that modifies either demethyl-PPT or PPT. The gene (*bar*) responsible for encoding this step confers resistance to bialaphos in the organism. The gene has been isolated and characterized and found to encode acetyl transferase activity which converts PPT to an acetylated metabolite that is non-phytotoxic (91).

Over the past few years, many vegetable and cereal crop species have been transformed with either the *bar* gene of *Streptomyces viridochromogenes* or the *pat* gene of *S. hygroscopicus* to impart resistance to PPT. Cloning of a PPT-resistant gene (*bar*) from *Streptomyces hygroscopicus* (166) and the transformation of PPT-resistant plants has been accomplished (34). A similar gene (*pat*) from *S. viridochromogenes* Tü 494, with the same function, was simultaneously isolated and introduced into various plant species (19, 37, 185). Presently there are more than 20 crop plant species transformed for resistance to PPT. Some of these genetically transformed plants are resistant to PPT at rates as high as 4 kg ai/ha (ca. 10 times the lowest normal field application rate) (166).

The anti-fungal activity of bialaphos and glufosinate was recently assessed on three pathogens *in vitro* and *in vivo* on PPT-resistant transgenic creeping bentgrass (*Agrostis palustris*), an important turf grass (99). Results show that bialaphos can simultaneously control weeds and fungal pathogens in this transgenic grass. Bialaphos has antibiotic activity against *Rhizoctonia solani* Kühn that causes rice sheath blight (35) and *Magnaporthe grisea* (Herbert) Barr (90) that causes rice blast disease. Substantial suppression of sheath blight symptoms occurred when bialaphos was applied to transgenic plants which had been infected with *R. solani* prior to herbicide treatment (168). Inoculated transgenic rice plants [bialaphos-resistant (*bar*) gene] had reduced lesions and symptoms of rice blast disease after bialaphos treatment (159). Thus, it may be possible to control these serious diseases by using *bar*-transgenic rice cultivars and bialaphos for weed and disease control. Glufosinate has also been successfully used to control the weed red rice (*Oryza sativa L.*, a conspecific weed of cultivated rice) in *bar*-transformed rice (141).

Since the discovery of PPT and its herbicidal properties, much research has been published on PPT and related compounds. Field trials have confirmed superior postemergence weed control and excellent crop safety with the use of glufosinate in four major transgenic (*pat* gene) crops: canola (*Brassica napus*), maize (*Zea mays*), soybean (*Glycine max*), and sugar beet *(Beta vulgaris* L.) (134). Major North American seed companies are developing glufosinate-tolerant varieties of a number of crops (133). Over 2 million acres of transgenic maize planted in the U.S. in 1998 (Dr. Rees, AgrEvo Corp., personal communication). Bialaphos and PPT are unique among commercial herbicides in that they have both antibiotic and herbicidal properties. This dual strategy will obviously be utilized more widely with the increasing availability of PPT-resistant crops. This is truly a remarkable success story for an agricultural product developed from a naturally occurring allelopathic compound with potent herbicidal activity .

4. Conclusions

It is expected that chemical herbicides, derived either synthetically or from natural products will remain the major method of weed control to produce the enormous quantities of food and fiber for peoples on the earth now estimated to be 8 billion by the year 2025 (170). Further research of other allelopathic compounds from plants or microbes may also result in the discovery of other novel compounds with superior herbicidal properties to help meet this challenge. With the increased interest in naturally occuring compounds as pesticides, allelopathic research should remain viable.

References

1. Abbas, H.K. and Boyette, C.D. (1992). Phytotoxicity of fumonisin on weed and crop species. *Weed Technology* **6**:548-552.
2. Abbas, H.K., Boyette, C.D. and Hoagland, R.E. (1995). Phytotoxicity of *Fusarium*, other fungal isolates, and of the phytotoxins fumonisin, fusaric acid and moniliformin to jimsonweed. *Phytoprotection* **76**:17-25.
3. Abbas, H.K., Boyette, C.D., Hoagland, R.E. and Vesonder, R.F. (1991). Bioherbicidal potential of *Fusarium monililforme* and its phytotoxin, fumonisin. *Weed Science* **39**:673-677.
4. Abbas, H.K., Duke, S.O. and Tanaka, T. (1993). Phytotoxicity of fumonisins and related compounds. *Journal of Toxicolicology and Toxin Reviews.* **12**:225-251.
5. Abbas, H.K., Vesonder, R.F., Boyette, C.D., Hoagland, R.E. and Krick, T. (1992). Production of fumonisins by *Fusarium moniliforme* cultures isolated from jimsonweed in Mississippi. *Journal of Phytopathology* **136**:199-203.
6. Acaster, M.A. and Weitzmann, P.D.J. (1985). Kinetic analysis of glutamine synthetase from various plants. *FEBS Letters* **189**:241-244.
7. Alston, T.A., Mela, L. and Bright, H.J. (1977). 3-Nitropropionate, toxic substance of *Indigofera*, is a suicide inactivator of succinate-dehydrogenase (rat liver mitochondria carbonium -N-5 flavin adducts 2-proton abstraction mechanism). *Proceedings of the National Academy of Science, USA* **74**: 3767-3771.
8. Amagasa, T., Paul, R.N., Heitholt, J.J. and Duke, S.O. (1994). Physiological effects of cornexistin on *Lemna pausicostata. Pesticide Biochemistry and Physiology* **49**: 37-52.
9. American Hoechst Corp. (1982). HOE-00661 *Technical Information Bulletin.* AMF 2464. Somerville, NJ
10. Arias, J.A. (1985). Secretory organelle and mitochondrial alteration induced by fusaric acid in root cell of *Zea mays. Physiological Plant Pathology* **27**:149-158.
11. Bacon, C.W., Hinton, D.M., Chamberlain, W.J. and Norred, W.P. (1994). *De novo* induction of adventitious roots in excised shoots of *Fusarium momiliforme. Journal of Plant Growth Regulation* **13**:53-57.
12. Bagchi, G.D., Jain, D.C. and Kuman, S. (1998). The phytotoxic effects of the artemisinin related compounds of *Artemisia annua. Journal of Medicinal and Aromatic Plant Science* **20**: 5-11.
13. Bartsch, K. and Tebbe, C. (1989). Initial steps in the degradation of phosphinothricin (glufosinate) by soil bacteria. *Applied and Environmental Microbiology* **55**:711-716.

14. Basset, W.L., Wiggins, J.R., Gleband, H. and Pressman, B.C. (1978). Monensin potentiation of potassium contracture in cat myocardium. *Journal of Pharmacology and Experimental Therapeutics* **207**:966-975.
15. Bayer, E., Gugel, K.H., Hägele, K., Hagenmaier, H., Jessipow, S., König, W.A., and Zähner, H. (1972). Phosphinothricin und phosphinothricyl-alanyl-alanin. *Helvetica Chimica Acta* **55**:224-39.
16. Behrendt, H., Matthies, M., Gildemeister, H., and Görletz, G. (1990). Leaching and transformation of glufosinate-ammonium and its main metabolite in a layered soil column. *Environmental Toxicology and Chemistry* **9**:541-549.
17. Bezuidenhout, S.C., Gelderblom, W.C.A., Gorst-Allman, C.P., Horak, R.M., Marasas, W.F.O., Spiteller, G. and Vleggaar, R. (1988). Structure elucidation of the fumonisins, mycotoxins from *Fusarium moniliforme*. *Journal of the Chemical Society Chemical Communications* **1984**:743-745.
18. Braunstein, A.E. (1973). Amino group transfer. In *The Enzymes*, 3rd ed., *Vol IX: Group transfer, Part B* : (Ed., P.D. Boyer), pp. 379-471. New York : Academic Press.
19. Broer, I., Arnold, W., Wohlleben, W., and Pühler, A. (1989). In *Proceedings Braunschweig Symposium of Applied Plant Molecular Biology*, p. 240. Galling, Braunschweig, Germany: Tech. Univ. Braunschweig
20. Brummell, D.A. and Hall, J.L. (1985). The role of cell wall synthesis in sustained auxin-induced growth. *Physiologia Plantarum* **63**:406-412.
21. Burmeister, H.R., Grove, M.D., Peterson, R.E., Weisleder, D. and Plattner, R.D. (1985). Isolation and characterization of two new fusaric acid analogues from *Fusarium moniliforme* NRRL-13,163. *Applied and Environmental Microbiology* **50**:311-314.
22. Carter, C.L. and McChesney, W.J. (1949). Hiptagenic acid identified as β-nitropropionic acid. *Nature* **166**:575-576.
23. Cawood, M.W., Gelderblom, W.C.A., Vleggaar, R., Behrend, Y., Theil, P.G. and Marasas, W.F.O. (1991). Isolation of the fumonisin mycotoxins: a quantitative approach. *Journal of Agricultural and Food Chemistry* **39**:1958-1962.
24. Chakrabarti, D.K. and Basuchaudhary, K.C. (1980). Correlation between virluence and fusaric acid production in *Fusarium oxysporum* f. sp. *carthami*. *Phytopathologia Zeitschrift* **90**:43-46.
25. Chelkowski, J., Zawadzki, M., Zajkowski, P., Logrieco, A. and Bottalico, A. (1990). Moniliformin production by *Fusarium* species. *Mycotoxin Research* **6**:41-45.
26. Chen, J., Miocha, C.J., Xie, W., Hogge, L., and Olson, D. (1992). Production of the mycotoxin fumonisin B_1 by *Alternaria alternata* f. sp. *lycopersici*. *Applied and Environmental Microbiology* **58**:3928-3931.
27. Chen, P.K. and Leather, G.R. (1990). Plant growth regulatory activities of artemisinin and its related compounds. *Journal of Chemical Ecology* **16**: 1867-1876.
28. Cole, R.J., Kirksey, J.W., Cutler, H.G., Doupnik, B.L., and Peckham, J.C. (1973). Toxin from *Fusarium moniliforme*, effects on plants and animals. *Science* **179**:1324-1326.
29. Coles, C.J.., Edmondson, D.E., and Singer, T.P. (1979). Inactivation of succinate dehydrogenase by 3-nitropropionate. *Journal of Biological Chemistry* **254**: 5161-5167.
30. Cutler, H.G., ed. (1988). *Biologically Active Natural Products: Potential Use in Agriculture*. American Chemical Society Symposium Series No. 380 : 483 pp. Washington, DC . American Chemical Society.
31. Cutler, H.G. (1988). Perspectives on discovery of microbial phytotoxins with herbicidal activity. *Weed Technology* **2**: 525-532.
32. D'Alton, A. and Etherton, B. (1984). Effects of fusaric acid on tomato root hair membrane potential and ATP levels. *Plant Physiology* **74**:39-42.
33. Deak, M., Donn, G., Feher, A., and Dudits, D. (1988). Dominant expression of a gene amplification-related herbicide resistance in *Medicago* cell hybrids. *Plant Cell Reports* **7**:158-61.
34. DeBlock, M., Botterman, J., Vanderwiele, M.,Dockz, J.,Thoen, C., Gossele, V., Movva, N.R., Thompson, C., van Montagu, J., and Leemans, J. (1987). Engineering herbicide resistance in plants by expression of a detoxifying enzyme. *EMBO Journal* **6**:2513-18.
35. De Datta, S.K. (1981). *Principles and Practices of Rice Production*. p. 443. New York : Wiley International.
36. DiTomaso, J.M. and Duke, S.O. (1991). Evaluating the effect of cinmethylin and artemisinin on polyamine biosynthesis as a possible primary site of action. *Pesticide Biochemistry and Physiology* **39**: 158-167.
37. Dröge, W., Broer, I., and Pühler, A. (1992). Transgenic plants containing the phosphinothricin-*N*-acetyl transferase gene metabolize the herbicide L-phosphinothricin (glufosinate) differently from untransformed plants. *Planta* **187**:142-51.
38. Duke, M.V., Paul, R.N., Elsohly, H.N., Sturtz, G., and Duke, S.O. (1994). Localization of artemisinin and artimisitene in foliar tissues of glanded and glandless biotypes of *Artemisia annua* L. *International Journal of Plant Science* **155**: 365-372.

39. Duke, S.O. (1991). Plant terpenoids as pesticides. In *Handbook of Natural Toxins;* Vol. 6, *Toxicology of Plant and Fungal Compounds.* (Eds., R.F. Keeler and A.T. Tu), pp. 269-296; Marcel Dekker, Inc., New York
40. Duke, S.O. (1986). Naturally occurring chemical compounds as herbicides. *Reviews of Weed Science* **2**:15-44.
41. Duke, S.O. (1986). Microbially produced phytotoxins as herbicides - A perspective. In *The Science of Allelopathy.* (Eds., A.R. Putnam and C.S. Tang), pp. 287-304. New York : John Wiley and Sons.
42. Duke, S.O. and Abbas, H.K. (1995). Natural products with potential use as herbicides. In *Allelopathy: Organisms, Processes and Applications*; American Chemical Society Symposium Series No. **582** : 348-362. (Eds., K.M. Dakshini, Inderjit and F.E. Einhellig). Washington, DC : American Chemical Society.
43. Duke, S.O., Vaughan, K.C., Croom, E.M., Jr., and Elsohly, H.N. (1987). Artemisinin, a constituent of annual worm wood (*Artemisia annua*), is a selective phytotoxin. *Weed Science* **35**: 499-505.
44. Einhellig, F.A. and Souza, I.F. (1992). Allelopathic activity of sorgoleone. *Journal of Chemical. Ecology* **18**:1-11.
45. Evidente, A., Capretti, P., Giordano, F. and Surico, G. (1992). Identification and phytotoxicity of 3-nitropropionic acid produced *in vitro* by *Melanconis thelebola. Experientia* **48**:1169-1172.
46. Felner, I. and Schenker, K. (1970). Total synthesis of the antibiotic anisomycin. *Helvetica Chimica Acta* **53**:754-63.
47. Fisher, H.-P. and Bellus, D. (1983). Phytotoxicants from microorganisms and related compounds. *Pesticide Science* **14**:334-346.
48. Fonné-Pfister, R., Chemla, P., Ward, E., Girardet, M., Kreuz, K.E., Honzatko, R.B., Fromm, H.J., Schär, H-P. , Grütter, M.G. and Cowan-Jacob, S.W. (1996). The mode of action and the structure of a herbicide complex with its target: binding of activated hydantocidin to the feedback regulation site of adenylosuccinate synthetase. *Proceedings of the National Academy of Science, USA* **93**: 9431-9436.
49. Fonvieille, J-L., Razki, A.,Touze-Soulet, J.M., Dargent, R. and Rami, J. (1991). Effect of monensin on the lipid composition of *Achyla bisexualis. Mycology* **95**:480-483.
50. Francey ,Y., Jacquet, J.P., Gairoli, S., Buchala, A.J. and Meier, H. (1989) The biosynthesis of β-glucans in cotton (*Gossypium hirsutum* L.) fibers of ovules cultured *in vitro. Journal of Plant Physiol*ogy **134**:485-491.
51. Gallina, M.A., and Stephenson, G.R. (1992). Dissipation of ^{14}C-glufosinate-ammonium in two Ontario soils. *Journal of Agriculture and Food Chem*istry **40**:165-68.
52. Gathercole, P.S., Thiel, P.G., and Hofmeyr, J. H. S. (1986). Inhibition of pyruvate dehydrogenase complex by moniliformin. *Biochemical Journal* **233**:719-723.
53. Gäumann, E. (1958). The mechanisms of fusaric acid injury. *Phytopathology* **48**:670-686.
54. Gäumann, E., Neaf-Roth, S. and Kobel, H. (1952). Über fusarunsaure, ein zweites welketoxin des *Fusarium lycopersici* Sacc. *Phytopathogische Zeitschrift* **20**:1-38.
55. Gerwick, B.C., Fields, S.S., Graupner, P.R., Gray, J.A., Chapin, E.L., Cleveland, J.A. and Heim, D.R. (1997). Pyridazocidin, a new microbial phytotoxin with activity in the Mehler reaction. *Weed Science* **45**:654-657.
56. Gilchrist, D.G. (1983). Molecular modes of action. In *Toxins and Plant Pathogenesis*, (Eds., J.M. Daly and B.J. Deverall), pp. 81-136. New York : Academic Press.
57. Giovanelli, J., Owens, L. and Mudd, S. (1971). Mechanism of inhibition of β-cystathionase by rhizobitoxine. *Biochemica et Biophysica Acta* **227**: 671-684.
58. Gonzalez, V.M., Kazimir, J., Nimbal, C., Weston, L.A. and Cheniae, G.M. (1997). Inhibition of a photosystem II electron transfer reaction by the natural product sorgoleone. *Journal of Agricultural and Food Chemistry* **45**: 1415-1421.
59. Goodrich, R. D., Garrett, J.D., Gast, D.R., Kirick, M.A., Lanson, D.A. and Meiske, J.C. (1984). Influence of monensin on the performance of cattle. *Journal of Animal Science* **58**:1484-1498.
60. Grayson , B.T., Williams, K.S., Freehauf, P.A., Pease, R.R., Ziesel, W.T., Sereno, R.L., and Reinsfelder, R.E. (1987). The physical and chemical properties of the herbicide cinmethylin (SD 95481). *Pesticide Sc*ience **21**:143-153.
61. Grote, R., Chen, Y., Zeeck, A., Chen, Z., Zähner, H., Mischnick-L©bbecke, P., and König, W.A. (1988). Metabolic products from microorganisms. 243 Pyridazocidin, a new antifungal antibiotic produced by *Streptomyces violaceoniges. Journal of Antibiotics* **41**:595-601.
62. Haas, P., and Müller, F. (1987) Behavior of glufosinate-ammonium in plants. *Brighton Crop Protection Conference - Weeds*, **10B**:1075-1082 .
63. Harned, R.L., Hidy, J. H., Corum, C.J., and Jones, K.L. (1951). Nigericin, a new crystalline antibiotic from an unidentified *Streptomycete. Antibiotic Chemistry* **1**: 594-596.
64. Harrington, P.M., Singh, B.K., Szamosi, I.T. and Birk, J.H. (1995). Synthesis and herbicidal activity of cyperin. *Journal of Agricultural and Food Chem*istry **43**: 804-808.

65. Heap, I.M. (1997). The occurrence of herbicide-resistant weeds worldwide. *Pesticide Science* **51**: 235-243.
66. Heim, D.R., Cseke, C., Gerwick, B.C., Murdoch, M.G. and Green, S.B. (1995). Hydantocidin: a possible proherbicide inhibiting purine biosynthesis at the site of adenylosuccinate synthetase. *Pesticide Biochemistry and Physiology* **53**:138-145.
67. Heisey, R.M. and Putnam, A.R. (1990). Herbicidal activity of the antibiotics geldanamycin and nigericin. *Journal of Plant Growth Regulation* **9**: 19-25.
68. Hershenhorn, J., Vurro, M., Stierle, A., and Strobel, G. (1993). *Septoria cirsii*, a potential biocontrol agent of Canada thistle and its phytotoxin beta-nitropropionic acid. *Plant Science* **94**:227-234.
69. Heupke, H.J. and Robinson, D. (1985) Intracellular transport of α-amylase in barley aleurone cells: evidence for the participation of the Golgi apparatus. *European Journal of Cell Biology* **39**:265-272.
70. Hoagland, R.E., ed. (1990). *Microbes and Microbial Products as Herbicides.* American Chemical Society Symposium Series No. 439; 341 pp. Washington, DC : American Chemical Society.
71. Hoagland, R.E. (1995). Phytotoxic effects of the microbial product 3-nitropropionate on several plant species. *Proceedings of the Southern Weed Science Society* p. 164.
72. Hoagland, R.E. (1996). Herbicidal properties of the antibiotic monensin. *Journal of the Science of Food and Agriculture* **70**:373-379.
73. Hoagland, R.E. (1999). Biochemical interactions of the microbial phytotoxin phosphinothricin and Analogues with plants and microbes. In *Biologically Active Natural Products: Agrochemicals*, (Eds., H.G. Cutler and S. Cutler), pp. 107-125. Boca Rotan, Florida, USA : CRC Press.
74. Hoagland, R.E. and Abbas, H.K. (1995). Phytotoxicity of *Fusarium* spp. and their natural products. *Proceedings 22nd Plant Growth Regulator Society of America* 109-114.
75. Hoagland, R.E. and Duke, S.O. (1982). Effects of glyphosate on metabolism of phenolic compounds. VIII. Comparison of the effects of aminooxy-acetate and glyphosate. *Plant and Cell Physiology* **23**: 1081-1088.
76. Hoerlein, G. (1994). Glufosinate (phosphinothricin), a natural amino acid with unexpected herbicidal properties. *Reviews of Environmental Contamination and Toxicology* **138**:73-145.
77. Hopkins, W.L. (1997). *Global Herbicide Directory*, 2nd edition. Indianapolis, IN, USA : Ag Chem Information Services.
78. Hoson, T. and Masuda, Y. (1991) The role of polysaccharide synthesis in elongation growth and cell wall loosening in intact rice coleoptiles. *Plant and Cell Physiology* **32**:763-769.
79. Hylin, J.W. and Matsumoto, H. (1960). The biosynthesis of 3-nitropropionic acid by *Penicillium atrovenetum*. *Archives of Biochemistry and Biophysics* **93**:542-545.
80. Jeannoda, V.L., Valisolaloa, J., Creppy, E.E., and Dirheimer, G. (1985). Identification of the toxic principle of *Cnestis glabra* as methionine sulfoximine. *Phytochemistry* **24**:854-55.
81. John, R.A., Charteris, A. and Fowler, L. J. (1978). The reaction of amino-oxyacetate with pyridoxylphosphate-dependent enzymes. *Biochemistry* **171**:771-779.
82. Kamikawa, T., Higuchi, F.,Taniguchi, M. and Asaka, Y. (1980). Toxic metabolites of an unidentified filamentous fungus isolated from zinnia leaves. *Agricultural Chemistry and Biology* **44**:691-692.
83. Kato, H., Nagayama, K., Abe, H., Kobayashi, R. and Ishihara, E. (1991). Isolation, structure, and biological activity of trialaphos. *Agricultural Biological Chemistry* **55**:1133-1134.
84. Kelsey, R.G. and Shafizadeh, F. (1980). Glandular trichomes and sesquterpene lactones of *Artemisia nova* (Asteraaceae). *Biochemal Systematics and Ecology* **8**: 371-377.
85. Kern, H. (1972). In *Phytotoxins in Plant Diseases*; (Eds., R.K.S. Wood, A. Ballio and A. Graniti), pp 35-48. New York : Academic Press.
86. Klayman, D.L. (1985). Qinghaosu (artemisinin): an antimalarial drug from China. *Science* **228**: 1049-1055.
87. Knight, T.J., Bush, D.R. and Langston-Unkefer, P.J. (1988) . Oat tolerant of *Pseudomonas syringe* pv. *tabaci* contains tabtoxinine-β-lactam-insensitive leaf glutamine synthetases. *Plant Physiology* **88**:333-339. .
88. Knight, T.J. and Langston-Unkefer, P.J. (1988) . Enhancement of symbiotic dinitrogen fixation by a toxin-releasing plant pathogen. *Science* **241**: 951-954.
89. Köcher, H. and Lötzsch, K. (1985). Uptake, translocation, and mode of action of the herbicide glufosinate-ammonium in warm climate weed species. *Proceedings of the Asian-Pacific Weed Science Society* **10**:193-98.
90. Kondo, Y., Shomura, T., Ogawa, Y., Tsuruoka, T., Watanabe, H., Totsukawa, K., Suzuki, T., Moriya, C., and Yoshida, J. (1973). Isolation and physiochemical and biological characterization of SF-1293 substance. *Meiji Seika Kaisha* **13**:34-44.
91. Kumada, Y., Anzai, H., Takano, E., Murakami, T., Hara, O., Itoh, R., Imai, S., Satoh, A. and Nagoaka, K. (1988). The bialaphos resistance gene (*bar*) plays a role in both self-defense and bialaphos biosynthesis in *Streptomyces hygroscopicus*. *Journal of Antibiotics* **41**:1839-1845.

92. Kutachera, U. and Briggs, W.R. (1987). Rapid auxin-induced stimulation of cell wall synthesis in pea internodes. *Proceedings of the National Academy of Sciences, USA* **84**:2747-2751.
93. Lakshminarayanan, K., and Subramanian, D. (1955). Is fusaric acid a vivotoxin ? *Nature (London)* **176**:697-698.
94. Langston-Unkefer, P.L., Robinson, A.C., Knight, T.J., and Durbin, R.D. (1987). Inactivation of pea seed glutamine synthetase by the toxin, tabtoxinine-β-lactam. *Journal of Biological Chem*istry **262**: 1608-1613.
95. Lea, P.J., Joy, K.W., Ramos J.L., and Guerro, M.G. (1984). The action of 2-amino-4-(methylphosphinyl)butanoic acid (phosphinothricin) and its 2-oxo-derivative on the metabolism of cyanobacteria and higher plants. *Phytochemistry* **23**:1-6.
96. Lea, P.J. and Ridley, S.M. (1989). Glutamine synthetase and its inhibition. In *Herbicides and Plant Metabolism.* (Ed., A.D. Dodge). Society for Experimental Biology Seminar Series No. **38** : pp. 137-170. Cambridge : University Press.
97. Leason, M., Cunliffe, D., Parkin, D., Lea, P.J., and Miflin, B.J. (1982). Inhibition of pea leaf glutamine synthetase by methionine sulfoximine. Phosphinothricin and other glutamate Analogues. *Phytochemistry* **21**:855-857.
98. Lieberman, M. (1979). Biosynthesis and action of ethylene. *Annual Review of Plant Physiology* **30**: 533-591.
99. Liu, C.A., Zhong, H., Vargas, J., Penner, D., and Sticklen, M. (1998). Prevention of fungal diseases in transgenic, bialaphos- and glufosinate-resistant creeping bentgrass (*Agrostis palistris*). *Weed Science* **46**:139-46.
100. Logusch, E.W., Walker, D.M., McDonald, J.F., and Franz, J.E. (1991). Inhibition of plant glutamine synthetase by substituted phosphinothricins. *Plant Physiol*ogy **95**:1057-1062.
101. Logusch, E.W., Walker, D.M., McDonald, J.F., Franz, J.E., Willafranca, J.J., DiIanni, C.L., Colanduoni, J.A. , Li, B., and Schineller, J.B. (1990). Inhibition of *Escherichia coli* glutamine synthetase by α- and γ-substituted phosphinothricins. *Biochemistry* **29**:366-372.
102. Lori, H.S., Leather, G.S., and Chen, P.K. (1994). Effects of two sesquiterpene lactones isolated from *Artemisia annua* on physiology of *Lemna minor*. *Journal of Chemical Ecol*ogy **20**: 967-978.
103. Manderscheid, R. and Wild, A. (1986). Studies on the mechanism of inhibition by phosphinothricin of glutamine synthetase isolated from *Triticum aestivum*. *Journal of Plant Physiology* **123**:135-142.
104. Marasas, W.F.O., Nelson, P.E., and Toussoun, T.A. (1984). Toxigenic *Fusarium* Species: *Identity and Mycotoxicology*. University Park, PA, USA : Penn. State University Press, 328 pp.
105. Mase, S. (1984). A new herbicide. *Japan Pesticide Inf*ormation **45**:27-30.
106. Mastalerz, P. (1959). Synthesis of phosphonic acids related structurally to glutamic acid. *Roczniki Akademii Rolniczej w. Poznaniu Fizyka Chemia* **33**:985-991.
107. Melroy, D. and Jones, R.L. (1986). The effect of monensin on intracellular transport and secretion of α-amylase isoenzyme in barley aleurone. *Planta* **67**:252-259.
108. Mio, S., Ichinose, R., Goto, K. and Sugai, S. (1991). Synthetic studies on (+)-hydantocidin (1): A total synthesis of (+)-hydantocidin, a new herbicidal metabolite from a microorganism. *Tetrahedron* **47**:2111-2120.
109. Mio, S., Shiruishi, M., Sugai, S., Haruyama, H. and Sato, S. (1991). Synthetic studies on (+)-hydantocidin (2): Alcohol addition approaches toward the stereoisomers of (+)-hydantocidin. *Tetrahedron* **47**:2121-2132.
110. Mio, S., Ueda, M., Hamura, M., Kitagawa, J. and Sugai, S. (1991). Synthetic studies on (+)-hydantocidin (4): Synthesis of stereoisomers of (+)-hydantocidin. *Tetrahedron* **47**:2145-2154.
111. Mollenhauer, H.H., Morré, J.D. and Proleskey, R.E. (1986). Monensin inhibition of growth of ryegrass seedlings. *Botanical Gazette* **147**:432-436.
112. Mollenhauer, H.H., Morré, D.J. and Norman, J.O. (1982). Ultrastructural observations of maize root tips following exposure to monensin. *Protoplasma* **12**:117-126.
113. Morré, D., Boss, J. W.F., Grimes, H. and Mollenhauer, H.H. (1983). Kinetics of Golgi apparatus membrane flux following monensin treatment of embryogenic carrot cells. *European Journal of Cell Biology* **30**: 25-32.
114. Morris, M.P., Pagan, C. and Warmke, H.E. (1954). Hiptagenic acid, a toxic component of *Indigofera endecaphylla*. *Science* **119**:322-323.
115. Muller, W.H. and Muller, C.H. (1964). Volatile growth inhibitors produced by *Salvia* species. *Bulletin Torrey Botanical Club* **91**: 327-330.
116. Murakami, T., Anzai, H., Imai, S., Satoh, A., Nagaska, K., Thompson, C. (1986). The bialaphos biosynthetic gene of *Streptomyces*: Molecular cloning and characterization of the gene cluster. *Journal of Molecular Genetics* **205**:42-50.
117. Nakajima, M., Itoi, K., Takamatsu, Y., Kinoshita, T. and Okazaki. T. (1991). Hydantocidin: a new compound with herbicidal activity from *Streptomyces hygroscopicus*. *Journal of Antibiotics* **44**:293-300.
118. Nakajima, M., Itoi, K., Takamatsu, Y., Sato, S., Furukawa, Y., Furuya, K., Honma, T., Kotani, Y., Kozasa, M.

and Haneishi, T. (1989). Cornexistin: a new fungal metabolite with herbicidal activity. *Journal of Antibiotics* **44**: 1065-1072.

119. Nakamoto, H., Ku, M.S.B. and Edwards, G.E. (1982). Inhibition of C_4 photosynthesis by benzamidooxy)acetic acid. *Photosynthesis Research* **3**: 293-305.

120. Nelson, P.E., Plattner, R.D., Shackelford, D.D. and Desjardins, A.E. (1992). Fumonisin B_1 production by *Fusarium* species other than *F. moniliforme* in section *Liseola* and by some related species. *Applied Environmental Microbiology* **58**:984-989.

121. Netzley, D.H. and Butler, L.G. (1986). Roots of sorghum exude hydrophobic droplets containing biologically active components. *Crop Science* **26**:776-778.

122. Niida, T., Inouye, S., Tsuruoka, T., Shomura, T., Kondo, Y., Ogawa, Y., Watanabe, H., Sekizawa, Y., Watanabe, T., and Igarashi, H. (1973). Antibiotic SF-1293 from *Streptomyces hygroscopicus*. Germ. Offen. DE 2 236 599. Meiji Seika Kaisha.

123. Nimbal, C.I., Yerkes, C.N., Weston, L.A. and Weller, S.C. (1996). Herbicidal activity and site of action of the natural product sorgoleone. *Pesticide Biochemistry and Physiology* **54**:73-83.

124. Nimbal, C.I. and Weston, L.A.(1996). Mode of action of sorgoleone, a natural product isolated from *Sorghum bicolor*. *Proceedings of the 2nd International Weed Control Congress*, **III**: 863-868.

125. Nimbal, C.I.,Pedersen, J., Yerkes, C.N., Weston, L.A. and Weller, S.C. (1996). Activity and distribution of sorgoleone in grain sorghum germplasm. *Journal of Agricultural and Food Chemistry* **44**:1343-1347

126. Omura, S., Hinotozawa, T., Imamura, N., and Marata, M. (1984). The structure of phosalacine, a new herbicide antibiotic containing phosphinothricin. *Journal of Antibiotics* **37**:939-940.

127. Omura, S., Murata, M., Hanaki, H., Hinotozawa, T., Oiwa, R., and Tanaka, H. (1984). Phosalacine, a new herbicidal antibiotic containing phosphinothricin. Fermentation, isolation, biological activity, and mechanism of action. *Journal of Antibiotics* **37**:829-835.

128. Omura, S., Murata, M., Imamura, N., Iwai, Y., Tanaka, H., Furusaki, A. and Matsumoto, T. (1984) Oxetin, a new antimetabolite from an actinomycete: Fermentation, isolation, structure and biological activity *Journal of Antibiotics* **37**:1324-1332.

129. Parker, R. and Williams, M.C. (1974). Factors affecting miserotoxin metabolism in timber milkvetch. *Weed Science* **22**:552-556.

130. Poland, B.W., Lee, S-F., Subramanian, M.V., Siehl, D.L., Anderson, R.J., Fromm, H.J. and Honzatko, R.B. (1996). Refined crystalline structure of adenylsuccinate synthetase from *Escherichia coli* complexed with hydantocidin 5'-phosphate, GDP, HPO_4^{2-}, Mg^{2+} and hadicidin. *Biochemistry* **35**:5753-5759.

131. Pruess, D.L., Scannell, J.P., Ax, H.A., Kellett, M., Weiss, F., Demny, T.C., and Stempel, A. (1973). Antimetabolites produced by microorganisms. VII. L-(N^5-phosphono)methionine-5-sulfoximinyl-L-alanyl-L-alanine *Journal of Antibiotics* **26**:261-266.

132. Rabie, C.J., Marasas, W.F.O., Thiel, P.G., Lübben, A. and Vleggaar, R. (1982). Moniliformin production and toxicity of different *Fusarium* species from southern Africa. *Applied Environmental Microbiology* **43**:517-521.

133. Rashe, E. (1995). In *Herbicide Resistant Crops in Pastures in Australian Farming Systems, Proceedings of Workshop*, (Eds., G.D. McLean and G. Evans), Parkes, Australia : Bureau of Resource Science, 253 pp.

134. Rasche, E. and Gadsby, M. (1997). Glufosinate-ammonium tolerant crops – International commercial developments and experience. *Brighton Crop Protection Conference - Weeds*, **9B**:941-946.

135. Ridley, S.M. and McNally, S.F. (1985). Effects of phosphinothricin on the isoenzymes of glutamine synthetase isolated from plant species which exhibit varying degrees of susceptibility to the herbicide. *Plant Science* **39**:31-36.

136. Robinson, D.G. (1981). The ionic sensitivity of secretion-associated organelles in root cap cells of maize. *European Journal of Cell Biology* **30**: 25-32.

137. Rodriguez, E., Towers, G.N., and Mitchell, J.C. (1976). Biological activities of sesquiterpene lactones. *Phytochemistry* **15**: 1573-1880.

138. Ronzio, R.A., Rowe, W.B., and Meister, A. (1969). Studies on the mechanism of inhibition of glutamine synthetase by methionine sulfoximine. *Biochemistry* **8**:1066-1075.

139. Ruff, M. D., Reid ,W.M. and Rahn, A.P. (1976) Efficacy of different feeding levels of monensin in the control of coccidiosis in broilers. *American Journal of Veterinary Research* **37**:963-967.

140. Rupp, W., Finke, M., Bieringer, H., and Langelueddeke, P. (1977). Herbicidal composition. Germ. Offen. DE 2 717 440. Hoechst AG.

141. Sankula, S., Braverman, M.P., Jodari, F., Linscombe, S.D., and Oard, J.H. (1997). Evaluation of glufosinate on rice (*Oryza sativa*) transformed with the BAR gene and red rice (*Oryza sativa*). *Weed Technology* **11**:70-75.

142. Sarhan, A.R.T. and Hegazi, A.M.A. (1988). Effect of nitrogen, phoshorus and potassium on sensitivity of pepper

plants to *Fusarium* wilt toxin. *Acta Phytopathology and Entomology of Hungary* **23**:57-60.
143. Sauer, H., Wild, A., and Rühle, W. (1987). The effect of phosphinothricin (glufosinate) on photosynthesis. II. The causes of inhibition of photosynthesis. *Zeitschrift fur Naturforschung.* **42c**:270-278.
144. Scannell, J.P., Pruess, D.L., Demney, T.C., Ax, H.A., Weiss, F., Williams T., and Stempel, A. (1972). L-(N^5-phosphono)methionine-*S*-sulphoximinyl-L-alanyl-L-alanine, an antimetabolite of L-glutamine produced by a *Streptomycete*. In *Chemistry and Biology of Peptides;* (Eds., R. Watler and J. Meinhafer), pp. 415-21. Ann Arbor, MI, USA : Ann Arbor Science Publishers.
145. Schnepf, E. (1983). Light-dependent, monensin-induced thylakoid swelling. *Naturwissenschaften* **70**: 260.
146. Sekizawa, Y. and Takematsu, T. (1983). How to discover new antibiotics for herbicidal use. In *Pesticide Chemistry, Human Welfare and the Environment, Vol. 2, Natural Products*; (Ed., J. Miyamoto), pp. 261-268. Oxford : Pergamon Press.
147. Shoji, J., Kozuki, S., Matsutani, S., Kubota, T., Nishimura, H., Mayama, M., Motokawa, K., Tanaka, Shimaoka, Y., N. and Otsuka, H. (1968). Studies on polyetherin A. I. Isolation and characterization. *Journal of Antibiotics* **21**:402-409.
148. Smith, A.E. (1988). Persistence and transformation of the herbicide [^{14}C]glufosinate-ammonium in prairie soils under laboratory conditions. *Journal of Agricultural and Food Chemistry* **36**:393-397.
149. Springer, J.P.,Clardy, J., Cole, R.J., Kirksey, J.W., Hill, R.K., Carlson, R.M. and Isidor, J.L. (1974). Structure and synthesis of moniliformin, a novel cyclobutane microbial toxin. *Journal of the American Chemical Society* **92**: 2267-2268.
150. Stermitz, F.R., Norris, F. A. and Williams, M. C. (1969). Miserotoxin, a new naturally-occuring nitro compound. *Journal of the American Chemical Society* **91**:4599-4600.
151. Sticher, L. and Jones, R.L. (1988). Monensin inhibits the secretion of α-amylase but notpolysaccharide slime from seedling tissues of *Zea mays*. *Protoplasma* **142**:36-45.
152. Stierle, A., Upadahyay, R. and Strobel, G. (1991). Cyperin, a phytotoxin produced by *Aschochyta cypericola*, a fungal pathogen of *Cyperus rotundus*. *Phytochemistry* **30**: 2191-2192.
153. Styer, C.H. and Cutler, H.G. (1984). Effects of moniliformin on mitosis in maize (*Zea mays*). *Plant and Cell Physiology* **25**:1077-1082.
154. Summers, L.A. (1980). *The Bipyridinium Herbicides*; pp. 316-916. London : Academic Press.
155. Siehl, D.L., Subramanian, M.V., Walters, E.W., Lee, S-F., Anderson, R.J. and Toschi, A.G. (1996). Adenylosuccinate synthetase: site of action of hydantocidin, a microbial phytotoxin. *Plant Physiology* **110**:753-758.
156. Sutko, J.L., Besch, J.R., Bailey, J.C., Zimmerman, G. and Watanabe, A.M. (1977). Direct effects of the monovalent cation ionophores monensin and nigericin on myocardium. *Journal of Pharmacology and Expermental Therapeutics* **203**: 685-700.
157. Tachibana, K. (1987). Herbicidal characteristics of bialaphos. *Pesticide Science and Biotechnology Proceedings, 6th (1986) International Congress of Pesticide Chemistry,* 145-148.
158. Tachibana, K., Watanabe, T., Sekizawa, Y., and Takematsu, T. (1986). Action mechanism of bialaphos. *Journal of Pesticide Science* **11**:27-31.
159. Tada, T., Kanzaki, H., Norita, E., Uchimiya, H., and Nakamura, I. (1996). Decreased symptoms of rice blast disease on leaves of *bar*-expressing transgenic rice plants following treatment with bialaphos. *Molecular Plant-Microbe Interactions* **9**:762-764.
160. Takahashi, S. (1994). Hydantocidin and cornexistin: new herbicidal antibiotics. In *Natural and Engineered Pest Management Agents.* Amer. Chem. Soc. Symp. Ser. No. **551** : 74-84. USA, Washington DC : American Chemical Society.
161. Takematsu, T., Konnai, M., Tachibana, K., Tsuruoka, T., Inouye, S., and Watanabe, T. (1979). Antibiotic SF-1293 as a herbicide. Jpn. Kokai Tokky Koho JP 79 067026. Germ. Offen. DE 2 848 224. Meiji Seika Kaisha.
162. Tartakoff, A.M. (1983). Perturbation of vesicular traffic with the carboxylic ionophore monensin. *Cell* **32**: 1026-1028.
163. Tartakoff, A.M. and Vassili, P. (1978). Comparative studies of intracellular transport of secretory proteins. *Journal of Cell Biology* **79**: 694-707.
164. Tate, S.S. and Meister, A. (1973). In *The Enzymes of Glutamine Metabolism*; pp. 77-86. New York: Academic Press.
165. Taylor, P.A., Schnoes, H.K. and Durbin, R.D. (1972). Characterization of chlorosis-inducing toxins from a plant pathogenic *Pseudomonas* sp. [*Phleum pratense*]. *Biochemistry Biophysics Acta* **286**: 107-117.
166. Thompson, C.J., Movva, N.R., Tigard, R., Crameri, R., Davies, J.E., Lauwereys, S.M., and Botterman, J. (1987).

Characterization of the herbicide-resistance gene BAR from *Streptomyces hygroscopicus*. *EMBO* Journal 6:2519-2523.
167. *The Pesticide Manual*, 6th Edition. (1979). Croydon, U.K. : British Crop Protection Council, p. 349.
168. Uchimiya, H., Iwata, M., Nojiri, C., Samarajeewa, P.K., Takamatsu, S., Ooba, S., Anzai, H., Christensen, A.H., Quail, P.H., and Toki, S. (1993). Bialaphos treatment of transgenic rice plants expressing a *bar* gene prevents infection by the sheath blight pathogen (*Rhizoctonia solani*). *BioTechnology* **11**:835-36 .
169. United States Patent Office No. 3, 162,525 (1964).
170. United States Bureau of Census (1998).
171. Vasil, I. K. (1996). Phosphinothricin-resistant crops. In *Herbicide-Resistant Crops*, (Ed., S.O. Duke), pp. 85-91. Boca Raton, FL, USA :CRC Press.
172. Venkatasubbaiah, P. , Van Dyke, C.G. and Chilton, W.S. (1992). Phytotoxic metabolites of *Phoma sorghina*, a new foliar pathogen of pokeweed. *Mycologia* **84**: 715-723.
173. Walworth, B.L. (1967). U.S. Patents 3 295 949 and 3 323 895.
174. Weber, H. A. and Gloer, J.B. (1988). Interference competition among natural fungal competitors: an antifungal metabolite from the coprophilous fungus *Preussia fleischhakii*. *Journal of Natural Products* **51**: 897-883.
175. Wendler, C, Barniske, M. and Wild, A. (1990). Effect of phosphinothricin and photorespiration of C_3 and C_4 plants. *Photosynthesis Research* **24**:55-61.
176. Weston, L.A., Nimbal, C.I. and Czarnota, M.A. (1997). Activity and persistence of sorgoleone, a long-chain hydroquinone produced by *Sorghum bicolor*. *Brighton Crop Protection Conference - Weeds*. **6B-6**:509-516.
177. Wild, A., and Mandersheid, R. (1994). The effect of phosphinothricin on the assimilation of ammonia in plants. *Zeitschrift fur Naturforschung* **39c**:500-504.
178. Wild, A. and Wendler, C. (1993). Inhibitory action of glufosinate on photosynthesis. *Zeitschrift fur Naturforschung* **48c**:369-373.
179. Wild, A. and Ziegler, C. (1989). The effect of bialaphos on ammonium-assimilation and photosynthesis. *Zeitschrift fur Naturforschung* **44c**:97-102.
180. Williams, M.C. (1981). Nitro compounds in foreign species of *Astralagus*. *Weed Science* **29**:261-269.
181. Williams, M.C. and Barneby, R.C. (1977). The occurrence of nitrotoxins in North American *Astragalus* (Fabaceae). *Brittonia* **29**:310-32.
182. Williams, M.C. and Gómez-Sosa, E. (1986). Toxic nitro compounds in species of *Astralagus* (Fabacea) in Argentina. *Journal of Range Management* **39**:342-344.
183. Williams, M.C., Norris, F. A., and VanKampen, K.R. (1970). Metabolism of miserotoxin to 3-nitro-1-propanol in bovine and ovine ruminal fluids. *American Journal of Veterinary Research* **31**:259-261.
184. Williams, M.C. and Ralphs, M. H. (1987). Effect of herbicides on miserotoxin concentration in Wasatch milkvetch (*Astralagus miser* var. *oblongifolius*). *Weed Science* **35**:746-748.
185. Wohlleben, W., Arnold, W., Broer, I., Hillemann, D., Strauch, E., and Pühler, A. (1988). Nucleotide sequence of phosphinothricin-*N*-acetyl-transferase gene from *Streptomyces viridochromogenes* Tue H 94 and its expression in *Nicotiana tabacum*. *Gene* **70**:25-37.
186. Yamada, O., Kaise, Y., Futatsuya, F., Ishida, S., Ito, K., Yamamoto, H. and Munakata, K. (1972). Studies on plant growth regulating activities of anisomycin and toyocamycin. *Agricultural and Biological Chemistry* **36**:2013-2015.
187. Yamamato, R., Sakurai, N. and Masuda, Y. (1981) Inhibition of auxin-induced cell elongation by galactose. *Physiologia Plantarum* **53**:543-547.
188. Yamamato, R. and Masuda, Y. (1984). Matrix polysaccharides of oat coleoptile cell walls. *Phytochemistry* **17**: 923-931.

7

Discovery of the triketone class of HPPD inhibiting herbicides and their relationship to naturally occurring β-triketones

CHRISTOPHER G. KNUDSEN*, DAVID L. LEE, WILLIAM J. MICHAELY, HSIAO-LING CHIN, NHAN H, NGUYEN, RONALD J. RUSAY, THOMAS H. CROMARTIE, REED GRAY, BYRON H. LAKE, TORQUIL E. M. FRASER[1] and DAVID CARTWRIGHT[1]
Zeneca Ag Products, 1200 S. 47th St., Richmond, CA 94804, USA

ABSTRACT

Derivatives of naturally occurring syncarpic acids were synthesized following the observation that few plants grew under the bottle brush plant (*Callistemon citrinus*, a member of the Myrtaceae family) which was known to produce these compounds. The compounds were tested in herbicide screening programmes and were shown to be herbicidal to many weeds both pre and post emergent albeit at rates too high to be of commercial value. Combining this work with concurrent efforts to discover new ACCase inhibitors led to the discovery of the 2-benzoylcyclohexane-1,3-diones, a novel class of bleaching herbicides whose mode of action was subsequently found to be the inhibition of the enzyme p-hydroxyphenylpyruvate dioxygenase. The structural similarities between the naturally occurring syncarpic acids and the new inhibitors of p-hydroxyphenylpyruvate dioxygenase were apparent and an attempt was made to combine the best features of both classes resulting in the synthesis of compounds in 2-benzoyl syncarpic acid class, which to date are the most effective herbicides known in either class. These compounds had persistence in the soil, in excess of what would be desirable for a commercial herbicide. The structure activity relationships of the natural syncarpic acids and 2-benzoyl-cyclohexane-1,3-dione classes of herbicides, their relationship to each other, their physical and environmental properties and the steps leading to the discovery of Mikado® are discussed.

Key words: Triketone, HPPD, hydroxyphenylpyruvate dioxygenase, leptospermone, syncarpic acid.

1. Discovery of the triketones

In 1977, Reed Gray, a biologist working for Zeneca Ag Products' Western Research Center (then Stauffer Chemical Company) made the observation that few plants were found growing under *Calistemon citrinus* growing in his back yard. *C. citrinus* commonly known as 'bottle brush plant', is an Australian native used extensively throughout California as an ornamental hedge. It is well adapted to winter rains and warm dry summers. The extent to which weeds were suppressed was out of proportion to what he expected from shading or leaf litter from the plant itself. Soil was taken from the area immediately beneath the plant and extracted and fractionated (Figure I) and the

---

*Correspondence author
[1] *Zeneca Ag Products, Jealott's Hill Research Station, Bracknell, Berkshire, RG42 6ET, UK*

S.S. Narwal et al. (eds.), Allelopathy in Ecological Agriculture and Forestry, 101–111.

resulting extracts sprayed on flats containing *Echinochloa crusgalli* (watergrass) at a rate of 100lb/acre.

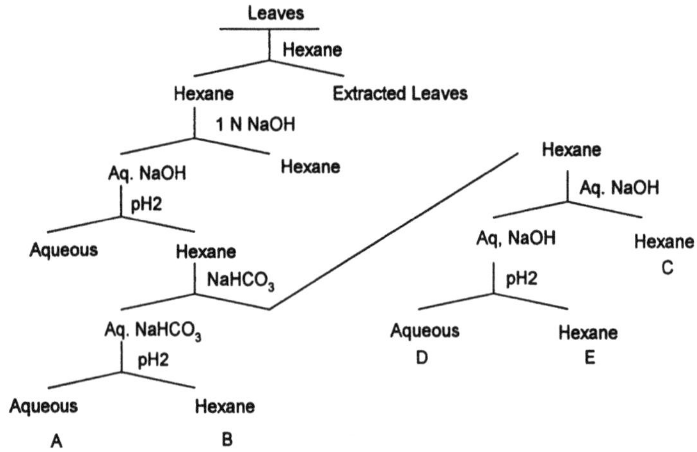

Figure 1. Flow chart for extraction of leptospermone from *Calistemon citrinus*

This experiment was repeated using fresh leaves of *Calistemon citrinus* and a similar result from corresponding fractions was obtained. The activity was found to reside in two fractions, the majority in Fraction B and a minor amount in Fraction E. This indicated that the compound responsible for the herbicidal activity was strongly acidic. These extracts (Fraction B and E) were then placed on a preparative TLC plate and developed. Seeds of *E. crussgalli* were placed on the plate and germinated. The seeds germinating in one section of the plate showed clear bleaching symptoms. The compounds at this band were isolated and a structure of one was determined using standard analytical techniques (1H NMR, IR and Mass Spec.). The first compound isolated and identified was 1-hydroxy-2-isovaloryl-4,4,6,6-tetramethyl cyclohexen-3,5-dione, leptospermone (Fig. 2. **I**) (5,6).

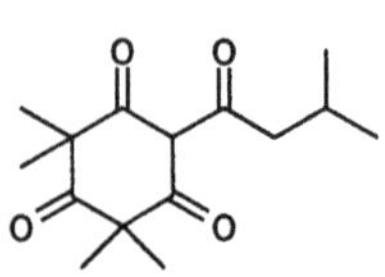

I

Figure 2. Leptospermone

Compound I was then synthesized using the route illustrated in Figure 3. Reaction of phloroglucinol II with isovaleroylnitrile III catalyzed by zinc chloride produced phluoroisovalerone imine IV which was then alkylated with iodomethane using sodium ethoxide in methanol to produce V. The intermediate V was then treated with aqueous hydrochloric acid to give isovaleroylsyncarpic acid I (leptospermone) (5).

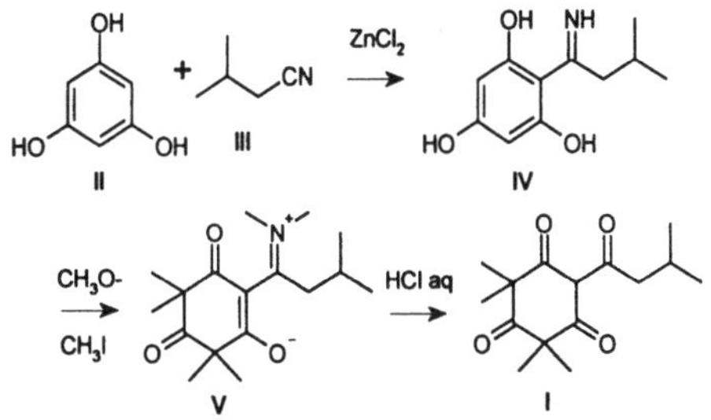

Figure 3. Synthesis of Leptospermone, I, from phloroglucinol, II.

Additional analogues were synthesized in exactly the same manner using acetonitrile, propionitrile, buteronitrile, and isobuteronitrile giving compounds II, III, and IV, respectively (Figure 4).

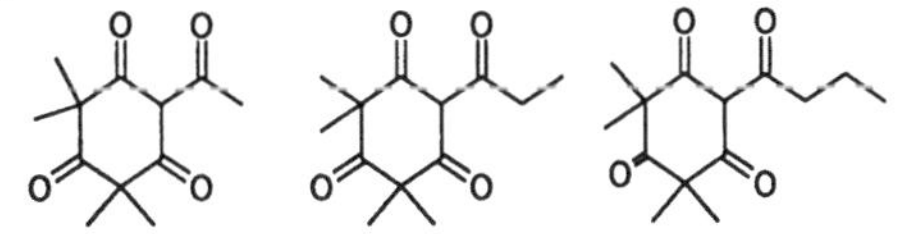

VI VII VIII

Figure 4. Analogues of leptospermone

These compounds were tested for phytotoxicity against weeds both pre-emergence and post-emergence. In the preemergence test, the seeds of seven different weed species were planted in individual rows using one species per row across the width of the flat. The species used were *Digitaria sanguinalis* (hairy crabgrass), *Setaria glauca* (yellow foxtail), *Echinochloa crusgalli* (watergrass), *Avena sativa* (California red oat), *Amaranthus reflexus* (redroot pigweed), *Brassica juncea* (Indian mustard) and *Rumex crispus* (curly dock). The flats were watered after planting. The compounds were weighed and dissolved in 50/50 acetone water with 1% polyoxyethlylene sorbitan monolaurate and then sprayed on the flat at a rate of 8lb/acre. Spray volume was 143 gal/acre. After treatment, the flats were placed in the greenhouse at a temperature of 70 to 85° F and watered by sprinkling. Two weeks after treatment the degree of injury or control was determined by comparison with untreated control plants of the same age. The injury rating from 0 to 100% was recorded for each species, with 0% representing no injury and 100% representing complete control.

In the post-emergence test, seeds of six plant species were used, including *Digitaria sanguinalis, Echinochloa crusgalli, Digitaria sanguinalis, Amaranthus reflexus, Brassica juncea*, and *Phaseolus vulgaris* (pinto bean). The plants were planted as above and placed in the greenhouse and watered daily for 10 to 14 days. When the primary leaves of the bean plant were almost fully expanded and the first trifoliate leaves were just starting to form, the plants were sprayed again at a rate of 8 lb/acre as described above. Spray volume was 238 gal/acre. The flats were returned to the greenhouse and injury was evaluated about 2 weeks after treatment using untreated controls as for the pre-emergence test. The results of the pre- and post-emergent treatments are presented in Table 1.

Table 1. Pre and Post-emergence activity of leptospermone analogues

Compound Number	Average Pre-emergence control	Average Post-emergence control
VI	53	82
VII	17	64
VIII	27	68
I	86	98

In 1980, Bill Michaely returned to this project in a round about way. He embarked upon a program to synthesize benzoyl analogues of the herbicide sethoxydim (4). Efforts to prepare analogue **IX** using the procedure of Rusay did not yield the desired imine **IX** but rather, after hydrolysis upon work-up, the triketone **X**. This triketone was examined in the primary herbicide screening and it showed no herbicidal activity. At that time Stauffer had an active program to identify antidotes for herbicides. The triketone, **X**, was shown to have interesting activity in these screens. Though no better than the antidote standards then available, it did inspire synthesis of analogues to improve activity. We embarked upon a programmme to improve upon this promising activity as an antidote. The first obstacle to this endeavour was that benzoyl cyanides are not easily synthesized and few are commercially available.

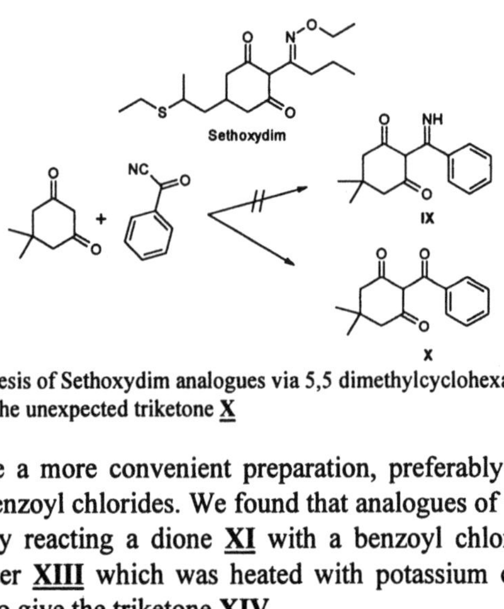

Figure 5. Attempted synthesis of Sethoxydim analogues via 5,5 dimethylcyclohexanedione and benzoyl cyanide yields the unexpected triketone **X**

We wished to have a more convenient preparation, preferably one using the more generally available benzoyl chlorides. We found that analogues of the triketone **X** could be easily prepared by reacting a dione **XI** with a benzoyl chloride **XII** to give the intermediate enol ester **XIII** which was heated with potassium carbonate to effect a Fries rearrangement to give the triketone **XIV**.

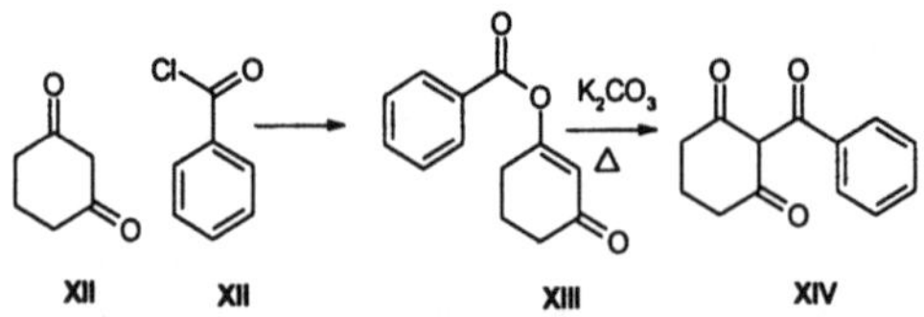

Figure 6. Synthesis of 2-benzoylcyclohexane dione **XIV** from cyclohexane dione **XI** and benzoyl chloride **XII** via the enol ester **XIII**

Unfortunately for the antidote programme, the first series of analogues not only showed no antidotal activity and many were in fact herbicidal. Herbicidal analogues are shown in Figure 7 and inactive compounds in Figure 8.

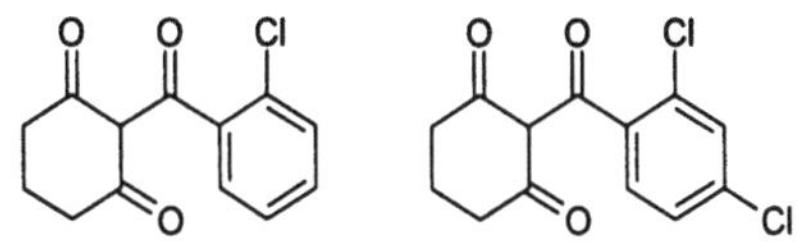

Figure 7. Ortho substituted herbicidal triketones

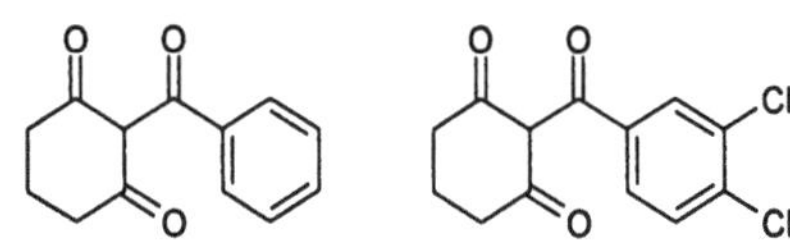

Figure 8. Herbicidally inactive triketones

A team was established to synthesize analogues within this series with sufficient activity and selectivity to compete with herbicides that were then on the market. Initial progress was good, with structural variations explored on both the dione and benzoyl portions of the molecule.

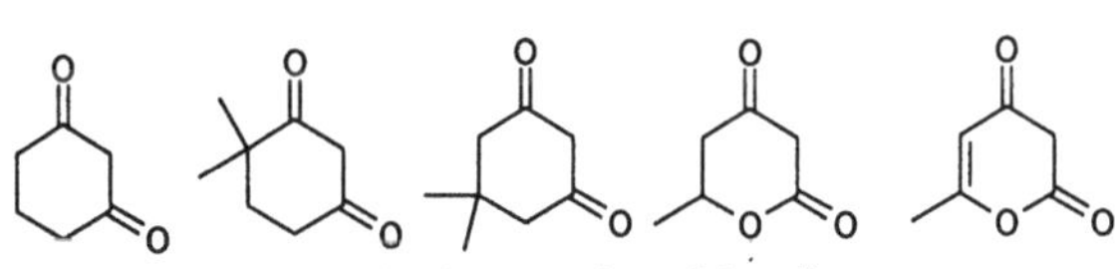

Figure 9. Structural variations explored in the dione portion of the triketone

Examples of diones synthesized are illustrated in Figure 9. Examples of the phenyl substitutions examined are illustrated in figure 10. We quickly determined that substitution at the 2 position of the benzoyl ring was absolutely essential. Compounds lacking any 2-substituent were inactive. Substitutions at the 3 and 4 positions were preferred. Substitutions at the 5 and 6 positions were tolerated but none demonstrated any special advantage. The initial triketone patent was published in 1986 (14). An analysis of the SAR of the 2 and 4 positions was published by David Lee *et al.* (9).

The work with the dione portion of the triketone series showed a clear preference for substitution in the 4 and 6 positions, with mono, di, tri and (as we later showed) tetra substitutions all accepted. Substitution in the 5 position was found to be more complicated, but disubstitution was clearly disfavored.

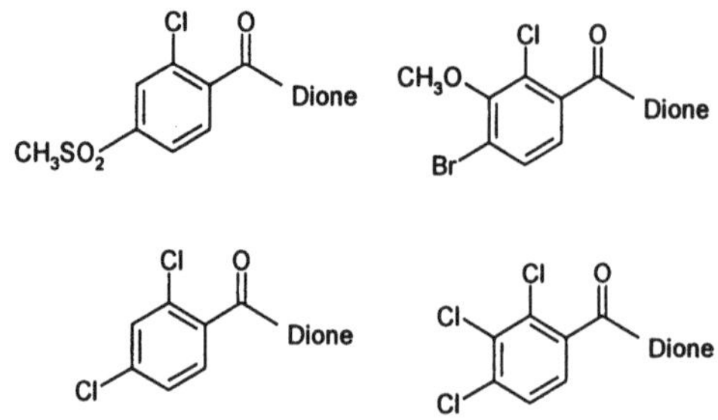

Figure 10. Examples of benzoyl substitutions which gave good herbicidal activity illustrating the scope available.

Until now we had been using the potassium carbonate mediated Fries rearrangement which limited us to thermally stable acid chlorides. We were not able to synthesize compounds with ortho-nitro substituents. This was solved with the invention of a cyanide promoted rearrangement of the enol esters (6). (Figure 11) The enol ester XV was stirred in acetonitrile with triethylamine as a base and a catalytic amount of cyanide in the form of acetone cyanohydrin. This invention offered us access to analogues hitherto unavailable, such as ortho-nitro triketones (e.g.XVI) and the phenyl analogues of the syncarpic acids.

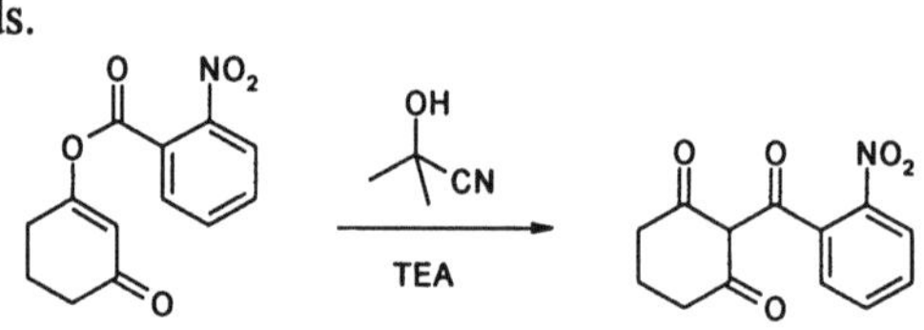

XV XVI

Figure 11. Synthesis of triketones via catalytic Fries rearrangement of an enol ester with acetone cyanohydrin.

We were surprised to discover that the syncarpic acid analogues with appropriately substituted 2-benzoyl substituent were extraordinarily active with LD90s observed at 0.001 kg/ha (greenhouse data XVII as opposed to 5 kg/ha for leptospermone I and its analogues XI, XII, and XIII. With such good activity, we felt it would be easy to find a molecule with sufficient activity and selectivity to be a commercial success. Unfortunately, this was not to be.

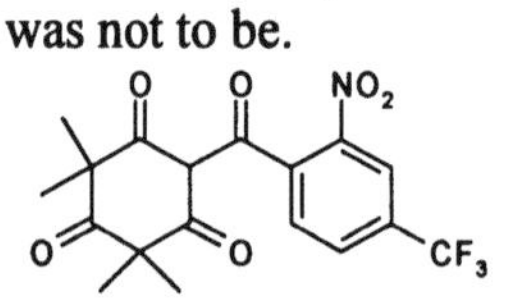

XVII

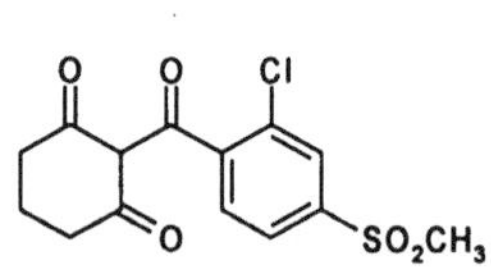

XVIII Mikado

In addition to offering farmers effective products that control weeds without harming crops, Zeneca believes in providing products with minimal impact on the environment. To that end, a target for our group was to discover a triketone with high activity, negligible phytotoxicity to the target crop and low environmental persistence. Our target was a half-life in soil of about two weeks. Examination of a few of the syncarpic acid analogues (e.g. **XVII**) showed that the half life in test soils was greater that 120 days. (Table 2) This may seem surprising considering that these analogues were closer to the natural product leptospermone **I**, than a 2-benzoyl cyclohexane dione, e.g. **XVIII**. We have not tested leptospermone in our soil persistence assays. Although a move closer to the natural product increased activity against weeds, that move took us further away from what we considered to be an environmentally acceptable compound.

Table 2. Soil persistence of triketone analogues

		$t_{1/2}$ Soil Degradation
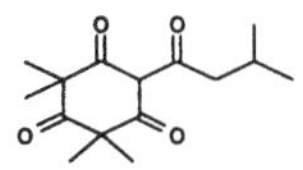	Leptospermone (I)	Not measured
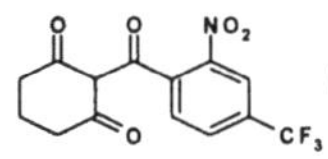	NTBC	Approx. 14 Days
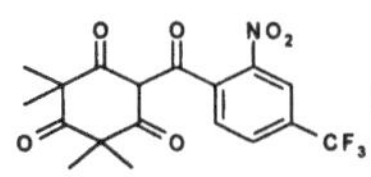	XVII	> 120 Days

2. Discovery of the mode of action

Upon publication of the first patents on this new class of herbicides there was speculation at Zeneca and in the literature about the mode of action. The most obvious symptom expressed by a treated plant is bleaching of the foliage, particularly when applied post emergent. It was shown that *in-vivo* phytoene levels do rise and this observation is consistent with the mode of action of herbicides such as Racer®, **XIX**, a typical bleaching herbicide.

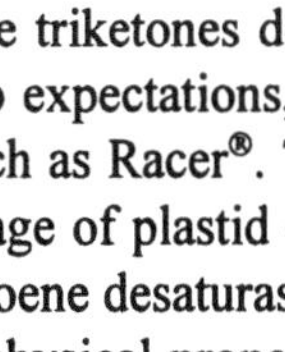

XIX

We were surprised to find that the triketones do not inhibit phytoene desaturase *in-vitro*, a result completely contrary to expectations, if indeed they share the same mode of action as bleaching herbicides such as Racer®. This result led to speculation that the triketones either act at a different stage of plastid development, or that some metabolic activation was required to elicit phytoene desaturase inhibition.

Based on our knowledge of the physical properties required of phytoene desaturase inhibitors, namely that they were typically high log P, planar molecules, we felt that the triketones must be acting at some other site. We began a search for alternatives, and found an article which described inhibitors of tyrosine hydroxylase (8). These authors showed that oudenone **XX** was a competitive with the pterin cofactor of tyrosine hydroxylase. It was speculated that ouedenone was not the inhibitor, but rather the triketone formed through hydrolysis (**XXI**) (Figure 12).

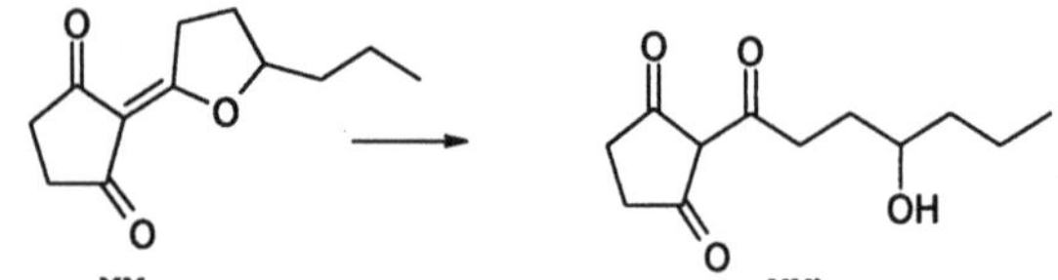

XX XXI

Figure 12. Structure of ouedenone and its hydrolysis product

These authors also described other inhibitors, in particular phenylacetylcyclohexane-1,3-dione as tyrosine hydroxylase inhibitors. Compounds similar to this had already been shown by us to be bleaching herbicides. At that time it was not clear to us how an inhibitor of tyrosine hydroxylase would cause bleaching. It was also anticipated that developing an assay, especially using a plant system, would be difficult. In contrast we felt that we could easily detect tyrosine build up in rats by treating them with a triketone and monitoring tyrosine levels in the urine. Rats were dosed with NTBC and the urine collected. Two colorimetric assays (ferric chloride and nitrosonaphthol) were used to compare the urine of treated rats with that from untreated controls (10). The urine of the treated animals tested positive in both assays. This indicated the presence of tyrosine or a tyrosine metabolite in the urine (Table 3).

Table 3 Results of two assays indicating the presence of tyrosine or its metabolites in rat urine following treatment with the triketone NTBC

Urine source	Ferric chloride assay	Nitrosonaphthol assay
NTBC-treated rats. Dose = 40 mg kg^{-1}	+	++
Negative control	-	-
Untreated	-	-

We then analyzed the urine directly using high pressure liquid chromatography and found the tyrosine level in the urine of treated animals to be substantially elevated compared to the untreated controls. The tyrosine levels were shown to be >2,000 n mol mL^{-1} even in animals dosed at only 1 mg kg^{-1}, compared to only 100 nmol mL^{-1} in the control animals. We were very surprised, then, to discover that NTBC does not inhibit tyrosine hydroxylase (2). It was clear that tyrosine metabolism was being interfered with thus we searched for other explanations. Our research led us to work demonstrating that tyrosine elevation occurs in the disease tyrosinemia type II, which is due to a natural deficiency of the enzyme tyrosine aminotransferase (17). This enzyme is not inhibited by NTBC (10). A nuclear magnetic resonance (NMR) analysis of the urine of NTBC-treated rats showed it contained substantial amounts of both *p*-hydroxyphenylpyruvic and *p*-hydroxyphenylacetic acids (2). This discovery was consistent with the colorimetric assays and indeed suggested that NTBC may interfere with an enzyme in tyrosine catabolism for which one of these acids is a substrate. *p*-Hydroxyphenylpyruvate is a substrate for the enzyme hydroxyphenylpyruvate dioxygenase (HPPD) (11). We showed that NTBC, as well as other triketones, were potent inhibitors of HPPD isolated from rat liver. We presumed that HPPD was also the target in plants and this was demonstrated (1,16,17). The metabolic pathway including tyrosine is illustrated below (Figure 13).

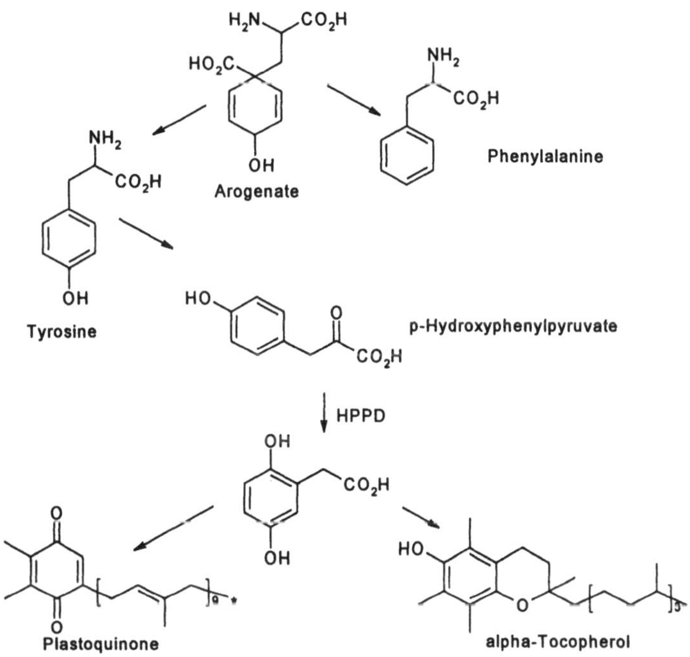

Figure 13. Biosynthesis of plastoquinones and α-tocopherol from tyrosine in plants

It is clear that inhibition of HPPD in plants should result in a decrease in the concentration of plastoquinone (PQ). Indeed this was shown to be the case. Plants treated with triketones (e.g. NTBC) showed an accumulation of tyrosine and a depletion of PQ in meristematic tissue (9). It was shown that PQ is a critical cofactor for the enzyme phytoene desaturase (12,13). Depletion of PQ leads to a depletion of carotenoids subsequent to phytoene in the carotenoid pathway which results in the bleaching of plant tissue. Thus, despite the radically different modes of action of the triketones (HPPD inhibitors) and bleachers such as Racer® (phytoenedesaturase inhibitors) the end result is the same. Plants are unable to convert phytoene to phytofluene which results in the destruction of photosynthetic pigments and the bleaching of plant tissues.

Conclusions

As with all research, the route from initial discovery of the herbicidal activity of leptospermone to a useful product, Mikado® was not direct. In retrospect we can easily measure the progress made. We started with leptospermone with an LD_{90} on broadleaf weeds of about 5 kg/ha, proceeded to increase herbicidal activity reaching an apex with the syncarpic acid analogue **XVII** (0.001 kg/ha, a 5000 fold improvement) and ended with Mikado® a triketone with and LD_{90} on broadleaf weeds of less than 0.1 kg/ha and an outstanding selectivity to maize. The path from leptospermone to Mikado®was not a traditional path of increasing phytotoxicity and crop selectivity. Of great concern to us was soil persistence and we found it necessary to sacrifice some activity to achieve the appropriate soil half life. Mikado was shown to have the best combination of good activity against broadleaf weeds, high selectivity to maize and excellent environmental profile.

References

1. Barta, I.C. and Boger, P. (1995). Benzolcyclohexanedione herbicides are strong inhibitors of purified *p*-hydroxyphenylpyruvic acid dioxygenase of maize. *Pesticide Science* **45**:286-287.
2. Ellis, Martin K., Whitfield, A.C., Gowans, T.R., Anton, Provan, W.M., Lock, E.A. and Smith , L. L. (1995). Inhibition of 4-hydroxyphenyl-pyruvate dioxygenase by 2-(2-nitro-4-trifluoromethylbenzoyl)-cyclohexane-1,3-dione and 2-(2-chloro-4-methanesulfonylbenzoyl)-cyclohexane-1,3-dione. *Toxicology and Applied Pharmacology* **133**:12-19.
3. Gray,Reed A.,Tseng,Chien K., Rusay, Ronald J.; 1-Hydroxy-2-(alkylketo)-4,4,6,6-tetramethylcyclohexen-3,5-diones. *United States Patent* 4,202,840
4. Heather, J. Brian, Milano Pamela D. (1986). Acylated diketonic compounds. *European Patent Application* EP 186,117.
5. Hellyer, R.O. (1968). The Occurrence of β-triketones in the steam-volitile oils of some myrtaceous Australian Plants. *Australian Journal of Chemistry* **21**: 2825.
6. Jain, A.C. and Seshadri, T.R. (1955). C-Methylation of phloroglucinol derivatives. *Proceedings of the Indian Academy of Science* **42A** : 279.
7. Knott, C.A. (1980). Post-emergence control of *Avena fatua* and *Agropyron repens* in peas using alloxydim-sodium and NP55. *Proceedings of the British Crop Protection Conference - Weeds* **15** : 39.
8. Koizumi, S., Nagatsu, T., Iinuma, H., Ohno, M., Takeuchi T. and Umezawa, H. (1982). Inhibition of phenylalanine hydroxylase, a pterin-requiring monooxygenase, by oudenone and its derivatives. *Journal of Antibiotics* **35** : 458.
9. Lee, David L., Knudsen, Christopher G., Michaely, William J., Chin,Hsiao-Ling, Nguyen, Nhan H., Carter, Charles G., Cromartie, Thomas H., Lake, Byron H., Shribbs, John M., Fraser, Torquil E. M. (1998). The Structure-Activity Relationships of the Triketone Class of HPPD Herbicides. *Pesticide Science* **54** : 377-384.
10. Lee, David L., Prisbylla, Michael P., Cromartie, Thomas H., Dagarin, Derek P., Howard, Stott W., Provan, W. McLean, Ellis, Martin K. Fraser, Torquil E.M., Mutter, Linda C. (1997). The discovery and structural requirements of inhibitors of p-hydroxyphenylpyruvate dioxygenase. *Weed Science* **45**:601-609.
11. Lindstedt, S. and Odelhog, B. (1987). 4-Hydroxyphenylpyruvate dioxygenasefrom human liver. *Methods in Enzymology* **142**:139-142.

12. Mayer, M. P., Beyer, P., Kleinig, K. (1990). Quinone compounds are able to replace molecular oxygen as terminal electron acceptor in phytoene desaturation in chromoplastsof *Narcissus psuedonarcissus. European Journal of Biochemistry* **191**:359-363 .
13. Mayer, M.P., Nievelstein, V. and Beyer, P. (1992). Purification and characterization of a NADPH dependent oxidoreductase from chromoplasts of *Narcissus psuedonarcissus:* a redox mediator possibly involved incarotene desaturation. *Plant Physiology and Biochemistry* **30**:389-398.
14. Michaely, W.J. and Kratz, G.W., Certain 2-(2-substituted benzoyl)-1,3-cyclohexanediones. US Patent 4,780,127
15. Natt, E., Kida, K., Odievre, M., Rocco, M.D. and Scherer, G. (1992). Point mutations in the tyrosine aminotransferase gene in tyrosinemia type II. *Proceeding of the National Academy of Science* **89**:9297-9301.
16. Prisbylla M. P., Onisko, B. C., Shribbs, J. M., Adams, D. O., Liu, Y., Ellis, M. K., Hawkes, T. R., and Mutter, L. C. (1993). The novel mechanism of action of the herbicidal triketones. *Proceeding of the British Crop Protection Conference - Weeds* **2**:731-738.
17. Secor, J. (1994). Inhibition of barnyardgrass 4-hydroxyphenylpyruvate dioxygenase by sulcotrione. *Plant Physiology* **106**:1429-1433 .

8

Phytotoxins as potential herbicides

FUMIO SUGAWARA

Department of Applied Biological Science
Research Institutes for Science and Technology,
Science University of Tokyo, Noda, Chiba 278-8510, Japan
sugawara@rs.noda.sut.ac.jp

Abstract

Plant pathogens are able to affect weedy plants, in part because of the phytotoxins produced by them. Phytotoxins of weedy pathogens are produced in culture media by most of the phytopathogenic fungi of weeds that we have studied. In many cases, the phytotoxins belong to a family of related compounds produced by the pathogen. These phytotoxins belong to a wide array of chemical substances including sesquiterpenoids, sesterterpenoids, diketopiperadines, peptides, spirocyclic lactams, isocoumarins and polyketides. Biological activity is usually observed in the range of 10^{-3} to 10^{-6}M concentration. The concept of using these molecules and related compounds as herbicides should be explored.

Key words : Allelochemicals, fungi, herbicides, microorganisms, novel chemical structure, pathogens, phytotoxins, wccds.

1. Introductiion

Plants are often infected with variety of diseases. The causes are non-biological (soil and weather conditions, environmental pollution, damages from agricultural chemicals and biological secondary metabolites) and pathogens (viruses, bacteria and fungi). In diseases caused by pathogens, various physical and biochemical interactions occur between the host plants and pathogens. In nature, host-parasite interactions are diverse and plants are infected with relatively few microorganisms. Furthermore, a given microorganism generally infects only a few plant species. This host specificity has been major area of research in plant pathology.

It is estimated that worldwide pests (pathogens, insects, nematodes and weeds) reduce crop yields by approximately 30% (1,13). Currently, pathologists and entomologists are examining the prospects of utilizing insects and pathogens in

S.S. Narwal et al. (eds.), Allelopathy in Ecological Agriculture and Forestry, 113–128.

biocontrol strategies to control weeds in agricultural crops. In particular, my group works not only on the isolation, identification and use of novel pathogenic fungi as weed control agents, but also on the phytotoxins produced by these organisms. The rationale behind this approach is, that plant pathogens, especially fungi and bacteria, induce disease symptoms in their respective host(s) due to the phytotoxins produced by them. These compounds may be candidates for novel herbicides or serve as models for herbicide development. Sometimes, these phytotoxins are useful tools for screening plants for toxin insensitivity (disease resistance) and as probes of normal physiological plant function. Except crop plants, the disease-inducing fungi and bacteria of many plants spp. have not been examined for their ability to produce phytotoxins. Potentially, there is a reservoir of novel biologically important substances, awaiting discovery in these organisms. Weed pathogens had millennia to evolve with their hosts and devise biochemical mechanisms to kill them or influence their physiology. Now we should take advantage of this, to devise new strategies for weed control. The first step in this direction is, an examination of the structure and function of phytotoxins from weed pathogens (13,14,43 and 44).

The genus *Helminthosporium* has been reclassified into three genera of *Bipolaris*, *Drechslera* and *Exselohilum* (1). Their perfect stages belong to *Cochliobolus*, *Pyrenophora* and *Setosphaeria*, respectively. These genera are wide spread in the fields and are recognized as major pathogens against both weeds and crops. It could be useful to know the nature of phytotoxins produced by these diverse genera. In this chapter, non-host specific phytotoxins produced by these genera are listed based on the structure and biogenetic relations.

2. Herbicidal phytotoxins

The phytotoxins with herbicidal potential have been divided into five classes, (a) polyketides, (b) terpenes, (c) amino acid derivatives, (d) mevalonates and polyketides and (e) amino acids, polyketides and miscellaneous.

2.1. POLYKETIDES

Polyketides compounds are intermediates in the pathway of fatty acids synthesis. Their structures have been depicted (Figs. 1) and bioactivity listed in Table 1. (-) Dihydropyrenophorin (I) is one tetraketide macrolide produced by *Drechslera avenae* (2) a causal agent of leaf blotch of oats (*Avena sativa*). While none of the plants tested (*Helianthus annuus, Hordeum vulgare, Sorghum vulgare, Glycine max, Triticum aestivum, Dactyloctenium aegyptium, Cynodon dactylon, Eleucine indica, Setaria glauca, Setaria viridis, Zea mays, Avena sativa, A. fatua*) responded to (-) dihydropyrenophorin (I) at concentrations much below 10^{-4} M. When applied to Johnson grass (*Sorghum halepense*) at 6.5 x 10^{-8} M, reddish flecks hypersensitivity were produced at the site of application.

Table 1. Phytotoxic Polyketides produced by *Bipolaris, Drechslera* and *Exselohilum*

Pathogen	Host Plant	Phytotoxin	Activity	Reference
		Tetraketide		
Drechslera avenae	Johnson grass	(-)-Dihydropyrenophorin (I)	Sunken lesion	47
		Pentaketides		
Cochliobolus lunata		Radicinin (II)		33
Bipolaris coicis		Radicinin (II)		21
Pyrenophora teres		Pyrenolides B (III)	Rice seedling	32
Drechslera indica	Purslane	Curvulin (IV)		16
	Spiny amaranth	*O*-Methyl curvulic acid (V)	Necrosis	
		Hexaketides		
Drechslera siccans	Ryegrasses	De-*O*-methyldiaporthin (VI)		11
Cochliobolus spicifer	Wheat	Spiciferone A (VII), C (VIII)		22-25
		Spiciferinone (IX), Spicifernine (X)		26, 27, 29, 30
		Heptaketides		
Exselohilum turticum	Johnson grass	Monocerin (XI)	Necrosis	8, 40
Exselohilum holmii	Crowfoot grass	Monocerin (XI)		51
Drechslera ravenelii		Monocerin (XI)		41
		Polyketides		
Drechslera phlei	Corn, Crabgrass	*trans*-resorcylide (XII)	Necrosis	10
Drechslera portulacae	Pursulane	Zeaenol (XIII)	Necrosis	46
		LL-Z1640-1 (XIV)		

Curvulin (IV) is a pentaketide produced by numerous fungi. It was isolated from *Drechslera indica* (16), a pathogen of common purslane (*Portulaca oleracea*) and spiny amaranth (*Amaranthus spinosus*). A related compound, *O*-methylcurvulic acid (V) was also found. At nanomolar concentrations, curvulin was selective against purslane and spiny amaranth. *O*-Methylcurvulic acid (V) was generally toxic, although a few plants were insensitive.

Drechslera siccans (11) (telemorph: *Pyrenophora lolii*) is a pathogen of perennial ryegrass (*Lolium perenne*) and oats (*A. sativa*). Its toxic extracts contained de-*O*-methyldiaporthin (VI), a family of isocoumarin. The toxin is selective and effective when applied in nanomolar amounts to leaves.

Cochliobolus spicifer (22-27, 29, 30), a pathogen of wheat, produced new hexaketide phytotoxins, which were tested against leaf protoplasts. Spiciferone A (VII) and spiciferinone (IX) are toxic, while, spicifernin (X) is not toxic against wheat and promoted rice seedling growth. The biosynthetic pathway of these compounds was suggested from the experiments using ^{13}C labeled acetate and methionine.

Monocerin (XI) has been isolated from *Exselohilum turticum* (8, 40), *Exselohilum holmii* (51) and *Drechslera. ravenelii* (41). The biosynthetic pathway of monocerin produced by *D. ravenelli* has been reported (41).

Resorcylides were first discovered in unidentified species of *Penicillium*. Two

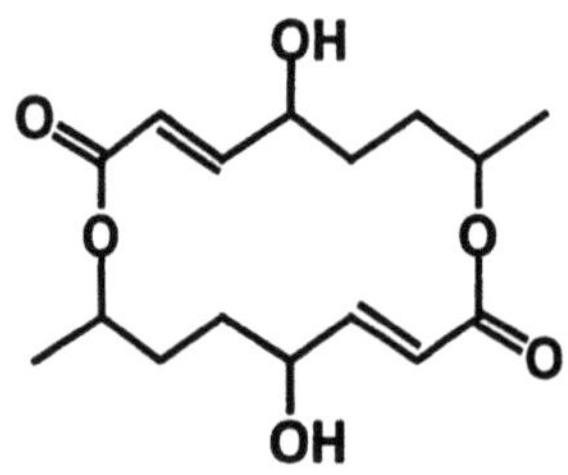

I: (-)-Dihydropyrenophorin

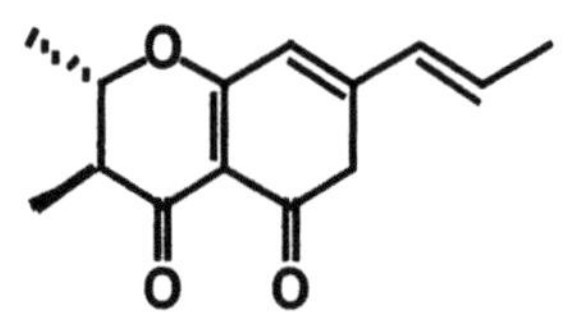

II: Radicinin

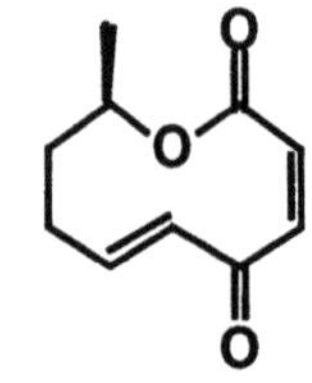

III: Pyrenolides B

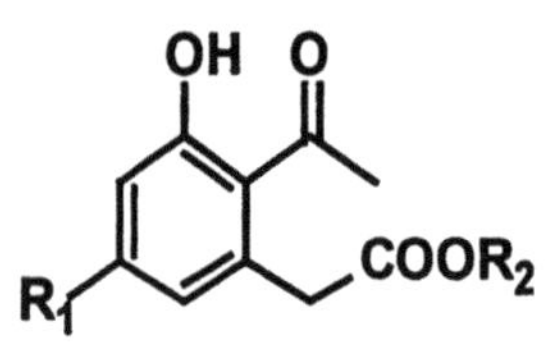

IV: Curvulin (R_1=H, R_2=C_2H_5)
V: *O*-methyl carvulic acid (R_1=OCH_3, R_2=H)

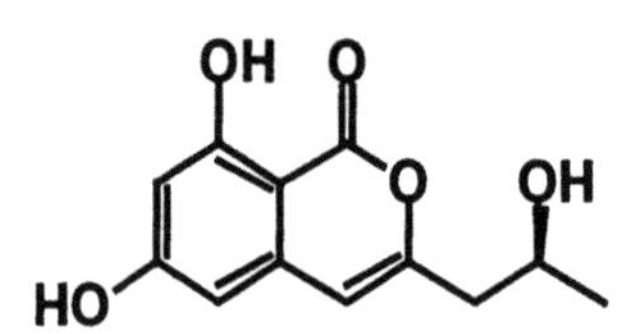

VI: De-*O*-methyldiaportin

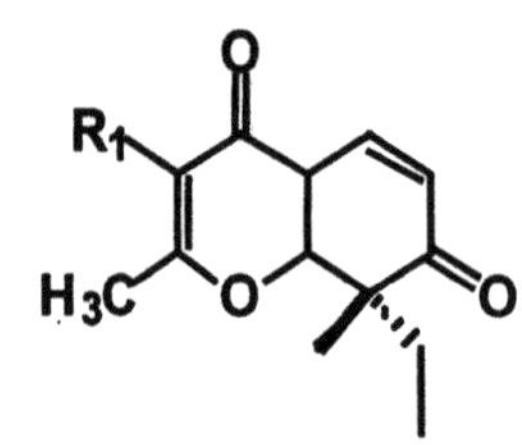

VII: Spiciferon A (R_1=CH_3)
VIII: Spiciferon C (R_1=CH_2OH)

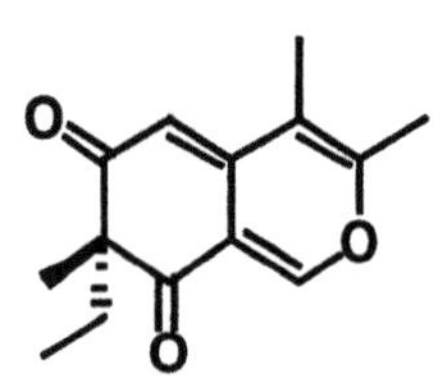

IX: Spiciferinon

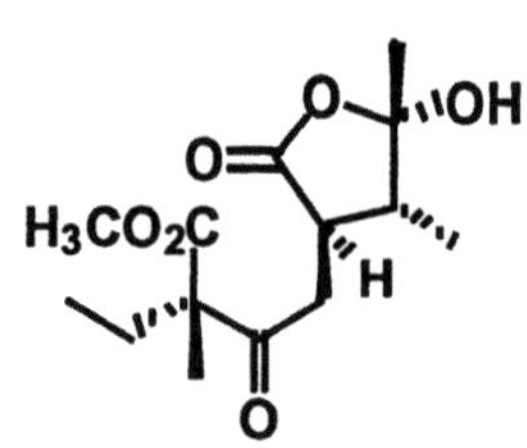

X: Spiciferinin

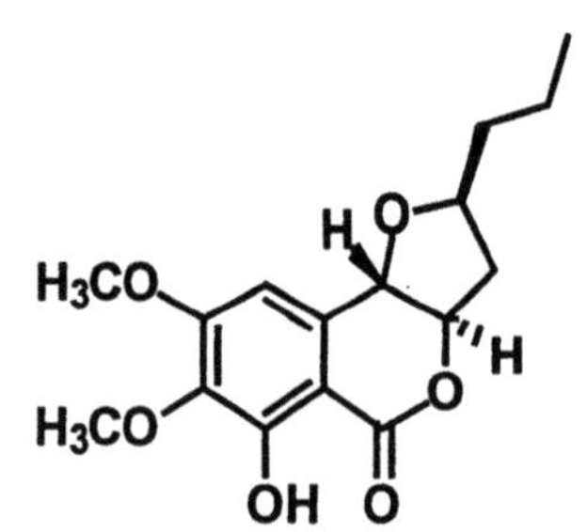

XI: Monocerin

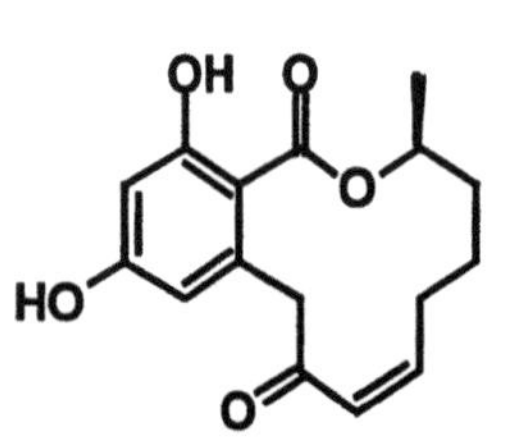

XII: *trans*-resorcylide

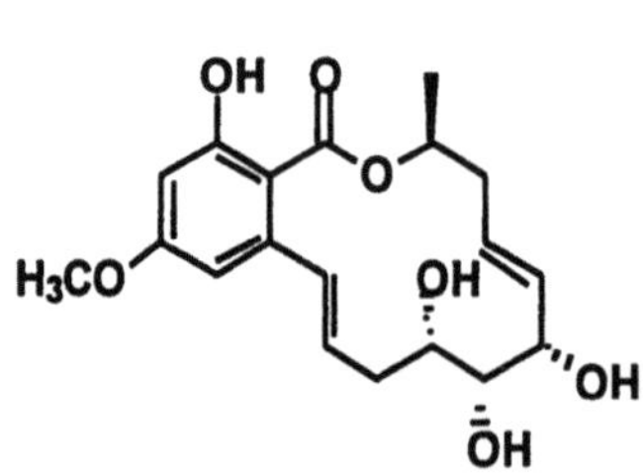

XII: Zeaenol

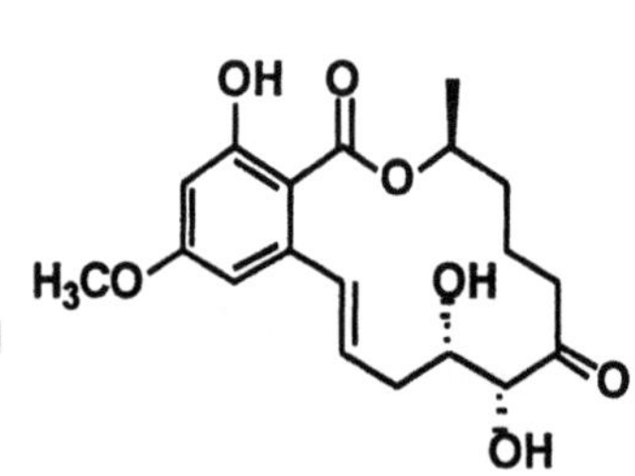

XIII: LL-Z1640-1

Fig. 1. Structures of phytotoxic polyketides

isomers, *cis* and *trans*, were identified. Trans-resorcylide (**XII**) is cytotoxic and antimicrobial and inhibits growth of roots and rice seedlings at 1 ppm concentrations, however, the *cis* isomer is inactive. These two isomers and a dihydro analogue have been identified in extracts of *Drechslera phlei* (10). The *trans* isomer caused necrosis in corn (*Z. mays*) and crabgrass (*Digitaria sanguinalis*) at 0.06 μg per leaf. At 2 μg per leaf, timothy (*Phleum pratense*), wild poinsettia (*Euphorbia heterophylla*) and sunflower (*Helianthus annuus*) were also sensitive.

Drechslera portulaca (46), a pathogen of leaf spotting disease of purslane (*Portulaca oleracea*), produced zealarenone analogues (**XIII**, **XIV**). When leaves of purslane were soaked in solutions of **XIII** and **XIV** at 10^{-6} to 10^{-5} M, necrotic regions developed. Both compounds also inhibited root elongation of cress (*Lepidium sativum*) seedlings.

2.2. TERPENES

Terpenes are the largest family of natural products and their biosynthetic pathway from acetate via mevalonate is well studied. Helminthosporol, eremophilane and ophioboline derivatives are often isolated from both weeds and crops pathogens (Table 2, Fig. 2-1 and 2-2).

Eremophilanes are bicyclic sesquiterpenoids and over 200 compounds of this group have been characterized in higher plants. Numerous eremophilanes have also been found in fungi. These include phomenone (**XXIII**) (*Phoma exigua*), phaseolinone (**XIXb**) (*Macrophoma phaseolina*), sporogen AO-1 (*Aspergillus oryzae*) and PR-toxin (*Penicillium roquefortii*). Phomenone (**XXIII**) and phaseolinone (**XIXb**) cause necrosis on dicots. PR-toxin is a dangerous mycotoxin, produced by the fermentative agent of blue cheese when cultured on corn (*Z. mays*) kernels. Phytotoxic eremophilanes were isolated from *Chochliobolus cynodontis* (anamorph: *Bipolaris cynodontis*) (45, 49), a fungal pathogen of Bermuda grass (*C. dactylon*). This pathogen makes two eremophilanes, biplaroxin (**XVIII**) and its dehydrate. Bipolaroxin (**XVIII**) showed some host selectivity and causes lesions on Bermuda grass at 38 μM, whereas, 0.7 mM concentration produces detectable symptoms on wild oats (*A. fatua*), sugarcane (*Saccharum officinarum*) and corn (*Z. mays*).

Drechslera gigantea (2, 3, 15), a pathogen of quackgrass (*Agropyron repens*), produces gray lesions on its hosts. Organic extracts of culture filtrates contain numerous eremophilanes, many of which have novel structures. The basic structure of eremophilanes from *D. gigantea* includes a rigid bicyclic system with a keto group at C8, unsaturation at C9-C10, a hydroxyl at C3 and *cis*-methyl groups at C4 and C5. Biogenic variability occurs via hydroxylation, epoxidation and dehydration at C6 through C13. When dicots were tested for sensitivity and selectivity, a necrotic reaction developed. When tested on monocots, however, gigantenone (**XIXa**), petasol (**XXa**) and phomenone (**XXI**) produced green islands, localized areas of chlorophyll retention in senescing tissues. Further studies revealed that, like cytokinin-induced green islands, eremophilane-induced green islands retain their photosynthetic ability. In addition, eremophilanes are as effective as cyclohexamide in inhibiting *in vitro* and *in vivo* protein synthesis, making their mode of action different from cytokinin. The differential response of monocots and dicots to compounds has indicated implications in agriculture, for selective herbicidal as well as growth-promotery effects.

Table 2. Phytotoxic Terpenoids produced by *Bipolaris, Drechslera* and *Exselohilum*

Pathogen	Host Plant	Phytotoxin	Activity	Reference
		Sesquiterpenoids Related to Helminthosporol		
Bipolaris sp	Johnson grass	Dihydroprehelminthosporol (**XV**)	Lesion	35
		Prehelminthosporol (**XVIa**)		
		Prehelminthosporolactone (**XVIb**)		
Bipolaris sp.		Victoxinine (**XVII**)	Necrosis	36
Bipolaris sorokiniana	Barley	Prehelminthosporol (**XVIa**)	Cell membrane	5,28,31
		Eremophilane-type Sesquiterpenoids		
Bipolaris cynodontis	Bermuda grass	Bipolaroxin (**XVIII**)	Zonate zone	45,49
Drechslera gigantea	Turf grass	Gigantenone (**XIXa**)	Protein synthesis	2,3,15
	Bermuda grass	Phaseolinone (**XIXb**)	Green island	
		Petasol (**XXa**)	Rhizogenesis	
		Phomenone (**XXI**)		
		Other eremophilanes (**XXb, XXII, XXIII**)		
		Ophiobolin-type Sesterterpenoids		
Drechslera oryzae	Rice	Ophiobolin A (**XXIV**),	Photosynthesis	53,58
		Ophiobolin B (**XXVIIIa**),		
		I (**XXIXa**), J (**XXXIa**)		
		3-Anhydroophiobolin A (**XXVII**)	Chlorosis, Lesion	
		25-Hydroxyophiobolin I (**XXIXb**)		
		6-Epiophiobolin I (**XXX**)	Lesion, CO_2 Fixation	
		8-Deoxyophiobolin J (**XXXIa**)		
Bipolaris oryzae	Rice	Ophiobolin A (**XXIV**)	Photosynthesis	53,57,58
		B (**XXVIIIa**), I (**XXIXa**)	Chlorosis, Lesion	
Helminthosporium sp.		Ophiobolin A (**XXIV**)	Photosynthesis	17
		3-Anhydroophiobolin A (**XXVII**)	Chlorosis, Lesion	
Bipolaris sorghicola	Johnson grass	Ophiobolin A (**XXIV**), I (**XXIXa**)	Photosynthesis	39,50
		6-Epiophiobolin A (**XXV**)	Chlorosis, Lesion	
		3-Anhydroophiobolin A (**XXVII**)		
		25-hydroxyophiobolin I (**XXIXb**)		
Drechslera maydis	Corn	Ophiobolin A (**XXIV**)	Photosynthesis	4,17,50
		C (**XXVIIIb**), I (**XXIXa**)		53,57,58
		6-Epiophiobolin A (**XXV**)	Chlorosis, lesion	
		25-Hydroxyophiobolin I (**XXIXb**)	Mitochondrial electron transport	
		3-Anhydro-6-epiophiobolin A (**XXVI**)		

Ophiobolins, a group of sesterterpenoids, have been implicated in two major disease epidemics, i.e., the Bengal rice famine of 1943 and the southern corn leaf blight epidemics in USA in 1972. In both diseases, the pathogen was a species of *Drechslera* (4, 17, 50, 53, 57, 58) or *Bipolaris* (37, 50, 53, 57, 58) (telemorph: *Cochliobolus*) known to produce a contingent of ophiobolins. 6-Epiophiobolin A (**XXV**), which

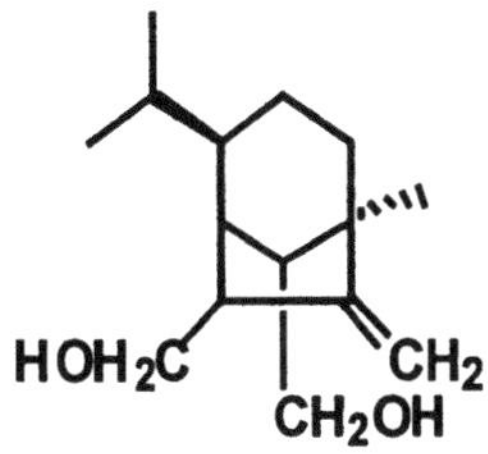

XV: Dihydrohydroprehelminthosporol

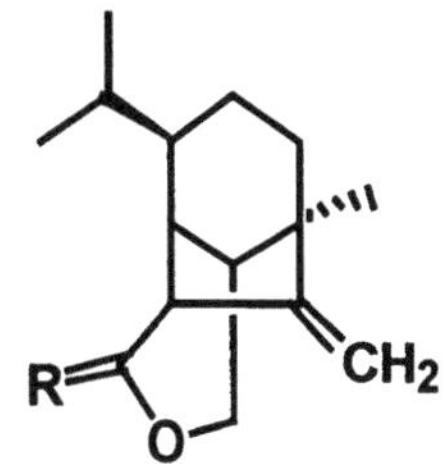

XVIa: Prehelminthosporol (R=H, OH)
XVIb: Prehelminthosporolacton (R=O)

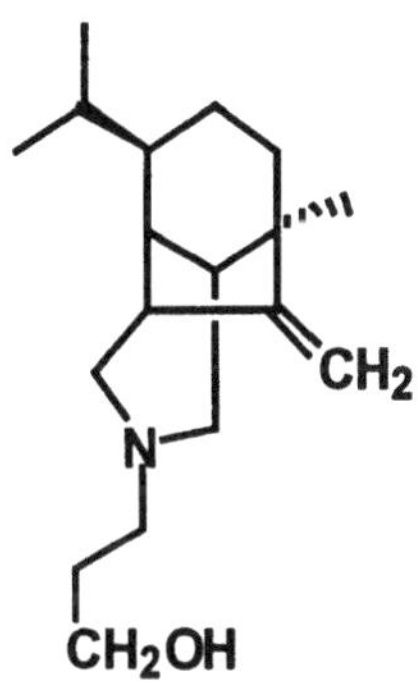

XVII: Victoxinin

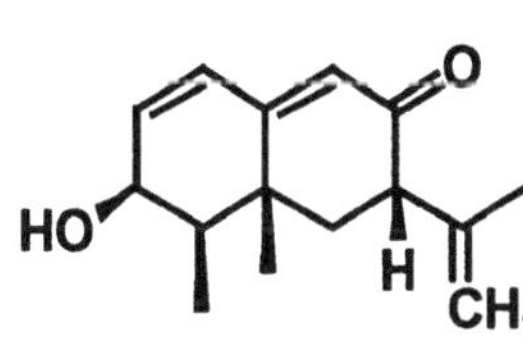

XVIII: Bipolaroxin

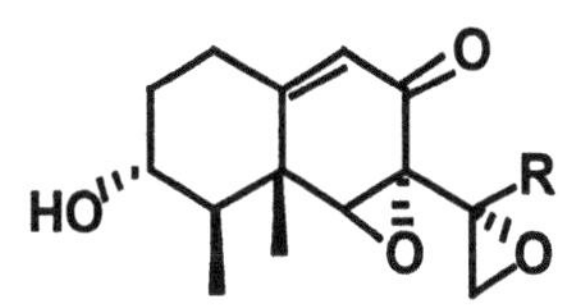

XIXa: Gigantenone (R=CH_3)
XIXb: Phaseolinone (R=CHO)

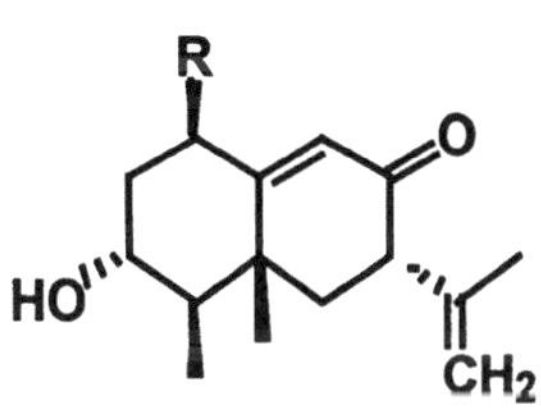

XXa: Petasol (R=H)
XXb: 1-Hydroxypetasol (R=OH)

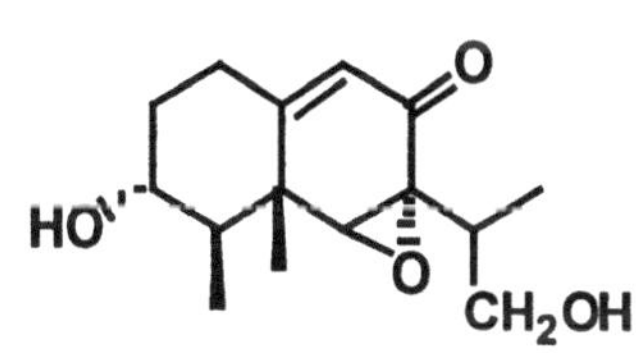

XXI: Gigantenone derivative

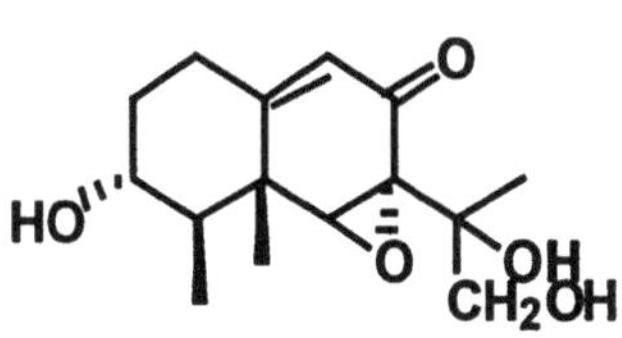

XXII: Gigantenone derivative

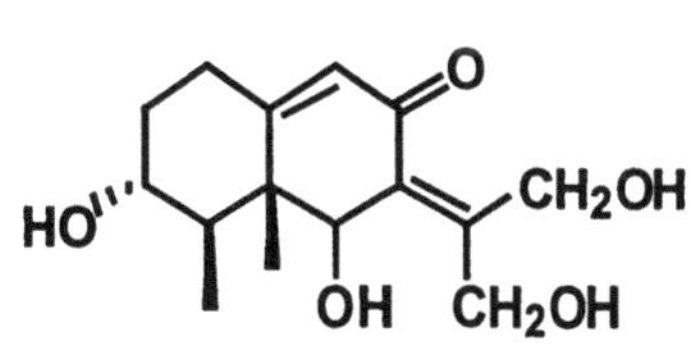

XXIII: Gigantenone derivative

Fig.2-1. Structures of phytotoxic terpenoids

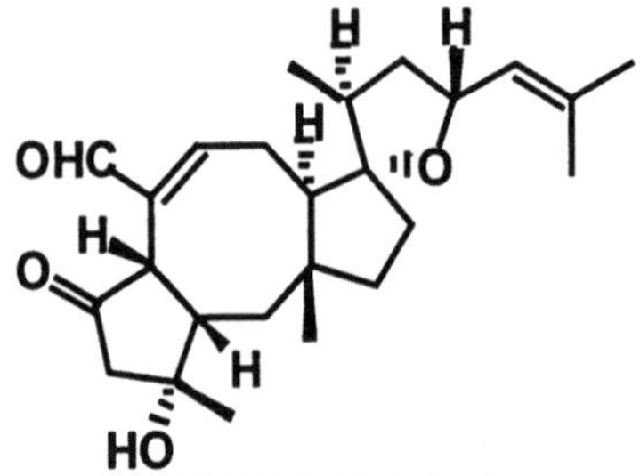

XXIV: Ophiobolin A

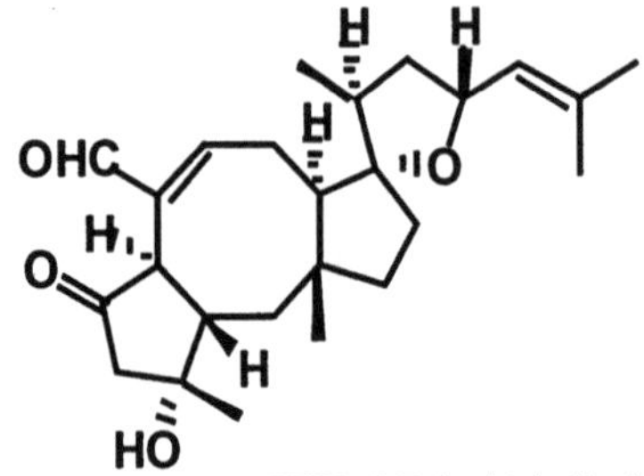

XXV: 6-Epiophiobolin A

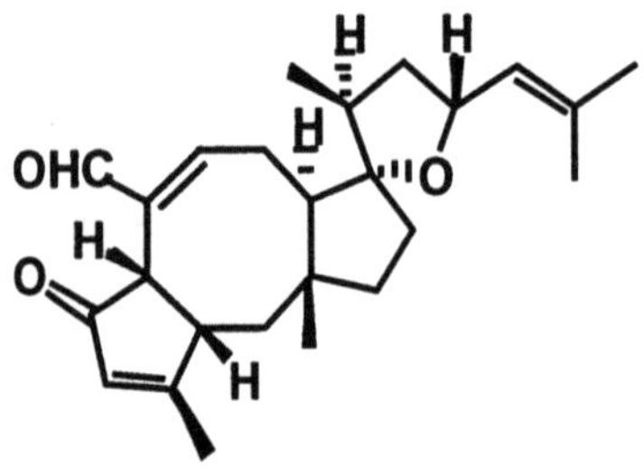

XXVI: 3-Anhydro-6-epiophiobolin A

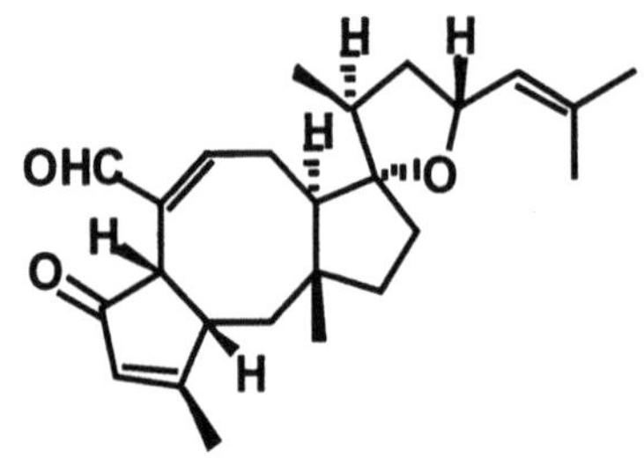

XXVII: 3-Anhydroophiobolin A

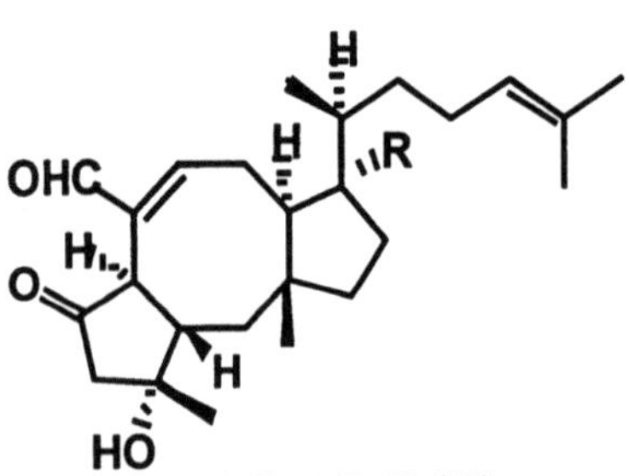

XXVIIIa: Ophiobolin B (R=H, OH)
XXVIIIb: Ophiobolin C (R=O)

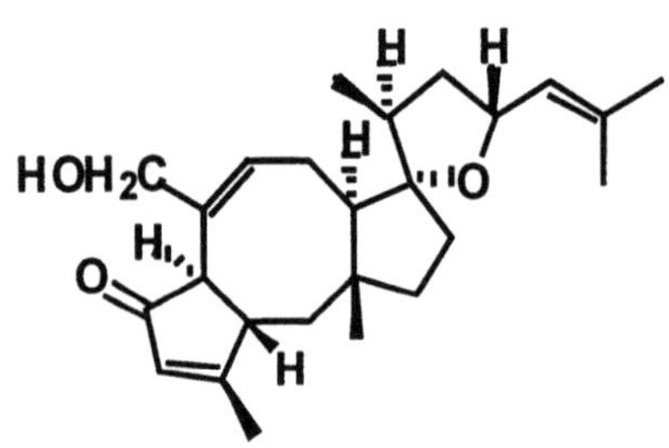

XXIXa: Ophiobolin I (R=CH_3)
XXIXb: 25-Hydroxyophiobolin I (R=CH_2OH)

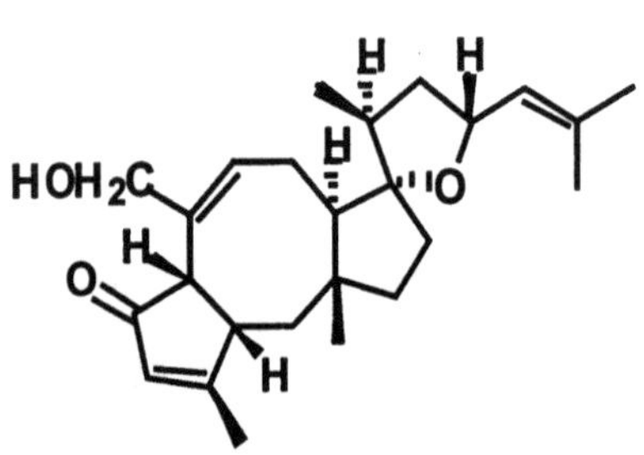

XXX: 6-Epiophiobolin I

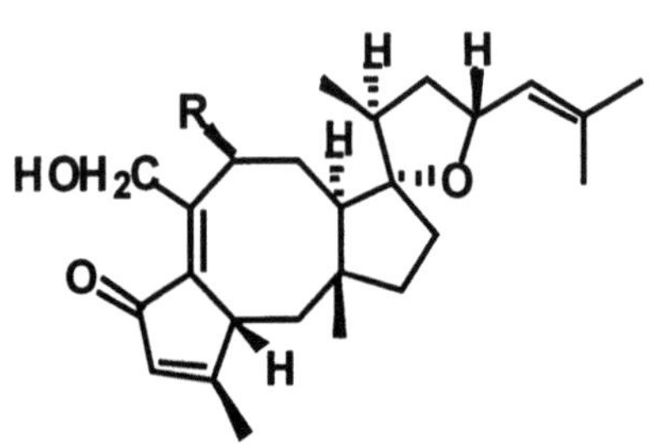

XXXIa: Ophiobolin J (R=OH)
XXXIb: 8-Deoxyophiobolin J (R=H)

Fig.2-2. Structures of phytotoxic terpenoids

differs from ophiobolin A (**XXIV**) only in the orientation of the proton on C6, can selectively inhibit CO_2 fixation in corn bearing TMS (Texas male sterile) cytoplasm at concentrations three orders of magnitude below that required for the same effect in normal corn cytoplasm. Mitochondria from normal corn cytoplasm are also selectively inhibited by 6-epiophiobolin A. Ophiobolin A (**XXIV**) is not discriminating in its toxicity towards these two corn germplasms. *D. heveae*, a pathogen of the rubber tree (*Hevea brasiliensis*), leafy spurge (*Euphorbia esula*) and wild poinsettia (*Euphorbia heterophylla*) produces a number of ophiobolins. Although 6-epiophiobolin A (**XXV**) is the most toxic, but all ophiobolins studied were phytotoxic to many grasses and dicots. *Bipolaris sorghicola*, a pathogen of Johnson grass (*S. halepense*), produces all of the ophiobolins that are also found in *Cochliobolus heterostrophus* (anamorph: *B. maydis*) and *Cochliobolus miyabeanus* (anamorph: *B. oryzae*), the causative agent of brown spot on rice.

2.3. AMINO ACID DERIVATIVES

Phytotoxins related to amino acids are depicted in Fig. 3, and listed in the Table 3. Amongst these, BZR-cotoxin I (**XXXVIII**), II (**XXXIX**), III, IV (**XXXX**) is quite impressive. BZR cotoxins were isolated from *B. zeicola* race 3 (54-56), a pathogen of corn (*Z. mays*). These compounds are not phytotoxic, if applied individually, however, their mixture shows strong phytotoxicity against rice (*Oryza sativa*) plants.

Tryptophol (**XXXII**) is a major metabolite in the culture filtrates of *D. nodulosum* (24), a pathogen of goose grass (*E. indica*). At 0.6 mM concentration, it was selectively toxic to young leaves of goose grass.

Exselohilum holmii (51) is a pathogen of crowfoot grass (*D. aegyptium*), an annual weed. This fungus produces two diketopiperazines, exserophilone (**XXXVI**) and its dihydrate. Both compounds showed non-selective toxicity towards several plant species and produce necrotic lesions.

Table 3. Phytotoxic amino acid derivatives produced by *Bipolaris, Drechslera* and *Exselohilum*

Pathogen	Host Plant	Phytotoxin	Activity	Reference
Drechslera nodulosum	Goose grass	Tryptophol (**XXXII**)	Necrosis, Chlorosis	48
Pyrenophora teres	Barley	*N*-(2-Amino-2-carboxyethyl) Aspartic acid (**XXXIII**)	Necrosis, Chlorosis	9
	Barley	Aspergillomarasmine (**XXXIV**) Anhydroaspergillomarasmine (**XXXV**)	Necrosis, Chlorosis	
Exselohilum holmii	Crowfoot grass	Exserohilone (**XXXVI**) 9,10-Dihydroexserohilone (**XXXVII**)	Necrosis	51
Bipolaris zeicola	Corn	BZR-cotoxin I (**XXXVIII**), II (**XXXIX**), IV (**XXXX**)	Rice seedlings	54-56

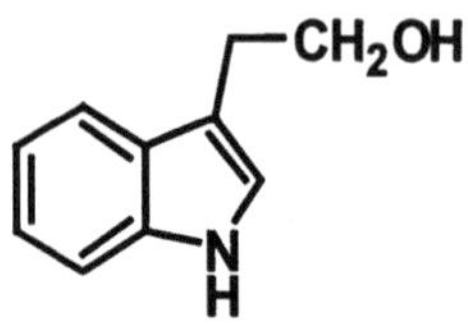

XXXII: Tryptophol

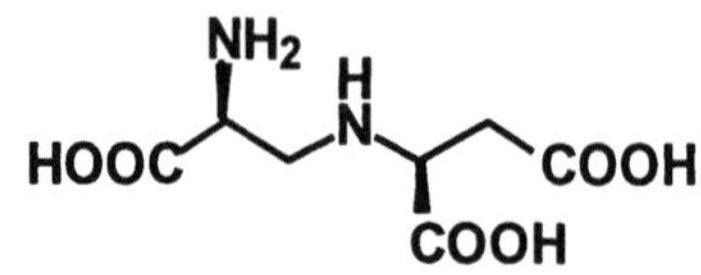

XXXIII: *N*-(2-Amino-2-carboxyethyl)-aspartic acids

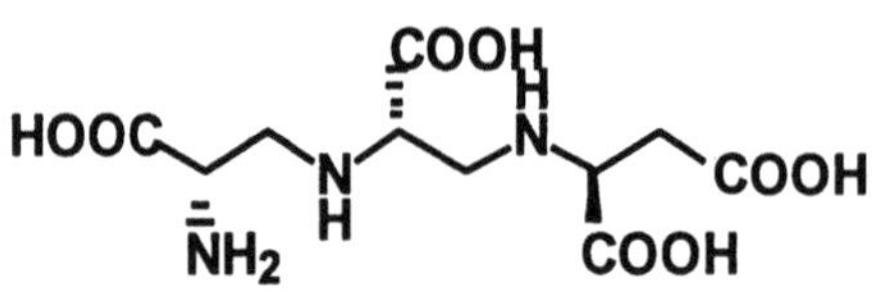

XXXIII: Aspartic acid

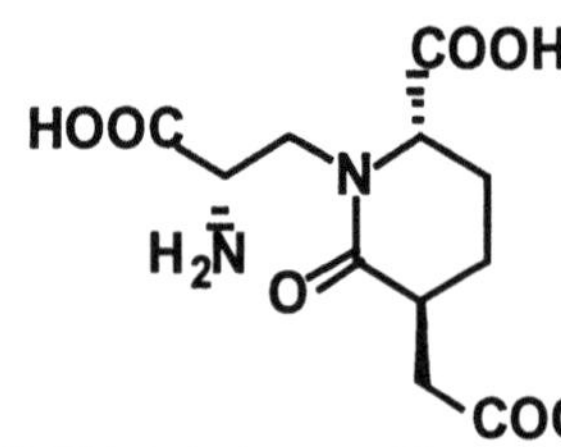

XXXV: Anhydroaspergillomarasmine

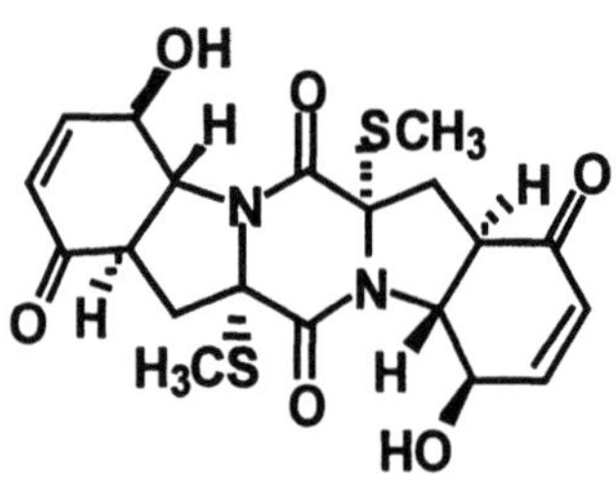

XXXVI: Exserohilone

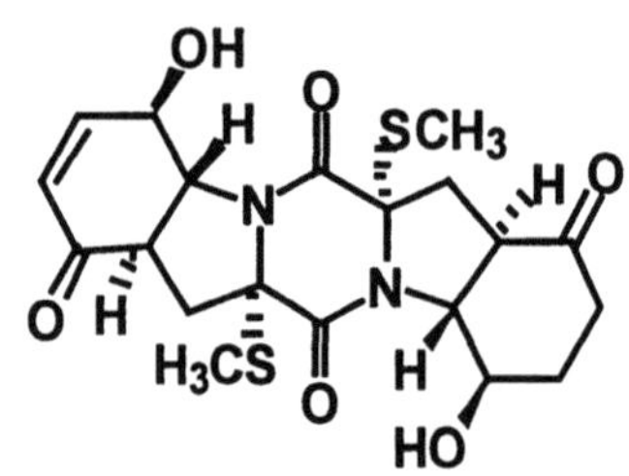

XXXVII: 9,10-Dihydroexserohilone

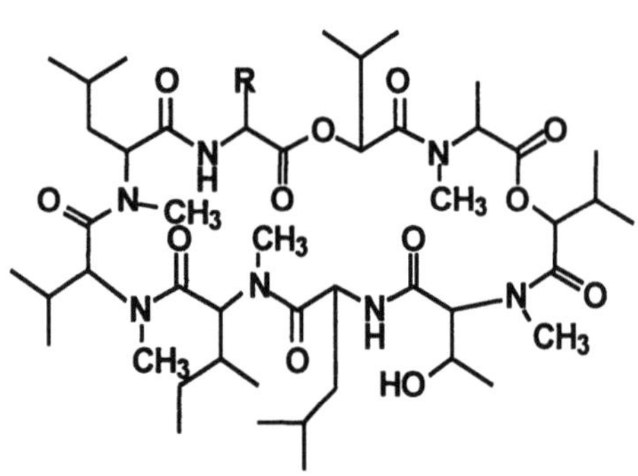

XXXVIII: BZR-Cotoxin I (R=CH_2OH)
XXXX: BZR-Cotoxin III (R=H)

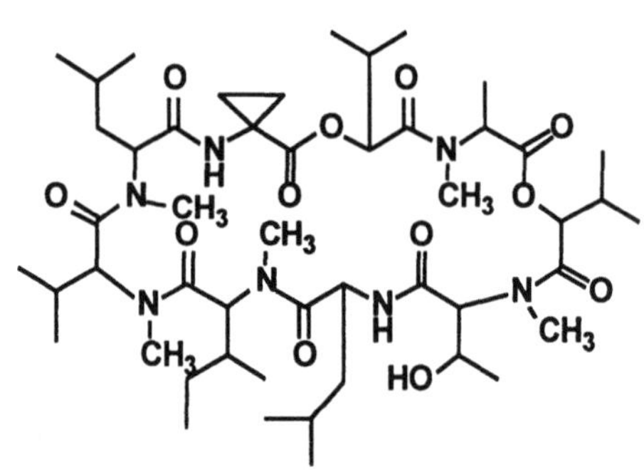

XXXIX: BXR-Cotoxin II

Fig. 3. Structure of phytotoxic aminoacid derivatives

2.4. MEVALONATE AND POLYKETIDES

Cochlioquinones were produced by *B. cynodontis* (20), a pathogen of finger millet (*Eleusine coracana*). The biosynthetic pathway of cochlioquinone A (**XXXXIa**) was constituted by the sesquiterpene part and the hexaketide part.

Table 4. Phytotoxic Mevalonate and Polykletide derivatives produced by *Bipolaris, Drechslera* and *Cochliobolus*

Pathogen	Host Plant	Phytotoxin	Activity	Reference
Bipolaris bicolor	Fingermillet	Cochlioquinone A (**XXXXIa**), B (**XXXXIb**)	Root inhibition	20
		Isocochlioquinone A (**XXXXII**) Stemphone (**XXXXIc**)	Diacyl glycerol kinase	
Drechslera saccharai	Sugarcane	Cochlioquinone A (**XXXXIa**), B (**XXXXIb**) Stemphone (**XXXXIc**)		18,34
Cochliobolus miyabeanus	Rice	Cochlioquinone A (**XXXXIa**), B (**XXXXIb**)		6

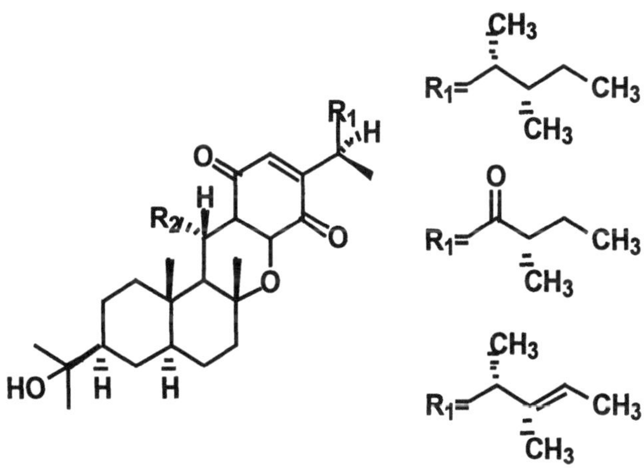

XXXXIa: Cochlioquinone A

XXXXIb: Cochlioquinone B

XXXXIc: Cochlioquinone C

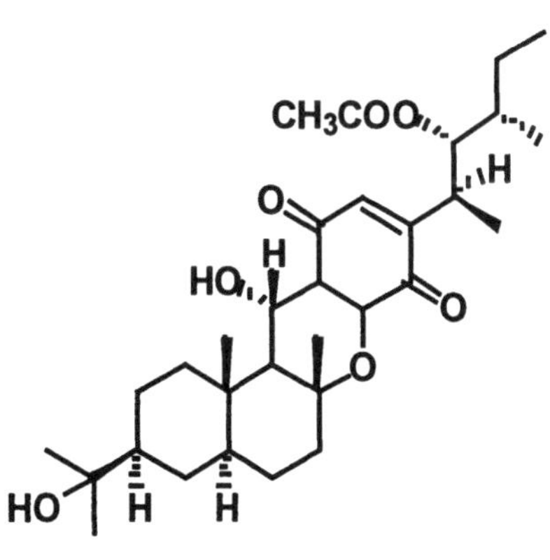

XXXXII: Isocochlioquinone A

Fig.4. Phytotoxic Mevalonate and Polykletide derivatives

2.5. AMINO ACID AND POLYKLETIDE DERIVATIVES AND MISCELLANEOUS

Triticones have been found in *Drechslera tritici-repentis* (12, 56) that causes tan spot on wheat (*T. aestivum*) and in *Curvularia clavata*, a pathogen of turf grass. Other several related species, causing diseases of monocotyledonous weeds, may also produce the triticones. Among the triticones, only triticone A (**XXXXIVa**) and B (**XXXXIVb**) are toxic. Triticone A causes necrosis on numerous plants, kills protoplasts of wheat and inhibits esterase activity and CO_2 fixation in wheat. Weed species are also sensitive, e.g. common lambsquarters (*Chenopodium album*), redroot pigweed (*Amaranthus retroflexes*) and dandelion (*Taraxacum officinale*). When tested on isolated chloroplasts of wheat, oats and spinach, triticone A (**XXXXIVa**) inhibits photosynthetic electron transport (PS-ETS). Triticone A (**XXXXIVa**) reacts stoichiometry with cysteine in less than 1 min. This affinity for sulfhydryl group is thought to be the reason for some of its biological activity- a premise reinforced by the ability of triticone A (**XXXXIVa**) to inhibit ficin, a known sulfhydryl protease. Thus, when assaying tissue or organelles, triticone A (**XXXXIVa**) appears to be non-selectively toxic. The biosynthetic pathway of triticones involved pentaketide, aspartic acid and S-methyl of methionine (personal communication from Professor Gary Strobel, Montana State University, Montana, USA).

Table 5. Phytotoxic amino acid and Polyketide derivatives produced by *Drechslera* and miscellaneous compounds by *Pyrenophra*

Pathogen	Host Plant	Phytotoxin	Activity	Reference
D. dematioidea		Cytochalasin B (**XXXXIII**)		19
D. tritici-repenti	Wheat	Triticone A (**XXXXIVa**), B (**XXXXIVb**) C (**XXXXVa**), D (**XXXXVb**)	Photosynthetic electron transport	12,51
P. teres	Barley	Pyreniliones A (**XXXXVIa**), B (**XXXXVIb**)	Lesion	7

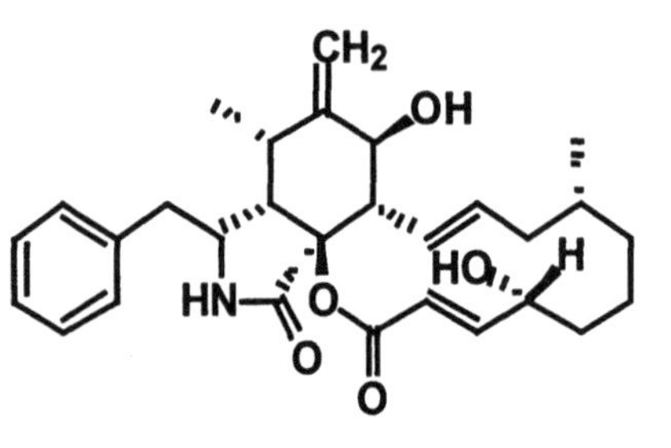

XXXXIII: Cytochalasin B

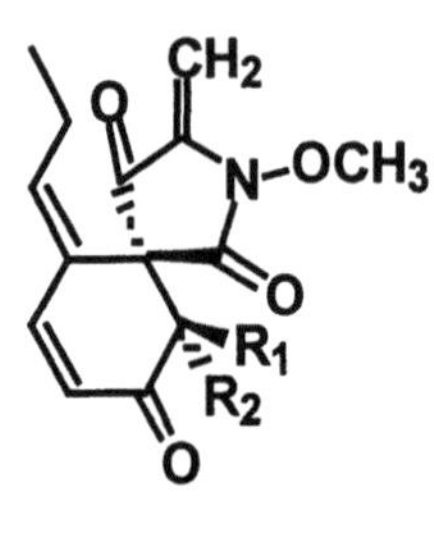

XXXXIVa: Triticone A (R_1=H, R_2=OH)
XXXXIVb: Triticone B (R_1=OH, R_2=H)

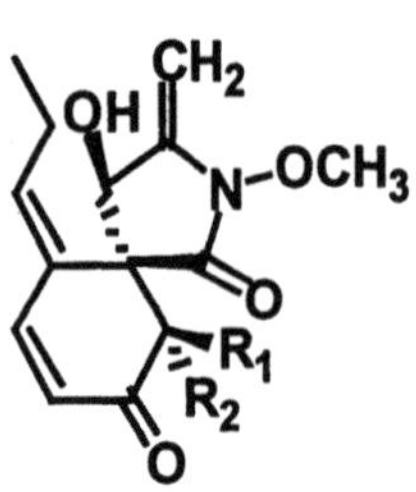

XXXXVIa: Pyreniliones A (R_1=H, R_2=OH)
XXXXVIb: Pyreniliones B (R_1=OH, R_2=H)

Fig.5. Structures of phytotoxic amino acid and Polyketide derivatives

3. Prospects and potential of structure research

The genus *Alternaria spp.* produces potential phytotoxins of weedy plants. For example, Maculosin (**XXXXVII**), is a cyclic dipeptide belonging to diketopiperadine group containing L-proline (42). Many related compounds have been discovered in isolates of *Alternaria alternata* from diseased spotted knapweed (*Centaurea maculosa*). Maculosin, cyclo (L-pro-L-Tyr), caused necrotic lesions on spotted knapweed at 10 μM concentration. Even at 1 mM concentration, maculosin did not affect test plants [lettuce (*Lactuca sativa*), lemon (*Citrus limon*), tomato (*Lycopercisicon esculentum*), apple (*Malus sylvestris*), sunflower (*H. anuus*), cucumber (*Cucumis sativus*), leafy spurge (*E. esula*), *Bidens pilosa*, dandelion (*T. officinale*), sagebrush (*Artemisia tridentata*), thistle (*Cirsium arvense*), crown of corns (*Euphorbia milli*), Johnson grass (*S. halepense*), quackgrass (*A. repense*), crabgrass (*Digitaria ischaemum*), corn (*Z. mays*)] or other *Centaurea* spp. Maculosin is the first phytotoxin produced by a weed pathogen that shows such a high degree of host specificity. It has been synthesized from *t*-BOC-L-proline and methyl ester of L-tyrosine (42). The synthetic product retains the host specificity of the natural product. Maculosin is the most active substance, but the other diketopiperadines were inactive. Understanding the mode of action of a host-specific toxin could provide insight into the development of plant pathogen-host plant interactions, also, the toxin receptor site might be discovered. ^{14}C-Maculosin was prepared using radiolabeled tyrosine and proline as starting materials in the chemical synthesis. After application of ^{14}C-maculosin to an unwounded leaf surface, there was little or no movement of radioactivity from the inoculation site. However, radioactivity was taken up with knapweed roots from the solution of ^{14}C-maculosin (38, 39). Further work could reveal the role of the receptor in production of toxin-induced symptoms.

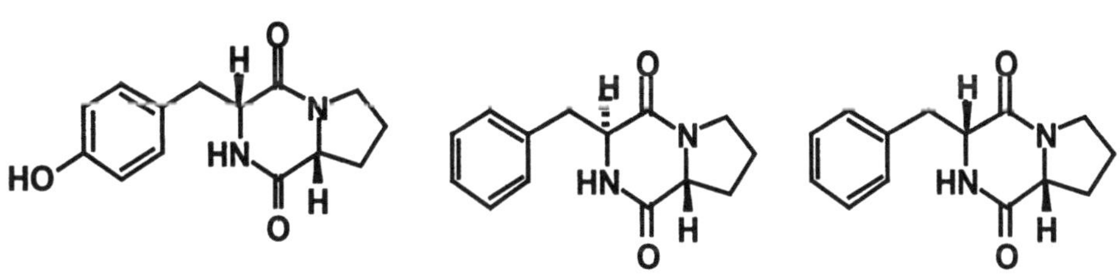

XXXXVII: maculosin **XXXXVIII** **XXXXVIV**

Fig.6. Structures of phytotoxic Diketopiperadine derivatives

4. Future areas of research

Studies on weed pathogens have shown that their phytotoxins are rich in novel structures and intriguing bioactivity. Even previously known chemicals such as curvulin (**IV**) have tremendous impact on the biology of organisms (16). The future of this field is exciting, especially in associations where one eukaryote, a fungus, produces toxins, which regulate or kill another eukaryote, a higher plant. These are pathogenic associations that offer tremendous potential to researchers for natural products for use as selective herbicides. Recent studies have discovered the herbicidal potential of both novel and previously known phytotoxins. Curvulin (**IV**) and maculosin (**XXXXVII**) can be synthesized and offer possibilities for direct application both as non-specific and as extremely selective herbicides. More complex organic compounds like the eremophilanes, resorcylides, triticones and ophiobolins are useful for understanding biochemical and physiological processes in plants and as selective, biologically rational agrochemicals. Many other microbes that attack weeds are candidates for further study.

References

1. Agrios, G. N. (1988). *Plant Pathology.* 3rd Edition. New York : Academic Press, USA.
2. Bunkers, G. J. and Strobel, G. (1991). Fate of petasol, a phytotoxin produced by *Drechslera gigantea* in plants. *Plant Science* 77 : 155-161.
3. Bunkers, G. J. and Strobel, G. (1991). A proposed mode of action for green island introduction by eremophilane phytotoxins produced by *Drechslera gigantea. Physiological and Molecular Plant Science* **38** : 313-323.
4. Canales, M. W. and Gray, G. R. (1988). 6-Epiophiobolin A and 3-anhydro- 6-epiophiobolin A- host specific phytotoxins of *Drechslera maydis* (*race* T). *Phytochemistry* **27** : 1653-1663.
5. Carlson, H., Nisson, P., Jansson, H.B. and Odham, G. (1991). Characterization and determination of prehelminthosporol, a toxin from the plant pathogenic fungus, *Bipolaris sorokiniana* using liquid chromatography/ mass spectroscopy. *Journal of Microbiological Methods* **13** : 259-69.
6. Canonica, L., Beretta, M. G., Colombo, L., Gennari, C., Ranzi, B. M. and Scolastico, C. (1980). Biosynthesis of cochlioquinones. *Journal of Chemical Society, Perkin Transactions* **1** : 2686-2690.
7. Coval, S. J., Hradil, C. M., M. Lu, H. S., Clardy, J., Satouri, S. and Strobel, G. (1990). Pyrenokine-A and -B, two new phytotoxins from *Pyrenophora teres. Tetrahedron Letters* **31** : 2117-2120.
8. Cuq, F., Herrmenn-Gorline, S., Klaebe, A., Rossignol, M. and Petitprez, M. (1993). Monocerin in *Exserohilum turticum* isolates from maize and a study of its phytotoxicity. *Phytochemistry* **34** : 1265-1270.
9. Friis, P., Olsen, C. E. and Moller, B. L. (1991). Toxin production in *Pyrenophora teres*, the Ascomycete causing the Net-Spot blotch disease of barley (*Hordeum vulgare* L.). *Journal of Biological Chemistry* **266** : 1332-1335.
10. Hallock, Y. (1988). *Isolation and Characterization of Novel Phytotoxins From Pathogenic Fungi.* Ph D. thesis. Ithaca, New York : Cornell University, U. S. A.
11. Hallock, Y. F., Clardy, J., Kenfield, D. and Strobel, G. (1988). De-*O*-methyldiaporthin, a phytotoxin from *Drechslera siccans. Phytochemistry* **27** : 3123-3125.
12. Hallock, Y. F., Liu, H. S. M., Clardy, J., Strobel, G., Sugawara, F., Samsoedin, R. and Yoshida, S. (1993). Triticones, spirocyclic lactams from the fungal plant pathogen *Drechslera tritici-repentis. Journal of Natural Products* **56** : 747-754.
13. Holm, L. G., Plucknet, D. L., Pancho, J. V. and Herberger, J. P. (1977). *The World Worst Weeds: Distribution and Biology.* Hawaii : University press.
14. Kenfield, D., Bunkers, G., Strobel, G. and Sugawara, F. (1988). Potential new herbicide- phytotoxins from plant pathogens. *Weed Technology* **2** : 519-524.
15. Kenfield, D., Bunkers, G., Woo, Y. H., Strobel, G., Sugawara, F., Hallock, Y. F. and Clardy, J. (1989). Gigantenone, a novel sesquiterpene phytohormone mimic. *Experientia* **45** : 900-902.
16. Kenfield, D. S., Hallock, Y. F., Clardy, J. and Strobel, G. A. (1988). Curvulin, and *O*-methylcurvulinic acid: phytotoxic metabolites of *Drechslera indica* which causes necrosis on purslane and spiny amaranthus. *Plant Science* **60** : 123-127.
17. Kim, J. M., Hyeon, S. B., Isogai, A. and Suzuki, A. (1984). Isolation of ophiobolin A and its analogues as inhibitor to photosynthesis. *Agricultural and Biological Chemistry* **48** : 803-805.

18. Lim, C.-H., Miyagawa, H., Tsurushima, T., Ueno, T. and Sato, M. (1996). Cochlioquinol: a new cochlioquinone derivative produced by the plant fungus *Bipolaris cynodontis*. *Bioscience, Biotechnology and Biochemistry* **60** : 724-725.
19. Lipski, K. M., McQuiggan, J. D., Loucy, K. J. and Fondy, T. P. (1987). Cytochalasin B: preparation, analysis in tissue extracts and pharmacokinetics after intraperiteneal bolus administration in mice. *Analytical Biochemistry* **161** : 332-340.
20. Miyagawa, H., Nagai, S., Tsurushima, T., Sato, M. and Ueno, T. (1994). Phytotoxins produced by the plant pathogenic fungus *Bipolaris bicolor* EI-I. *Bioscience, Biotechnology and Biochemistry* **58** : 1143-1145.
21. Nakajima, H., unpublished data.
22. Nakajima, H., Fujimoto, H., Kimura, Y. and Hamasaki, T. (1993). Importance of ketone function for the phytotoxicity of spiciferone A, a phytotoxin produced by the fungus *Cochliobolus spicifer*. *Bioscience, Biotechnology and Biochemistry* **57** : 1938-1939.
23. Nakajima, H., Fujimoto, H., Matsumoto, R. and Hamasaki, T. (1993). Biosynthesis of spiciferone A and spicifenin. Bioactive metabolites of phytopathogenic fungus, *Cochliobolus spicifer*. *Journal of Organic Chemistry* **58** : 4526-4528.
24. Nakajima, H., Fukuyama, K., Fujimoto, H., Baba, T. and Hamasaki, T. (1994). Absolute stereochemistry of spiciferones and spicifinin, bioactve metabolites of the fungus *Cochliobolus spicifer*. *Journal of Chemical Society, Perkin Transaction* **1** : 1865-1869.
25. Nakajima, H., Hamasaki, T., Kohno, M. and Kimura, Y. (1991). Spiciferon B and C, minor phytotoxin from the fungus, *Cochliobolus spicifer*. *Phytochemistry* **30** : 2563-2565.
26. Nakajima, H., Hamasaki, T., Maeta, S., Kimura, Y. and Takeuchi, Y. (1990). A plant growth regulator produced by the fungus, *Cochliobolus spicifer*. *Phytochemistry* **29** : 1739-1743.
27. Nakajima, H., Hamasaki, T. and Kimura, Y. (1989). Structure of speciferone A, a novel γ-pyrone plant growth inhibitor produced by fungus *Cochliobolus specifer* Nelson. *Agricultural and Biological Chemistry* **53** : 2297-2299.
28. Nakajima, H., Isomi, K., Hamasaki, T. and Ichinoe, M. (1994). Sorokinianin: a novel phytotoxin produced by phytopathogenic fungus *Bipolaris sorokiniana*. *Tetrahedron Letters* **35** : 9597-9600.
29. Nakajima, H., Kimura, Y. and Hamasaki, T. (1992).. Speciferone, an azaphilone phytotoxin produced by the fungus *Cochliobolus specifer*. *Phytochemistry* **31** : 105-107.
30. Nakajima, H., Matsumoto, R., Mimura, Y. and Hamasaki, T. (1992). Biosynthesis of spicifernin, a unique metabolite of the phytopathogenic fungus, *Cochliobolus spicifer*. *Journal of Chemical Society Chemical Communication* 1654-1656.
31. Nisson, P., Accession, H., Jansson, H. B. and Odham, G. (1993). Production and release of the phytotoxin prehelminthosporol by *Bipolaris sorokiniana* during growth. *FEMS Microbiological Ecology* **102** : 91-98,.
32. Nukina, M., Ikeda, M. and Sassa, T. (1980). Two new pyrenolides, fungal morphogenic substances produced by *Pyrenophora teres* (Diedicke) Drechsler. *Agricultural and Biological Chemistry* **44** : 2761-2762.
33. Nukina, M., and Marumo, S. (1977). Radicinol: a new metabolite of *Cochliobolus lunuta* and absolute stereochemistry of radicinin. *Tetrahedron Letters* **18** : 3271-3272.
34. Ogawara, H., Higashi, K., Machida, T., Takashima, J., Chiba, N. and Mikawa, T. (1994). Inhibitors of diacylglycerol kinase from *Drechslera sacchari*. *Journal of Antibiotics* **47** : 499-501.
35. Pena-Rodriguez, L. M., Armingeon, N. A. and Chilton, W. S. (1988). Toxins from weed pathogens. I. Phytotoxins from *Bipolaris* pathogen of Johnson grass . *Journal of Natural Products* **51** : 821-828.
36. Pena-Rodriguez L. M. and Chilton, W. S. (1989). 3-Anhydroophiobolin A and 3-anhydro-6-*O*-ophiobolin A, phytotoxic metabolites of the Johnson grass pathogen *Bipolaris sorghicola*. *Journal of Natural Products* **52** : 1170-1172.
37. Pena-Rodriguez L. M. and Chilton, W. S. (1989). Victoxinin and prehelminthosporolactone, two major phytotoxic metabolites produced by *Bipolaris* sp., a pathogen of Johnson grass. *Journal of Natural Products* **52** : 899-901.
38. Park, A. H., Stierle, A. C. and Strobel, G. A. (1994). Metabolism of maculosin, a host specific phytotoxin produced by *Alternaria alternata* on spotted knapweed (*Centaura maculosa*). *Phytochemistry* **35** : 101-106.
39. Park, A. H. and Strobel, G. A. (1994). Cellular protein receptors of maculosin, a host specific phytotoxin of spotted knapweed (*Centaura maculosa*). *Biochimica et Biophysica Acta* **1199** : 13-19.

40. Robeson, D. J. and Strobel, G. (1982). Monocerin, a phytotoxin from *Exserohilum turticum* (=*Drechslera turticum*). *Agricultural and Biological Chemistry* **46** : 2681-2683.
41. Scott, F. E., Simpson, L. A., Trimble, L. A. and Vederas, J. C. (1984). Biosynthesis of monocerin. Incorporation of ^{2}H-, ^{13}C- and ^{18}O-labeled acetates by *Drechslera ravenelii*. *Journal of Chemical Society, Chemical Communication* : 756-758.
42. Stiele, A. C., Cardellina II, J. H., and Strobel, G. A. (1988). Maculosin, a host specific phytotoxin for spotted knapweed (*Centaureaa maculosa*). *Proceedings of National Academy of Science, U. S. A.* **85** : 8008-8011.
43. Strobel, G., Sugawara, F. and Clardy, J. (1987). Phytotoxins from plant pathogens of weedy plants. In *Allelochemicals : Role in Agriculture and Forestry*, (Ed., G.R. Waller). ACS Symposium Series **330** : pp. 516-523. Washington DC : American Chemical Society.
44. Strobel, G., Sugawara, F. and Hershenhorn, J. (1992). Pathogens and other products affecting weedy plants. *Phytoparasitica* **20** : 307-323.
45. Sugawara, F., Hallock, Y. F., Bunkers, G., Kenfield, D. S., Strobel, G. and Yoshida, S. (1993). Phytoactive eremophilanes produced by the weed pathogen *Drechslera gigantea*. *Bioscience, Biotechnology and Biochemistry* **57** : 236-239.
46. Sugawara, F., Kim, K., Kobayashi, K., Uzawa, J., Yoshida, S., Murofushi, A., Takahashi, N. and Strobel, G. (1992). Zearalenone derivatives produced by *Drechslera portulacae*. *Phytochemistry* **31** : 1987-1990.
47. Sugawara, F. and Strobel, G. (1986). (-)-Dihydropyrenophorin, a novel and selective phytotoxin produced by *Drechslera avenae*. *Plant Science* **43** : 1-5.
48. Sugawara, F. and Strobel, G. (1987). Tryptophol, a phytotoxic metabolite of *Helminthosporium nodulosum* on weedy plants. *Phytochemistry* **26** : 1349-1351.
49. Sugawara, F., Strobel, G., Fisher, L. F., Van Duyne, G. D. and Clardy, J. (1985). Bipolaroxin, a selective phytotoxin produced by *Bipolaris cynodontis*. *Proceedings of National Academy of Science U. S. A.* **82** : 8291-8294.
50. Sugawara, F., Strobel, G., Strange, R. N., Siedow, J. N., Van Duyne, G. D. and Clardy, J. (1987). Phytotoxins from the pathogenic fungi *Drechslera maydis* and *Drechslera sorghicola*. *Proceedings of National Academy of Science U. S. A.* **84** : 3081-3085.
51. Sugawara, K., Sugawara, F., Strobel, G., Fu, Y., Cun-Heng, H. and Clardy, J. (1985). Exserohilone, a novel phytotoxin produced by *Exserohilum holmii*. *Journal of Organic Chemistry* **50** : 5631-5933.
52. Sugawara, F., Takahashi, N., Strobel, G., Strobel, S., Liu, H. S. M. and Clardy, J. (1988). Triticones A and B, novel phytotoxins from the plant pathogenic fungus *Drechslera tritici-repentis*. *Journal of American Chemistry Society* **110** : 4086.
53. Sugawara, F., Takahashi, N., Strobel, G., Yun, C. H., Gray, G., Fu, Y. and Clardy, J. (1988). Some new phytotoxic ophiobolins produced by *Drechslera oryzae*. *Journal of Organic Chemistry* **53** : 2170-2172.
54. Ueda, K., Xiao, J. Z., Doke, N. and Nakatsuka, S. (1992). Structure of BZR-cotoxin II produced by *Bipolaris zeicola* race 3, the cause of leaf spot disease in corn. *Tetrahedron Letter.* **33** : 5377-5380.
55. Ueda, K., Xiao, J. Z., Doke, N. and Nakatsuka, S. (1994). Structure of BZR-cotoxin I produced by *Bipolaris zeicola* race 3, the cause of leaf spot disease in corn. *Tetrahedron Letters* **35** : 7033-7036.
56. Ueda, K., Xiao, J. Z., Doke, N. and Nakatsuka, S. (1995). Isolation and structure of BZR-cotoxin IV produced by *Bipolaris zeicola* race 3, the cause of leaf spot disease in corn. *Tetrahedron Letters* **36** : 741-743.
57. Xiao, J. Z., Tsuda, M., Doke, N. and Nishimura, S. (1991). Phytotoxins produced in spores of *Bipolaris oryzae*. *Phytopathology* **81** : 58-64.
58. Yun, C. H., Sugawara, F. and Strobel, G. A. (1988). The phytotoxic ophiobolins produced by *Drechslera oryzae*, their structures and biological activity on rice. *Plant Science* **54** : 237-243.

9

Evaluation of selected pharmaceuticals as potential herbicides: bridging the gap between agrochemicals and pharmaceuticals

STEPHEN J. CUTLER*, ROBERT E. HOAGLAND[1] and HORACE G. CUTLER
Natural Products Discovery Group, Southern School of Pharmacy,
Mercer University, 3001 Mercer University Drive, Atlanta, Georgia 30341, USA

Abstract

Much has been written in the field of agrochemicals about the application of phenoxy compounds as herbicidal agents. The illustrious history of these agents includes their accelerated development by the Allies for use during World War II. Among the most famous of these agents are 2,4-dichlorophenoxyacetic acid (2,4-D) and its congener, 2,4,5-trichlorophenoxyacetic acid (2,4,5-T). Because of their high specific activities, these compounds became templates for the production of a number of plant growth regulators such as Fruitone and Tomato Bloom. Interestingly, the pharmaceutical industry was developing a group of phenoxy derivatives which proved to be useful in the treatment of type III hyperlipoprotenemia, a condition where triglycerides and total cholesterol levels are elevated. The most recognized agent in this pharmaceutical class is clofibrate which inhibits *sn*-glyceryl-3 phosphate acyltransferase, the enzyme responsible for the conversion of acetate to mevalonate. Furthermore, its mechanism of action also includes inhibition of 3-hydroxy-3-methylglutaryl-CoA (HMGCoA) reductase. In addition, the aftermath of World War II served as a catalyst in the development of the benzodiazepines, a well recognized pharmaceutical class possessing sedative/hypnotic and tranquilizing properties. It is of interest to note that these agents are actually allelopathic natural products whose chemical structures were ignored until the development of Valium as a pharmaceutical agent. Had these natural product chemical templates not been relegated to library book shelves, they may have served a useful purpose in their development as medicinals. Recently, some benzodiazepines have been shown to possess agrochemical properties including inhibition against *Phytophthora infestans*, late blight of potato. The separate, but distinct, agrochemical and pharmaceutical development of various classes of agents will be discussed.

Key words: Agrochemicals, bioassay, herbicides, pharmaceuticals, weed, wheat.

---

*Correspondence author
[1]USDA-ARS, Southern Weed Science Research Unit, Stoneville, Mississippi, 38776, USA.

S.S. Narwal et al. (eds.), Allelopathy in Ecological Agriculture and Forestry, 129–137.

1. Introduction

When reading the scientific literature, it is readily apparent that the two disciplines, agrochemicals and pharmaceuticals, operate in a scientific vacuum relative to each other. Only in rare instances do the disciplines come together with a common denominator applicable to both fields. In September 1997, the American Chemical Society=s Division of Agrochemicals supported a three day symposium on ABiologically Active Natural Products: Agrochemicals and Pharmaceuticals≅ in order to bridge the two disciplines. This has resulted in the publication of two companion volumes which are dedicated to merging these areas of natural products (10, 11). Even before this ACS Symposium, the United States Congress indicated an interest in natural products research. In 1992, the United States Senate Committee on Appropriations (US Senate Report No. 102-119) issued a directive to the United States Department of Agriculture to develop a plan on the beneficial uses of agricultural commodities in the manufacture of pharmaceuticals, biological and biosafe pesticides and other natural products [Agricultural Research Services (ARS) Working Conference on Natural Products for Control of Agricultural Pests, Richard B. Russell Center, Athens, GA, October 19-20, 1993]. It was apparent from the workshop that there were many relationships between agrochemicals and pharmaceuticals than might have been previously anticipated. Two examples are: the benzodiazepines, a pharmaceutical class of antianxiety agents which may also have agrochemical use as antifungals and plant growth regulators; and the phenoxy derivatives of 2,4-dichlorophenoxyacetic acid, an agrochemical used as a broadleaf herbicide which is structurally similar to the fibrates, a pharmaceutical class of cholesterol lowering agents.

One of the most successful classes of pharmaceutical agents are the benzodiazepines which were first reported to possess tranquilizing properties in 1957 (16). The first compound, chlordiazepoxide (Librium), was introduced into clinical use in 1960 (Figure 1). The success of chlordiazepoxide led to the development of Diazepam (Valium) in 1959 with its clinical launch in 1963. The valuable clinical properties of diazepam led to an explosion in the development of benzodiazepines as a class of anxiolytics, anti-depressants, sedative-hypnotics and muscle relaxants. Within a few years, literally thousands of benzodiazepine derivatives had been synthesized and evaluated for their pharmacological properties. With this large arsenal, it was not long before the structure-activity-relationships (SAR) of the benzodiazepines were developed (17). It was noted that substitutions on the C-7 position played a significant role in the compound=s potency. However, the most significant derivatives are those possessing rings which join the #1 and #2 positions. The Adiazole-≅ and Atriazolebenzodiazepines≅ are therapeutically the most potent of the benzodiazepines.

However, to suggest that the discovery of the benzodiazepines came about with the serendipitous discovery of chlordiazepoxide is erroneous. In fact, the structural elucidation of the first benzodiazepine was unequivocally determined several years earlier (2). The group was successful in the isolation and structural characterization of cyclopenin (Figure 2), the nitrogen containing secondary metabolite of *Penicillium cyclopium*. However, due to its limited antibiotic activity against *Micrococcus pyogenes* var *aureus* and *Escherichia coli*, the publication was ignored. Later, cyclopenol (Figure 2), a second naturally occurring benzodiazepine, was isolated from *Penicillium viridicatum* (1, 14). Even so, the scientific community displayed little interest in the discovery of this novel heterocyclic class of agents.

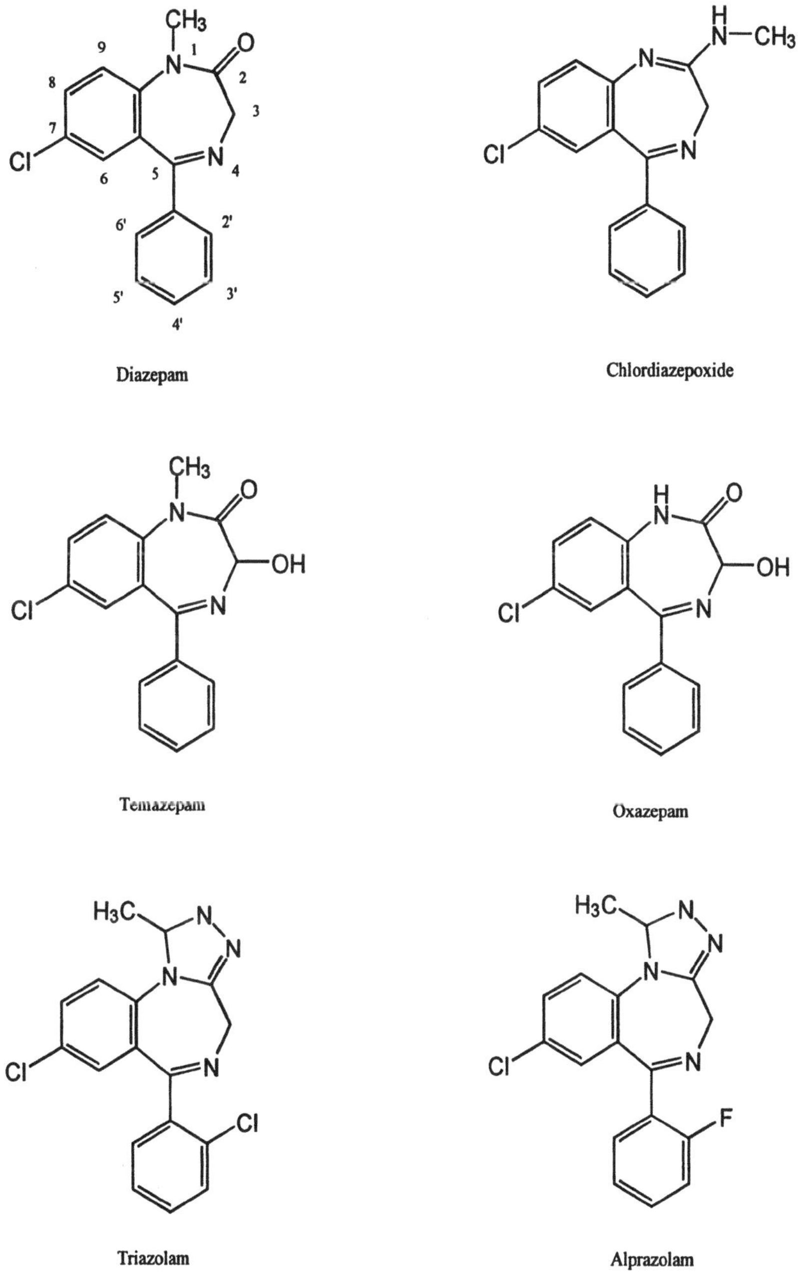

Figure 1. Chemical structures of benzodiazepines

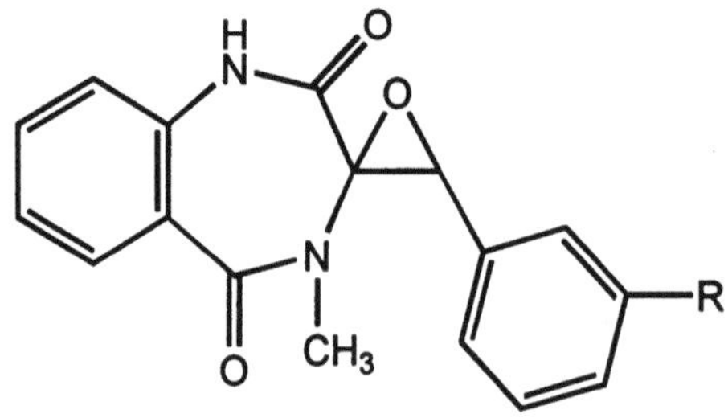

R = H Cyclopenin

R = OH Cyclopenol

Figure 2. Chemical structures of naturally occurring benzodiazepi

After two decades, a complete report on the agrochemical and pharmaceutical properties of cyclopenin and cyclopenol was made (7, 8). Cyclopenin and cyclopenol were found to significantly inhibit the growth of etiolated wheat coleoptiles. Furthermore, cyclopenin was found to induce malformations in the first set of trifoliate leaves of nine-day-old bean (*Phaseolus vulgaris* L. cv Black Valentine) seedlings at treatments of 10^{-2}M and to cause stunting and necrosis in one-week-old corn (*Zea mays* L. cv Norfolk Market White) seedlings. Cyclopenol had no visible effects on bean or corn but was effective against *Phytophthora infestans*, late blight of potato (*Solanum tuberosum* L.). This organism, which was responsible for the Great Famine in Ireland in 1846, is a continuing problem in the potato-growing areas of the world (19). The number of applications of fungicides to control this fungus can be as high as 18 per growing season in the United States. Additionally, cyclopenin, but not cyclopenol, was shown to have tranquilizing properties which confirms ring substitutions play an important role.

Another group of compounds which has been independently and separately developed as agrochemicals and pharmaceuticals is phenoxyacetic acids. In the field of agrochemicals, the most famous are 2,4-dichlorophenoxyacetic acid (2,4-D) and 2,4,5-trichlorophenoxyacetic acid (2,4,5-T) (Figure 3) (13, 15). Interestingly, 2,4-D was

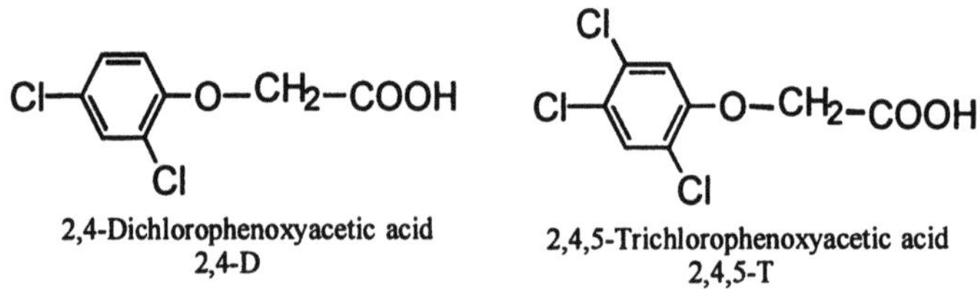

2,4-Dichlorophenoxyacetic acid
2,4-D

2,4,5-Trichlorophenoxyacetic acid
2,4,5-T

Figure 3. Structure of phenoxy acetic acids

developed during World War II with the disclosure of its synthesis in 1941 by Pokorny (15). Because of the war, the Official Secrets Act prevented any earlier report by the British. As a result of their high profile of activity, phenoxyacetic acids became templates for the development of a number of agrochemicals (18). Fruitone® [2-(3-chlorophenoxy)-propanoic acid] and Tomato Bloom® (4-chlorophenoxyacetic acid) resulted from these efforts. Other phenoxyacetic acid derivatives include: Hoelon® (diclofop), Fusilade® (fluazifop-butyl), Verdict® (haloxyfop-methyl), Gallant® (haloxyfop-methyl) and Whip® (fenoxaprop-ethyl).

Independently, the phenoxyacetic acids were developed as pharmaceutical agents or, as they are more commonly referred to, the fibric acids. Clofibrate, 2-(4-chlorophenoxy)-2-methyl propionate represents the first of the fibric acids (Figure 4) introduced into clinical practice. The clinical use of the fibrates stems from their ability to induce hypolipoprotenemia, a condition where serum levels of triglycerides and cholesterol are reduced. The mechanism of action for the fibrate derivatives is through inhibition of 3-hydroxy-3-methylglutaryl-CoA (HMGCoA) reductase as well as *sn*-glycerol-3-phosphate acyltransferase (3). Based on the clinical results of clofibrate, several fibric acids have been developed for use in the treatment of coronary heart disease.

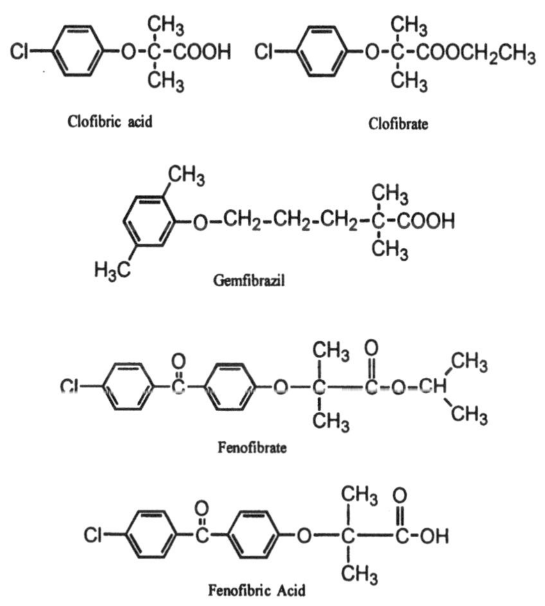

Figure 4. Fibric acid derivatives

Due to the chemical similarity of the fibric acids to the phenoxyacetic acids and the synthetic benzodiazepines to naturally occurring benzodiazepines, it was of interest to evaluate the agrochemical properties of the fibric acids and the synthetic benzodiazepines.

2. Material and methods

Wheat seeds (*Triticum aestivum* L., cv. Wakeland) were grown in moist vermiculite in a plastic tray. The tray was covered with aluminum foil and placed in total darkness for four days. The temperature was maintained at 22$\pm$1°C (6, 12). After the wheat had germinated and grown for 4 days, the seedlings were removed from the trays, the roots and caryopses removed and the apices were fed into a Van der Weij guillotine so that the first 2 mm could be removed and discarded under a green safe-light (540 nm) (5). The next 4 mm of each seedling was used for the bioassay. Ten segments were placed in a test tube containing the test compound and 2 ml of phosphate-citrate buffer (pH 5.6) which had been supplemented with 2% sucrose. Using a roller tube apparatus, the test tubes were rotated at 0.25 rpm for 18 h. The sections were removed, blotted dry, placed on a glass plate and magnified three-fold with a photographic enlarger. The coleoptile lengths were measured, recorded and statistically analyzed.

Intact plant bioassays included either corn (*Zea mays* L., cv. Norfolk Market White) or beans (*Phaseolus vulgaris* L., cv. Black Valentine) plants (6, 7). The plants were grown in clay pots outdoors for 1 week before a solution of the test compound was applied to the leaves. The test solution was prepared by dissolving the test compound in 0.5 ml of acetone and 4.5 ml of 1% Tween 20 dissolved in water. In beans, 1 ml of test solution was added to each of the plants, while the corn plants were individually treated with 100 µl, placed in each sheath. There were four plants per pot, and experiments were triplicated. Visual effects were recorded. Controls received a foliar application of the solution vehicle.

Compounds were obtained from Sigma-Aldrich Chemical Company, Milwaukee, Wisconsin. Fenofibric acid was prepared through the hydrolysis of fenofibrate. This was accomplished by dissolving fenofibrate in a mixture of ethanol:KOH (18:1) followed by heating at 85EC for approximately 1.5 h. The ethanol was evaporated *in vacuo*, water (pH=2) added, and the mixture extracted with diethyl ether. The ether was dried over anhydrous sodium sulfate, the sodium sulfate removed by filtration and the resulting ether evaporated to yield a fluffy white product. This material was found to be analytical pure based on FT-IR, NMR, TLC, and elemental analysis.

3. Results and discussion

3.1. AGROCHEMICAL PROPERTIES OF BENZODIAZEPINES

The biological properties of cyclopenin, the first reported benzodiazepine, were studied 20 years after its discovery (8). In that study, cyclopenin (10^{-4} M) inhibited the growth of etiolated wheat coleoptiles by 33%. However, at 10^{-3} M, cyclopenin inhibited the growth of wheat coleoptile by 100%. Unfortunately, its analogue, cyclopenol only inhibited coleoptile growth by 20% at 10^{-3} M as compared with the control group (9). Nevertheless, it appeared that the benzodiazepine structure possessed interesting biological properties and stimulated an interest in the evaluation of additional agents possessing this heterocyclic system. Several pharmaceutical benzodiazepines were obtained and evaluated for their potential use as agrochemicals. The first compounds to be evaluated were those initially synthesized by Sternbach (16). They were chlordiazepoxide and diazepam and both exhibited significant inhibition of wheat coleoptile growth. Chlordiazepoxide at 10^{-3} M inhibited coleoptile growth by 40 %. However, at the same concentration (10^{-3} M), diazepam caused 100% inhibition. The metabolites of diazepam include oxazepam and temazepam which were developed for use in patients with compromised liver enzymes. In the coleoptile bioassay, temazepam had a greater activity; i.e., 100% inhibition at 10^{-3} M, while oxazepam had approximately 35% at that concentration. At 10^{-4} M, neither of these agents possessed inhibitory properties in the coleoptile bioassay. Lorazepam, a chloride derivative of oxazepam, was found to have a higher level of inhibitory effects than oxazepam; i.e., 100% inhibition at 10^{-3} M. Furthermore, at 10^{-4} M, lorazepam caused 34 % growth inhibition of the wheat coleoptile while oxazepam had no effects at this concentration. It is clear that the heavy atom on position 2N imparts a higher effect than the un-substituted derivative. The condensation of an additional ring on position 1 and 2 of the diazepine portion yields an interesting group of therapeutic agents. These Atriazole≅ and Aimidazole≅ benzodiazepine derivatives are among the most potent pharmaceutical agents available for anxiety. It was of interest to evaluate a few of these compounds in the coleoptile bioassay, to determine if there was a correlation between the two systems. Triazolam and alprazolam, which differ from each other by the substitution of a 2N-chloro and a 2N-fluoro, respectively, were found to inhibit coleoptile growth by 42% at 10^{-3} M, while having no inhibitory properties at 10^{-4} M. Although it was evident from this study that there is not a direct correlation with plant and mammalian systems, the inhibitory activity in the wheat coleoptile bioassay showed that this bioassay is capable of identifying biologically active compounds which have potential use as anti-anxiety, sedative-hypnotic, muscle relaxant and anti-convulsant agents. Furthermore, this bioassay has been successfully used to detect immunosuppressants anti-microbials and anti-amoebic agents.

In the whole plant bioassay, only chlordiazepoxide induced necrosis when applied to week-old corn plants at 10^{-2} M. After one week, necrosis was no longer evident and the plants resumed normal growth (9). There were no visual effects from any of the benzodiazepines tested on one-week-old bean plants.

3.2. AGROCHEMICAL PROPERTIES OF FIBRIC ACIDS

The wheat coleoptile bioassay was used as the primary bioassay in determining the potential use of pharmaceutical fibrates as agrochemicals. The agrochemical, 2,4-dichlorophenoxy acetic acid (2,4-D) was also tested in this bioassay. 2,4-D completely inhibited coleoptile growth at 10^{-3} M and at 10^{-4} growth was significantly reduced. At concentrations of 10^{-5} M and lower, 2,4-D showed stimulatory effects with a promotion of growth over the control group. Clofibric acid inhibited the growth of etiolated wheat coleoptiles by 60% at 10^{-4} and 100% at 10^{-3} M. The ethyl ester form of clofibric acid, clofibrate, was found to induce 100% inhibition at 10^{-3} and 10^{-4} M. Interestingly, this is the preferred derivative for treatment of hyperlipoprotenemia. Gemfibrazol was found to significantly inhibit wheat coleoptile growth at 10^{-3} (100%) and 10^{-4} M (64%). Although fenofibrate was inactive in the coleoptile bioassay, its free acid, fenofibric acid, elicited a 40% inhibition at 10^{-2} M.

In whole plant bioassays, clofibric acid had no effect on week-old corn plants at 10^{-2} M. However, in week-old bean plants clofibric acid inhibited growth by 50% at 10^{-2} M. In fact,10 days after treatment, the plants appeared as if they had been treated with 2,4-D in that the trifoliates became leathery and malformed. A similar effect on one-week-old bean plants was observed with the use of clofibrate, with growth inhibition of 50% at 10^{-2} M. In addition, clofibrate exhibited necrotic effects on corn within 48 h of 10^{-2} M. After 10 days, these same corn plants were inhibited 50% in their growth relative to controls. Gemfibrazol significantly inhibited wheat coleoptile growth by 100% and 60% at 10^{-3} and 10^{-4} M, respectively. Although gemfibrazol had no effects on bean, it did cause slight necrosis on corn at 10^{-2} M. Fenofibrate was completely inactive in the coleoptile bioassay and in the bean and corn seedling tests. However, fenofibric acid was active at 10^{-2} M with significant inhibition (40%) of coleoptile growth, but fenofibric acid was not found to be active on corn or bean seedlings.

Using the wheat coleoptile bioassay in conjunction with whole plant systems, it is possible to evaluate the agrochemical properties of pharmaceutical agents which are structurally similar to certain herbicides, plant growth regulators, and other pesticides . This study demonstrates the importance not only of the use of pharmaceuticals as agrochemicals, but the agricultural use of agents which have been thoroughly evaluated for mammalian toxicity.

References

1. Birkenshaw, J.H., Luckner, M. Mohammed, Y.S. Mothes, K. and Stickings, C.E. (1963). Studies on the biochemistry of microorganisms. 114. Viridicatol and cyclopenol, metabolites of *Penicillium viridicatum* Westling and *Penicillium cyclopium* Westling. *Journal of Biochemistry* **89**: 196-202
2. Braken, A., Pocker, A. and Raistrick, H. (1954). Studies in the biochemistry of microorganisms. 93. Cyclopenin, a nitrogen-containing metabolite product of *Penicillium cyclopium* Westling. *Journal of Biochemistry* **51**: 587-595
3. Cocolas, G.H. (1998). Cardiovascular Agents. In *Wilson and Gisvold=s Textbook of Organic Medicinal and Pharmaceutical Chemistry*, (Eds., J.N. Delgado and W.A. Remers), pp. 616-617. 10th edition. Philadelphia: Lippincott.
4. Crosby, D.G., Moilanen, K.W. and Wong, A.S. (1973). Environmental generation and degradation of dibenzodioxins and dibenzofurans. *Environmental Health Perspectives* September pp. 259-266
5. Cutler, H.G. (1984). A fresh look at the wheat coleoptile bioassay. *Proceedings of the Eleventh Annual Meeting of the Plant Growth Regulator Society of America*, 1
6. Cutler, H.G. (1986). Isolating, characterizing, and screening mycotoxins for herbicidal activity. In *The Science of Allelopathy*, (Eds., A.R. Putnam and C-S. Tang), pp.147-170. New York: Wiley Interscience
7. Cutler, H.G., Ammerman, E. and Springer, J.P. (1988). Diverse but specific biological activities of four natural products from three fungi. In *Biologically Active Natural Products in Agriculture*, (Ed., H.G. Cutler), ACS Symposium Series No. 380 : 79-90. Washington, DC: American Chemical Society.

8. Cutler, H.G., Crumley, F.G., Cox, R.H., Wells, J.M. and Cole R.J. (1984). The biological properties of cyclopenin and cyclopenol. *Plant Cell and Physiology* **25**: 257- 263
9. Cutler, H.G. and S. J. Cutler. (1999). Agrochemicals and pharmaceuticals: The connection. In *Biologically Active Natural Products: Agrochemicals* (Eds., H.G. Cutler and S.J. Cutler), pp 1-15. Boca Raton, Florida: CRC Press.
10. Cutler, H.G. and Cutler, S.J. (1999). *Biologically Active Natural Products: Agrochemicals*, (Eds., H.G. Cutler and S.J. Cutler). Boca Raton, Florida: CRC Press.
11. Cutler, S.J. and Cutler, H.G.(1999). *Biologically Active Natural Products: Pharmaceuticals*, (Eds., S.J. Cutler and H.G. Cutler), Boca Raton, Florida: CRC Press.
12. Hancock, C.R., Barlow, H.W. and Lacey, H.J. (1964). The East Malling coleoptile straight-growth test method. *Journal of Experimental Botany* **15**: 166-176
13. Holden, C. (1979). Agent orange furor continues to build. *Science* **205**: 770-772
14. Mohammed, Y.S. and Luckner, M. (1963). The structure of cyclopenin and cyclophenol, metabolic products *from Penicillium cyclopium Westling and Penicillium viridicatum Westling. Tetrahedron Letters* **28**: 1953-1958
15. Pokorny, R. (1941). Some chlorophenoxyacetic acids. *Journal of the American Chemical Society* **63**: 1768
16. Sternbach, L.H. (1979). The benzodiazepine story. *Journal of Medicinal Chemistry* **22**: 1-7
17. Sternbach, L.H., Randall, L.O., Banziger, R. and Lehr, H. (1967). Drugs affecting the central nervous system. In *Medicinal Research Series*, (Ed., A. Burger), Vol.**1**: 237-264. New York: Marcell Dekker.
18. Thompson, W.T. (1992). *Agricultural Chemicals, Book III, Miscellaneous Agricultural Chemicals*. Fresno, CA: Thompson Publications. **83**: 86
19. Wood, J.L. (1953). Three billion dollars a year. In *Plant Diseases: the Yearbook of Agriculture*, pp 1-9. Washington, DC: United States Department of Agriculture.

SECTION III

MULTIPLE CROPPING SYSTEMS

10

Allelopathic interactions in multiple cropping systems

S. S. NARWAL
Department of Agronomy,
CCS Haryana Agricultural University, Hisar – 125 004, India

Abstract

In western countries (Europe, USA, Canada etc.) very severe winters restrict the crop production to only summer season and in the remaining months land remains fallow. The decomposition of residues of former crop and release of allelochemicals occur during the fallow period. Hence, there is no effect of preceding crop on the succeeding crop. While in tropical and subtropical countries, climate is suitable for round the year cropping. In assured irrigation areas, farmers grow 2-4 crops in the same field in a calendar year. It has been found that as the cropping intensity increases, the total quantum of crop residues incorporated annually into the soil increases and on the other hand the duration of fallow period decreases and in multiple cropping systems (MCS) of 300 or 400% intensity, it becomes negligible or nil. Hence under such situations, the decomposition of crop residues of previous crop and release of allelochemicals coincides with the germination and early seedling growth of following crops. Thus allelochemicals influence the establishment and growth of next crop. The recently introduced sunflower crop proved most harmful to succeeding summer crops and drastically reduced the germination growth and yields. In crop rotations, there are numerous such examples in tropical and subtropical countries. On the other hand, the crop yields decline in monocultures due to autotoxicity or soil sickness owing to allelopathic interactions.

In all tropical and subtropical countries, crop mixtures/intercropping of 2-7 crops are practised to overcome the risk of crop failure. Under such situations, allelopathic interaction are inevitable, hence, the available information has been reviewed in this chapter. In South American countries several crop mixtures e.g. bean + corn + squash etc. are highly productive and provides pests (weeds, insects, nematodes, diseases) control due to release of allelochemicals.

To feed the increasing population, the intensification of multiple cropping systems is likely to increase in future, hence, more research is needed on their allelopathic effects.

1. Introduction

The nature of cropping systems adopted in a given region, depends upon the climatic conditions and availability of water for raising of crops. Unlike temperate regions, the

S.S. Narwal et al. (eds.), Allelopathy in Ecological Agriculture and Forestry, 141–157.

climatic conditions in tropical and sub-tropical regions of the World are favourable for round the year cropping. In those areas, where assured supply of irrigation/rain water is available, the farmers grow more than 2 crops in a Calendar year. Therefore, allelopathic interactions play a major role in such areas. While in arid and unirrigated areas, farmers grow single crop in a year due to moisture deficiency for crop growth. On the other hand, in temperate regions, very low temperatures during winter season allow growing of a single crop in a year. In single cropping (100% cropping intensity) areas, allelopathic interactions do not play a major role. Because during the long fallow period of 200-260 days, the soil incorporated crop residues are decomposed completely and allelochemicals released during the decomposition are transformed to non-phytotoxic compounds. Besides, allelochemicals released into the soil through root exudates from the previous crop, perhaps also become non-toxic to succeeding crop. Hence, the preceding crop has little or no allelopathic influence on the succeeding crop.

In cropping system, mainly two types of allelopathy related problems viz., soil sickness and autotoxicity have been reported. In general, cropping systems are of three types viz., monocultures, crop rotations and crop mixtures or intercropping. Allelopathy plays a significant role in all these cropping systems. It decreases the crop growth and yields in continuous monocultures, due to build up of phytotoxins or harmful microbes or both in the growing medium and gives rise to autotoxicity and soil sickness. Crop rotations are practised to overcome such harmful effects of monocultures, the preceding crops may be inhibitory or stimulatory to the succeeding crops. Allelopathy also plays a great role in crop mixtures or intercropping systems due to inhibitory or stimulatory effects of one plant on another.

2. Monoculture

According to Francis (16), "monoculture is growing of same sole crop on the same land" and this definition has been used in text. Under optimal conditions of growth, monocultures of annual crops do not show inhibitory effects on yields but it may not be true in their continuous monocultures or in perennial crops. This is due to accumulation of phytotoxins or build up of harmful microbes or both, which leads to autotoxicity and/or soil sickness.

2.1 AUTOTOXICITY

Autotoxicity limits the yields of perennial crops and of annual crops in continuous monocultures. It results from the accumulation of phytotoxins from decomposing plant residues, root exudates and from multiplication of harmful pathogens in the soil. The autotoxicity has been observed in alfalfa (48,64), sugarcane (91), rice (9), asparagus (26, 27), pigeonpea (28), sunflower (12) and summer squash (30) (Table 1). Similar problems of yield decline occur in wheat and corn in areas of stubble mulch farming.

Preceding alfalfa crop released phytotoxins in soil which adversely affected the re-establishment of alfalfa crop, however, saponins had no such effect (48). In germination chamber experiments, water soluble compounds present in methanol extracts of soil cropped to alfalfa or barley decreased seedling root lengths of test crops viz., alfalfa 'L-720', wheat 'Nugaines', radish, 'Crimson giant'. Five days after

Table 1. Crops exhibiting autotoxicity and soil sickness in monocultures

Donor crop	Source of phytotoxicity		Recipient crop		Reference
	Compounds	Soil microbes	Name	Inhibitory effect	
AUTOTOXICITY					
Sorghum	ND	-	Sorghum	Inhibited crop growth	6
Rice	p-hydroxybenzoic, syringic, vanillic, ferullic, propionic, o-hydrophenylacetic and butyric acids H_2S gas	Putida	Rice	Suppressed root growth, tillering, panicle initiation and low yield	7,8,9,91
Sugarcane	p-hydroxybenzoic, p-coumaric, syringic, ferulic, vanillic, formic, acetic, oxalic, melonic, tartaric, malic and fusaric acids	*Fusarium oxysporum*	Sugarcane	Poor germination and growth	86,91
Asparagus	ND	*Fusarium* spp.	Asparagus	Poor growth and yield	26,37
Alfalfa	ND	-	Alfalfa	Poor re-establishment	48
Sunflower	ND	-	Sunflower	Reduced plant stand and yield attributes	12
Summer squash	ND	-	Summer squash	Reduced seedling growth	30
Pigeonpea	Terpenoids, polyphenols	-	Pigeonpea	Declined growth and yield	28
SOIL SICKNESS					
Alfalfa	Saponin	-	Alfalfa	Reduced seedling vigour, retarded root development and nodules, poor plant stand of short spindly yellowish plants and lower herbage yields	36,38,88
Alfalfa	Saponin	-	Cotton	Decreased yields	49
Wheat	ND	33 fungal spp. from *Penicillium*, *Fusarium*, *Mortierella* genera. *Gaemannomyces graminia* fungus and nematodes (*Pratylenchus pratensis* and *Tylenchohynchus dobius*)	Wheat	Decreased shoot and root growth and tillering	1,24,74 85,89
Sugarbeet	ND	*Haterodera schachtii* and *Pratylenchus napus* nematodes increased, Azotobactor and nitrifying bacteria decreased	Sugarbeet	Decreased germination, attack or black root rot	19,74
Corn	ND	*Pratylenchus pratensis*, and *Helicotylenchus dihystera*	Corn	Increased soil toxicity and reduced crop growth	33
Lupine, potato	ND	*Penicillium*, *Aspergillus* fungi	Lupine, potato		52
Cowpea	ND	-	Cowpea	Inhibited crop growth	71
Rice	ND	*Fusarium moniliforme*	Rice	Reduced growth and yield	83
Mungbean	ND	*Rotylenchulus*, *Meloidogyne*	Mungbean	Reduced growth and yield	83
Cowpea	ND	*Meloidogyne*	Cowpea	Reduced growth and yield	83

ND = Not determined.

germination, seedling dry weights of alfalfa and radish in alfalfa soil extract were lower than wheat or redclover, 'Kenland'. The soil extracts from cropped soil had no harmful effect on growth of red clover. The extracts of crop residues from soil cropped to alfalfa or barley reduced root lengths of seedlings, however, the extracts of alfalfa residues proved more inhibitory than barley (66). Sugarcane yields were declined in subsequent ratoons partly due to phytotoxicity of decomposing residues (under waterlogged conditions) or microbial toxins (90,91). Autotoxicity is one of the major reasons of low yields of second or third crop of rice in paddy growing areas of the World and it has been investigated by Chou and co-workers in Taiwan. The phytotoxins produced from the decomposing residues of preceding rice crop under waterlogged or submerged or saturated soil conditions suppressed root growth, panicle initiation and grain yields of rice (7,91,100). Asparagus may give high yields upto 25-30 years after planting but its autotoxicity markedly declined the yields after few years of successive growth in the same field. The phytotoxic compounds produced from decomposing residues of asparagus and from pathogenic fungi Fusarium spp. were toxic to its own growth (26,37,42,80). The extracts reduced hypocotyl growth in lettuce, shoot growth in asparagus and inhibited radicle elongation in barley, lettuce and asparagus (27). Hartung *et al.* 26) reported that allelochemicals from asparagus have direct physiological and biochemical effects on its plant that predisposes them to Fusarium diseases. Sorghum autotoxicity was observed only in sandy soils of Senegal, West Africa, due to low absorption capacity of the soil and inability of the soil microflora to detoxify the soil phytotoxins before the sowing of next sorghum crop (6). Continuous monoculture of pigeonpea declined the growth and yield of subsequent crops owing to phytotoxicity from terpenoids, polyphenols and other allelochemicals (28). The autotoxins produced from the decay of prairie sunflower (*Helianthus rigidus*) residues reduced the plant population and inflorescence numbers and size in the centre of clone (12). Itulya (30) reported that the leachates of summer squash reduced its own growth due to autotoxicity.

2.2. Soil sickness

In tropical and subtropical regions of the World, crops may be grown round the year if irrigation water is not limiting. But when a crop is grown continuously in the same soil for long periods, subsequent plantings often grow poorly compared with similar planting in soil never planted to the species concerned. The condition is species specific and other crops grow well in the same soil (Table 1). In sick soils phytotoxic materials liberated from the plants or plant residues may accumulate in soil through physical sorption (73). Clover crop failed in the field where it had been grown constantly owing to soil sickness (9) caused by root exudates and may be corrected by crop rotations having an interval of 7-8 years between the clover crops (14,73). According to Schreiner and associates soil sickness is caused by single cropping (70,71). Extracts of cowpea sick soil inhibited cowpea growth but after the extraction of cowpea sick soil inhibited cowpea growth but after the extraction of inhibitors from the soil it did not inhibit cowpea growth. Continuous cropping of oat, wheat and barley caused soil sickness owing to accumulation of toxins (61). Soil sickness may be caused by (a) unbalanced use of nutrients, micronutrient deficiencies and excess of fertilization, (b) destruction of soil structure and physical properties of soil, (c) evolution of phytopathogenic microflora, (d) disproportionate development of several groups of microflora, (e) increased production of pests and weeds, (f) changes in soil pH and (g) accumulation of phytotoxins in the soil (24,25,36,58,83,85). However, in this chapter only the reasons having phytotoxic effects (c, d and g) have been discussed. In the early phases of soil sickness, most of the toxins are native metabolites of higher or lower plants but in the later phases humid substances are dominant such as phenolic acids and closely related aromatic compounds.

2.2.1. *Alfalfa*

Areas in Alberta which once produced good crops of alfalfa now grow consistently poor crops having short, spindly and yellowish plants which do not respond to fertilization and irrigation. However, these soils are suitable for growing of alsike clover and red clover (88). Soils under monocultures of alfalfa suppressed the production of herbage roots and nodules. Alfalfa sown after corn or sorghum produced higher plant population and yield compared with continuous alfalfa (Table 2).

Table 2. Plant population and yield of alfalfa (after 6 years under various cropping sequences) Illinois, Urbana

Cropping sequence	Plants/ft^2	Dry matter yield (t/ha)
Corn-alfalfa	15.0	3.78
Corn-soybean-alfalfa	12.5	3.48
Continuous alfalfa	6.5	1.90

Source: 38.

Autophytotoxins were found in continuously alfalfa cropped soils, which inhibited the seedling establishment and growth and not in the root exudates of alfalfa (38). According to Kehr *et al.* (36), the relative yields of alfalfa for land with no known history of alfalfa cropping were 26, 40 and 35% higher in first, second and third year after sowing, respectively, compared with the same land area resown to alfalfa (Table 3).

Table 3. Average plant stand and forage yield of alfalfa in the year of seedling and in subsequent years at Mead, Nebraska

Year seeded	Plant stand (%)			Forage yield (t/ha)		
	Year of seeding	Subsequent I	Years II	Year of seeding**	Subsequent I	Years* II
1970	95	95	90	8.25	17.15	16.46
1976	96	96	86	***	12.42	12.22
1977	97	82	-	6.59	11.97	-
1978	98	0	-	-	6.52	-

*Three cuttings per year, ** Four cuttings per year, *** Comparable data not available.
Source : 36.

The stand of alfalfa became poor in subsequent years of continuous alfalfa crop owing to phytophthora and anthracnose diseases which caused soil sickness. Plant stand and yield declined rapidly under continuous alfalfa than in alfalfa-soybean rotation. Soil fatigue caused premature death of red clover plants owing to choking of their root xylem vessels with brown mass (5).

In germination chamber experiments, water soluble compounds present in methanol extracts of soil cropped to alfalfa or barley decreased seedling root lengths of test crops viz., alfalfa 'L-729', wheat 'Nugaines', radish, 'Crimson giant'. Five days after germination, seedling dry weights of alfalfa and radish in alfalfa soil extract were lower than wheat or red clover 'Kenland'. The soil extracts from cropped soil had no harmful effect on growth of red clover. The extracts of crop residues from soil cropped to alfalfa or barley reduced root lengths of seedlings, however, the extracts of alfalfa residues proved more inhibitory than barley (66). The soil sickness from alfalfa may also decrease yields of other crops which follow alfalfa. Cotton yields were decreased when it was grown after 3-4 years old lucerne crop owing to excretion of saponins in the soil (49).

2.2.2. *Wheat*

Inhibitory compounds in the aqueous extracts of soil under wheat monoculture with conventional and no tillage inhibited the root growth of wheat seedlings by 32 and 10% and shoot growth by 19 and 21% respectively (85). Continuous monocultures of wheat decreased tillering and crop yields of wheat (24) owing to accumulation of allelochemicals in the soil (3). The allelochemicals may change the hydrophillic nature of the soil colloids and thus directly inhibit the germination of crops. Maximum soil toxicity occurred during the kernel formation stage of wheat. Continuous monocultures of wheat accumulated 1.4 – 1.7 times more allelochemicals (n-coumarin, n-oxybenzoic, ferulic and vanillic acids) and resulted in the build up of harmful microbes in the soil than that under crop rotations (3). The extracts of these phytotoxic microbes were found to decrease the seed germination of radish by 50-80%. Golovko (2) isolated and identified 33 species of fungi mainly from *Penicillium*, *Fusarium* and *Mortierella* genera which made the soil sick and their toxicity ranged from 20 to 100%. Continuous monocultures of wheat also enhanced the accumulation of fungus *Gaeumannonyces araminis* responsible for root rot (89) and parasitic namatodes such as *Pratylenchus pratensis* and *Tylencohynchus dobina* (74).

2.2.3 *Other crops*

Continuous monocultures of corn on soddy podzolic soil increased the soil toxicity owing to build up of toxic forms of bacteria, fungi and parasitic nematodes such as *Pratylenchus pratensis* and *Helicotylenchus dihysteria* (33,74). However, Ventura *et al.* (85) did not observe soil sickness from 8-years of corn and sorghum monocultures compared with rice, mungbean and cowpea. Growth inhibition in monocultures of rice, mungbean or cowpea began with second crops and increased with succeeding crops. Soil sickness was not due to the nutrient status of the soil but was owing to the toxins produced from the decomposing crop residues and from build up of fungi in rice and nematodes in legumes. Soil sickness was observed in continuous monoculture of rice and mungbean for 8 years but not in rice-mungbean rotation (Table 4). Continuous cultivation of sugarbeet lowered its seed germination from 77 to 33%, decreased the availability of nitrogen and phosphorus to crop plants owing to reduction in the population of N and P compounds transforming bacteria and that of Azotobacter by 5-10 times (19). It also lead to the build up of nematodes (*Heterodera schachti* and *Pratylenchus napus*) which increased crop susceptibility to black root rot (74). Continuous monocultures of tobacco increased the incidence of root diseases caused by soil borne pathogen *Thielaviopsis basicola* compared to rye grass – corn – tobacco rotation (60). Because, fungitoxic substances produced from decomposing residues of corn or rye grass inhibited the germination of Conidia or Chlaymydospores of *T. basicola* and thus reduced its population and root rot diseases. Continuous cultivation of lupines or potato increased the population of toxic fungi (*Penicillium* and *Aspergillus*) compared to rye-lupine or rye-potato rotations (52). The continuous monocultures of cabbage for eight crop seasons accumulated amino acids in growing substrate of perlite (22).

Table 4. Effect of previous cropping condition on yield of upland rice

Crop	Previous crop treatment	Grain yield (t/ha)
Rice 'IR 2061-464-2-4'	3 continuous rice crops	0.36 b
	15 month continuous fallow	1.75 a
Rice 'IR 52'	16 continuous rice crops	0.85 b
	2 years continuous fallow	2.35
Rice 'IR 747-B2-6—3'	4 continuous rice crops	0.95 b
	4 continuous cowpea crops	3.02 a
Mungbean 'MG 50-10A'	16 continuous mungbean crops	0.15 b
	8 continuous sorghum crops	0.53 a
	6 continuous sorghum crops	1.21
	2 Mungbean + 1 year fallow	0.60 a
Cowpea 'EG green pod'	5 continuous cowpea crops	0.60 b
	5 continuous maize crops	1.64 a
Mungbean 'MG 50-10A'	16 continuous mungbean crops	0.42 a
	16 continuous cowpea crops	0.47 a
Sorghum 'Cosor 2'	4 continuous sorghum crops	3.45 a
	4 continuous rice crops	3.71 a
	16 continuous sorghum crops	2.95 a
	16 continuous cowpea crops	3.30 a
Maize 'DMR 2'	5 continuous maize crops	3.63 b
	5 continuous cowpea crops	4.47 a

In crops the same cropping period means followed by a common letter are not significantly different at 5% level
Source : 85.

2.3. YIELD DECLINE IN SUGARCANE RATOON CROP

Ratooning is a major practice in sugarcane and is done for a period of 2 to 8 years. However, in every successive ratoon, the plant stand is decreased by about 10% and this decline in crop stand considerably reduces the cane yields in successive ratoons. Studies in Taiwan and Australia have attributed this to allelopathy (8,46). In a ratoon crop, the decomposition of old roots may produce phytotoxic compounds, which may affect the root and shoot growth of new tillers. Besides, decomposition of sugarcane leaves under waterlogged conditions produced p-hydroxybenzoic, p-coumaric, syringic, ferulic, vanillic, formic, acetic, oxalic, malonic, and malic acids; which at 50 ppm in water culture inhibited the growth of sugarcane roots. Also, the rhizosphere soils from the fields of poor ratoon crops contained a high population of *Fusarium oxysporum*, which produced fusaric acid. This acid inhibited the shot and root growth of sugarcane (90). The retardation in the sugarcane growth is due to the inhibitory metabolites of the fungi *F. oxysporum* and *Trichoderma harzianum* which fluorished in sugarcane residue added soils (34). No such allelopathic studies have been done in India; however, such studies if conducted may help to overcome the yield decline in successive ratoons and to determine the causes of varietal variability for ratooning capacity.

3. Crop rotations

In crop rotations, one crop may harm another one through depletion of plant nutrients, phytotoxins released from decomposing residues may injure the plant roots in contact with them and toxins in root exudates may be harmful to subsequent crop. Since all the plants are not equally sensitive to allelochemicals, therefore, ideal crop rotations avoids negative effects of allelochemicals and capitalize on their stimulatory effects.

Good crop rotations maintain the soil productivity, control the insects and diseases and overcome the soil sickness or autotoxicity in crops. Crop rotations remove the ill effects of soil sickness arising from continuous cropping of tobacco (60), lupine (52), potato (52), alfalfa (38, 36), rice (83) and mungbean (81) and autotoxicity caused by pangola grass (7). Crop rotations overcome the major causes of soil sickness such as (i) accumulation of phytotoxins in the soil (61), (ii) multiplication of phytopathogenic soil microflora (21,33,52), (iii) increased incidence of pests (74) and diseases (36,60) and (iv) destruction of soil physical properties (24). Numerous studies have been done on the crop rotations but very few dealt with allelopathic aspects and their causative agents. Cereals (wheat, oat, barley, rice, corn, sorghum) based rotations have been studied most compared with rotations of other crops. Inhibition or stimulation of crop yields have been observed in rotations of cereals-cereals, cereals-legumes and other crops.

Allelopathy has been implicated in yield decline in crops when sown after sesame, sorghum(10,78), corn (18), rice (83,90), pea (55), mungbean (83) and sunflower (69). Besides allelopathic effects also reduced crop yields in continuous monocultures of sorghum (6), barley (51), rice (83), tobacco (60) and mungbean (83). Sorghum was one of the first crops to be recognized for having detrimental effects on the succeeding crops of corn, wheat and oat (10,68). Sorghum decreased the growth and yield of following sorghum crop in sandy soils of Senegal, West Africa due to low absorption capacity of the soil microflora to detoxify the soil toxins before the sowing of next sorghum crop (6). Aqueous extracts of sorghum field soils, decaying mulch and its residues reduced seed germination, radicle growth and dry matter of sorghum, pearlmillet, maize and foxtail millet by creating water deficit in these crops (29). In sorghum/corn/soybean-oat rotations, heavy residues of sorghum and corn inhibited growth of oat plants, therefore, oat following soybean produced bold and heavy seeds compared with those after corn or sorghum (90). Su and Fang (78) studied sorghum/corn-rape/barley/peas, sorghum/millet-wheat and corn/groundnut-wheat rotations. Corn and groundnut proved better preceding crops of wheat than sorghum or millets. While previous crop of sorghum was favourable to barley and pea but harmful to rape and reverse hold true for preceding corn. Wheat, peas and rape were good previous crops for maize as were wheat, barley and oat for sorghum. In corn-soybean rotation, the corn was benefited from soybean effect, owing to release of 30-60 Kg N/ha and to allelopathy (50). In barley/potato-barley rotations, barley sown after barley gave 5-18% less yield than that after potato. Because phenols from the decomposing residues of preceding barley reduced the root growth of succeeding barley crop at heading stage (51).

In rice-soybean rotation, soil incorporation of rice straw decreased the grain yields of soybean compared with burning of rice straw owing to inhibition of biological nitrogen fixation in soybean by the phytotoxins from the decomposing rice straw (67). In 8-years study, rice-mungbean rotation maintained high yields over their monocultures (Table 5), because, soil, fungi and root infecting nematodes caused soil sickness in monocultures of rice and mungbean, respectively. While, rice-mungbean rotation mitigated the soil sickness problem by suppressing the population of these microbes (85). Similarly, corn-rye grass-tobacco rotation controlled the root rot disease of tobacco monocultures caused by *Thielaviopsis basicola* (60). Alfalfa grown in corn-alfalfa rotation gave 50-60% higher yields than its monocultures (38).

Table 5. Effect of previous crops on plant population and yield of mungbean 'MG 50-10A' in field

Crop history	Plant population ($x10^3$/ha)			Grain yield (kg/ha)
	4 week	6 week	8 week	
15 continuous mungbean crops	221a	98c	64c	61.5d
2 year fallow	263a	185b	158b	419.6a
1 rice crop-1 mungbean crop-1 rice crop	284a	263a	228a	473.1a
1 year mungbean and 1 season fallow	271a	210b	190ab	514.7

In a column means followed by a common letter are not significantly different LSD, P = 0.05
Source: 85.

Sesame reduced the yield of following cabbage due to the injurious effects of its residues (75) Preceding pea crop inhibited the germination and growth of succeeding rye crop because of release of inhibitory free radicles from the decomposing pea stubbles (55). Allelochemicals released from the sunflower plants and residues decreased yields of subsequent sorghum, through interference in water balance (67). Sugarbeet decreased the stem and leaves growth of subsequent soybean crop in dry type volcanic ash soil in District Hokkaido, Japan. It was speculated to be caused by some allelopathic compounds (46). Potatoes gave better yields when grown after tobacco than after corn, while small grains did better following potatoes than following tobacco (18). In tobacco/corn-rice rotations, tobacco leaf litter added 5.32 kg nicotine/ha which stimulated the radicle and plumule growth of corn but inhibited that of rice (66). Pangola grass-water melon rotation increased the growth of pangola grass by 40% over its monocultures, because water melon eliminated the phytotoxins of pangola grass residues (7).

Narwal *et al.* (57) investigated the effects of preceding sunflower crop in field studies and bioassays on eight summer crops viz., sorghum 'HC 1', maize 'Safed', pearlmillet 'HHB67', clusterbean 'HG27', Cowpea 'FS68', greengram 'K851', cotton 'H777' and sunflower 'Jawala Mukhi'. The preceding sunflower crop decreased the growth (plant height, dry matter) and yield of all subsequent crops as compared to those sown in fallow plots. The maximum reduction in growth, yield attributes and grain yield occurred in cotton and sunflower and moderate in legumes (clusterbean, cowpea, greengram) (Table 6). The sunflower straw management practices also decreased the growth and yield of test crops and followed the order : soil incorporation of straw + stubble (root) > straw burnt > straw removed > straw + stubble removed. The phytotoxicity of sunflower biomass persisted in soil upto 9 weeks after soil incorporation. The aqueous extracts of sunflower stem and leaf delayed and inhibited the germination of all test crops.

Table 6. Inhibitory effect of sunflower straw management on test crops

Straw management treatment	Succeeding test crops yield (% inhibition over fallow/control)							
	Grain				Fodder crops			
	Pearlmillet	Green-gram	Sun-flower	Cotton	Sorgh-um	Maize	Cluster-bean	Cowpea
Straw burnt	22.84	49.93	78.18	80.21	25.69	27.71	52.46	48.96
Straw removed	20.39	45.58	80.16	76.10	23.37	25.38	48.29	45.32
Straw + root removed	15.07	34.11	78.00	71.00	18.15	20.06	41.00	38.04
Straw added to soil	36.22	62.90	82.10	86.29	39.18	41.16	68.91	65.60

Source : Narwal et al. (1999a).

The further pot and biomass decomposition studies (58) showed that sunflower infested soil inhibited the germination, growth and grain yields of all test crops. In germination, the maximum reduction occurred in cotton and cereals (sorghum, maize, pearlmillet) and moderate in legumes (soybean, cowpea, blackgram, greengram, clusterbean and pigeonpea). The amendment of sunflower infested soil with sunflower straw increased the seedling mortality of test crops (Table 7). The addition of sunflower roots caused maximum mortality in seedlings of test crops. The seedling mortality followed the order : cowpea > clusterbean > greengram > maize > pearlmillet > sorghum i.e. legumes were most sensitive to decomposing biomass of sunflower. Under labortoryconditions, biomass decomposition of sunflower took about 9 weeks.

3.1. INHIBITION OF NITRIFICATION AND BIOLOGICAL NITROGEN FIXATION

Phytotoxins produced during the decomposition of crop residues inhibit the nitrification process in the soil and biological nitrogen fixation in legumes. The maintenance of corn residues on the soil surface increased the concentration of nitrification inhibitors (ferulic and *p*-coumaric acids) in the soil, which decreased the population of nitrosomonas and thus increased the concentration of NH_4^+ over NO_3^- compared with the soil without corn residues (45). In south Taiwan, soybean following rice, yielded higher when rice residues were burnt than when decomposed in the field (1), because phenolics produced from decomposing rice residues inhibited the growth of N fixing bacteria (*Rhizobium japonicum*), reduced nodule number and thus decreased biological nitrogen fixation in soybean (69). Similarly, soil incorporation of vines and storage root residues of sweet potato reduced the nodulation of nitrogen fixation in cowpea (86).

3.2. REDUCED NUTRIENT UPTAKE

Ion uptake by plants is important for the growth, development and yield and many allelopathic agents affect their uptake. Sugarbeet alters the zinc status of the soil to the

Table 7. Effect of sunflower infested soil amended with sunflower biomass on seedling mortality of test crops

Test crops	RS (Rhizosphere soil)				RS+R (Rhizosphere soil + roots)				RS+ FP (Rhizosphere soil + full plant biomass)			
	1996		1997		1996		1997		1996		1997	
	Seedling mortality (%)	Total	Seedling mortality (%)	Total	Seedling mortality (%)	Total	Seedling mortality (%)	Total	Seedling mortality (%)	Total	Seedling mortality (%)	Total
Sorghum	0	0	0	0	9.1(7), 10.0 (8)	19.1	0	0	6.7(11)	6.7	0	0
Pearl-millet	0	0	0	0	20.0(7), 20.0(11)	40.0	6.6(12)	6.6	20.0(10.), 8.3(11)	28.3	0	0
Maize	0	0	13.2(6)	13.2	18.2(5), 20.0(12)	38.2	33.2(6)	33.2	0	0	0	0
Cotton	12.5(10)	12.5	6.6(12)	6.6	12.5(8)	12.5	0	0	8.3(6), 26.7(8), 9.1 (10)	44.1	0	0
Cowpea	8.3(6), 9.1(8), 10.0(10)	27.4	0	0	69.2(7), 15.4(8), 7.7(9)	92.3	6.6(6)	6.6	33.3(7), 33.3(9)	66.6	0	0
Green-gram	0	0	6.6(12), 13.2(13)	19.8	7.7(5), 50.0(6)	57.7	6.6(6)	6.6	11.1(7)	11.1	6.6(12)	6.6
Cluster-bean	14.3(10)	14.3	6.6(12)	6.6	23.1(6), 16.7(9), 8.3(11)	48.1	13.2(12)	13.2	10.0(8)	10.0	0	0
Black-gram	-	-	13.2(6), 13.2(13)	26.4	-	-	13.2(11)	13.2	-	-	6.6(13)	6.6
Soybean	-	-	13.2(6)	13.2	-	-	6.6(7), 13.2(12)	9.8	-	-	6.6(5)	6.6
Pigeon-pea	-	-	6.6(12), 6.6(13)	13.2	-	-	20.6(6), 6.6(10)	26.6	-	-	6.6(12), 33.2(13)	39.8

Figure in parenthesis indicates days after sowing (DAS); - indicates crops were not sown.

extent that succeeding corn and beans are severely deficient (4). Sugarbeet did not make the soil zinc deficient but it added toxins to the soil that interfered with zinc uptake by other crops. All phenolic acids inhibited P_{32} and K uptake due to an increase in the membrane permeability to inorganic ions (20). Ferulic acid at 0.5 and 1.0 mM inhibited P_{32} uptake by soybean seedlings (49). Juglone was the most inhibitory phenolic acid with 79% inhibition of K absorption (2). Generally, flavonoids were more inhibitory than phenolic acids at 10^{-4}M concentration.

4. Crop mixtures

Productivity of crop mixtures or intercropping systems may be decreased or increased depending on the inhibitor or stimulatory effects of component crops on each other provided growth resources (light, water, nutrients and space) are not limiting. Much of the research on agrophytocenosis has been conducted in the USSR and major part of this section is based on their english summaries. In crop mixtures or intercropping systems the growth and yield of component crops increases due to greater nutrient absorption, better weed control etc. than in pure crops, yet the mechanism have not been fully understood. Root exudates play a major role in the productivity of crop mixtures as they may improve growth (45,65) and yield (35,63) of component crops, through improved ion exchange (65), greater nutrient uptake (35,41,81) and partial weed control (32) compared with pure crops.

Several legumes have been shown,to increase maize growth in mixtures (15,45). To improve the productivity of mixtures, new varieties of legumes stimulatory to the non-legume component have been developed, for example, late maturing varieties of soybean for corn+soybean mixture (45). Sokolova and Mikryukev (77) studied white mustard + pea/barley mixtures. In white mustard + pea mixture, the root exudates of white mustard improved the phosphorus absorption and proportion of pea plants in the crop mixture, therefore, produced higher yield in mixtures than their sole crops. In pea + barley mixture, barley had no positive effect on pea but the ratio of barley in mixture increased in drought years compared with wet years. In white mustard + barley mixture, strong development of xylem occurred with a consequent decrease in the parenchyma of barley.

White mustard proved a better component of crop mixture with pea than with barley. In crop mixtures, legumes improved the phosphorus and potassium uptake of cereals by increasing the availability of soil phosphorus (81). For example, in oat + lupine mixture, phosphorus uptake of oat was greater than in its sole crop (64). Maize + horsebean mixtures produced high yields compared with sole crops due to greater exchange of root exudates between the component crops (63). In mustard + broccoli intercropping, broccoli gave very high yields because mustard stimulated the growth and yield parameters (number of harvestable heads, diameter of the inflorescence and total biomass) of broccoli (Table 8) (31). In buckwheat + mustard mixture the root exudates of buckwheat inhibited the growth of mustard plants (82). In oat + alfalfa mixture, alfalfa cultivar "MN allelopathy' reduced grain yield of oat cultivars compared with other alfalfa cultivars (56). In corn + water melon intercroping, corn pollens decreased the respiration and growth of watermelon seedlings (11). On the other hand, addition of 1-2 kg seed/ha of wild heliotrope (*Heliotropium europeum*) to the legume mixtures reduced their weeds (30-70%) and pests (23).

Table 8. Effect of mustard intercropping on yield attributes and yield of broccoli

Treatment	Head			
	No. of plants/plot	Diameter (cm)	Weight (g)	Yield (t/acre)
Broccoli pure	73	13.55a	353.6a	4.41
2 mustard plants/m^2 + brocccoli	89ab	14.16a	332.8a	5.02
4 mustard plants/m^2 + broccoli	101b	14.62a	358.35a	6.25
8 mustard plants/m^2 + broccoli	99b	15.83a	406.35b	6.74

Mean with the same letters are not significantly different, P = 0.05.
Source : 31

Root exudates play a vital role in productivity of crop mixtures, besides volatiles from crop shoots may also influence nutrient uptake. Tomato + cucumber mixture excreted amino acids, aliphatic organic acids, phenolics and several unidentified compounds in nutrient culture (17). Exudates of carrot roots and from leaves of beet, tomato and sweet potato contained volatile compounds (acetaldehyde, propionic aldehyde, acetone, methanol, ethanol and other unidentified compounds) but propionic aldehyde greatly reduced the growth of test species (13). In wheat + rye mixtures, wheat growth was inhibited when its proportion was small, while the growth of rye was stimulated in all proportions. Since, the degree of stimulation or inhibition depends on the concentration of unidentified substances secreted from the root exudates of both species (40). In oat + alfalfa mixture, the root exudates of oat reduced the growth of alfalfa but alfalfa exudates slightly increased the oat growth (39). In general, root exudates may inhibit or stimulate the absorption, migration of ions and growth of plants. In maize + pea mixture, as the proportion of maize decreases its inhibition by pea increases and vice-versa (42). Kaurov (35) reported greater dry matter production in bird's foot trefoil + yellow lupine mixture than their pure crops, because lupine improved the mineral nutrition of bird's foot trefoil plants. In buckwheat + lupine/mustard mixture, buckwheat yields were 12-35% higher than the pure crop, as the root exudates of mustard and lupine stimulated the growth and development of buckwheat. While in buckwheat + oat/lupine mixture, the root exudates of oat were inhibitory to buckwheat, conversely, buckwheat root exudates increased the productivity of oat by 10-20% but decreased the growth and yield of lupine (94).

In maize + pea mixture, root exudates may inhibit or stimulate the growth, absorption and ion accumulation of plants, however, total N, P, K uptake was greater than in sole crops and the nutrient uptake of maize was higher than pea (41,42). Similarly in bird's foot trefoil + yellow lupine mixture lupine improved the mineral nutrition of bird's foot trefoil and bird's foot trefoil absorbed 32^P more actively than lupine. The rate of 32^P migration from lupine to bird's food trefoil was greater than that from bird's foot trefoil to bird's foot trefoil plants (35). According to Rakhteenko (65), the rate of exchange of root exudates in pea + oat and wheat + white mustard mixtures was 1.5 to 2 times and 3 to 7 times greater, respectively, than their pure crops. Substances in root exudates of pea and vetch stimulated the photosynthesis, absorption of 32^P, uptake of N, K and Ca by barley and oat plants. On the contrary, substances in root exudates of oat and barley inhibited these processes in pea and vetch plants. In pea + white mustard mixture, mustard root exudates improved the phosphorus absorption of pea plants than in mixture of pea with other crops (74). Volatile substances from the shoots of soybean, chickpea and bean reduced the 32^P uptake of maize plants, but combined volatile substances secreted from their roots and shoots stimulated its uptake and improved the maize yields.

Refereences

1. Asian Vegetables Research and Development Centre (1978). *Soybean Report*, 1976. Shanhua, Taiwan, China: AVRDC.
2. Barke, N. (1977). *Inhibition of Ion Absorption in Avena sativa Roots by Diethyestibesterol and other Phenolic Compounds*. Ph.D. Dissertation. West Lafayette, Indiana: Purdue University.
3. Berestetskiy, O.A. and Kalmykova, N.A. (1974). *Mikobiologicheskiy Zhurnal* **36**: 726-732.
4. Boawn, I.C. (1965). Sugarbeet induced zinc deficiency. *Agronomy Journal* **57**: 509-511.
5. Bogdan, G.P., Shpilevoy, B.E. and Maistrnko, O.U. (1982). Ultrastructural alterations in the clover root xylem under soil fatigue. *In Role Allelopatii V Rastenie Vodstve*. Pp. 33-42. Kiev: Naukova Dhunka.
6. Burgeos-Leon, W., Ganry, F., Nicon, R. and Chopart, J.L. (1980). *Agronomy Tropical* **35**: 319-334.
7. Chou, C.H. (1986). The role of allelopathy in sub-tropical agroecosystems in Taiwan. In *The Science of Allelopathy* (Eds., A.R. Putnam and C.S. Tang) pp. 57-73. New York: Wiley Interscience.
8. Chou, C.H. (1987). Allelopathy in subtropical vegetation and soils in Taiwan. *In Allelochemicals: Role in Agriculture and Forestry* (Ed., G.R. Waller). ACS Symposium Series No. **330**: 102-117. Washington, D.C.: American Chemical Society.
9. Chou, C.H. and Chiou, S.J. (1979). *Journal of Chemical Ecology* **5**: 839-859.
10. Cole J.S. and Halstead, A.L. (1922). Methods of Winter Wheat Production at Fort Hyes Research Station. *USDA Bulletin No.* 1094.
11. Cruz, O.R., Anaya, A.L. and Ramos, L. (1988). *Journal of Chemical Ecology* **14**: 71-86.
12. Curtis, J.T. and Cottam, G. (1950). Antibiotic and autotoxic effects in prairve sunflower. *Bulletin of The Torrey Botanical Club* **77**: 187-191.
13. Dadykin, V.P., Stepanov, L.N. and Rykhova, B.E. (1970). On importance of volatile plant-secretions under the development of closed systems. *In Physiological-Biochemical Basis of Plant Interactions in Phytocenosis* (Ed., A.M. Grodzinsky) **1**: 118-124. Kiev: Naukova Dumka.
14. DeCandolle, M.A. (1832). *Physiologie vegetable* Vol. **3**: 110.Paris: Bechet, Juene, Library, Faculty of Medicine.
15. Dzybenko, N.N. and Petrenko, N.I. (1971). On biochemical interaction of cultivated plants and weeds. In *Physiological-Biochemical Basis of Plant Interactions in Phytocenosis* (Ed., A.M. Grodzinsky) **1**: 118-124. Kiev: Nankova Dumka.
16. Francis, C.A. (1989). Biological efficiencies in multiple croppping systems. *Advances in Agronomy* **42**: 1-43.
17. Gaidamak, V.M. (1971). Biologically active substances in nutrient solutions after cucumbers and tomatoes grown in pure and multiple used broken brick. In *Physiological-Biochemical Basis of Plant Interactions in Phytocenosis* (Ed., A.M. Grodzinsky) **2**: 55-60. Kiev: Naukova Dumka.
18. Garner, W.W., Lunn, W.M. and Brown, D.E. (1925). *Journal of Agricultural Research* **30**: 1095-1132.
19. Geller, I.A., Kalmykova, N.A. and Petrusha, L.K. (1977). *Sakharnaya Svelka* **1**: 31-32.
20. Glass, A.D.M. (1975). Inhibition of phosphate uptake in barley roots by hydroxybenzoic acids. *Phytochemistry* **14**: 2127-2130.
21. Golovko, E.A., Elanskaya, I.A. and Kostroma, E.Y. (1981). Allelopathic soil fatigue and phytotoxic properties of soil microbial fungi. In *Allelopatiya vestesvennykhi isknstvennykh fitolesenozakh. Collection of Scientific Papers*. Pp. 86-95. Kiev: Naukova Dumka.
22. Gorobets, S.A. (1981)., In *Allelopatiya Vestesvennykh I isknsstvennykh fitosenozakh. Collection of Scientific Papers*. pp. 152-159. Kiev: Naukova Dumka.
23. Grechkanov, D.M. and Radionov, V.I. (1971). Interaction of the components of nector fodder mixtures with wild heliotrops. *In Physiological-Biochemical Basis of Plant Interactions in Phytocenosis* (Ed., A.M. Grodzinsky) **2**: 88-100. Kiev: Naukova Dumka.'
24. Grodzinsky, A.M. and Golovko, E.A. (1983). Allelopathic problems in soil fatigue. *Soviet Soil Science* **15**: 54-62.

25. Grummer, G. (1955). *Die gegenseitigebonerer Pflanzen Allelopathie*. Germany, Jena: Fischer pp. 162.
26. Hartung, A.C. and Stephens,C.T. (1983). *Journal of Chemical Ecology* **9**: 1163-1183.
27. Hazebroek, J.P., Garrison, S.A.and Ianfagna, T.G. (1989). *Journal of American Horticultural Science* **114**: 152-158.
28. Happerly, P.R. and Diaz, M. (1983). The allelopathic potential of pigeonpeas in puerto Rico. *Journal of Agricultural University*, Puerto Rico **67**: 453-463.
29. Hussain, F. and Gaddon, M.A. (1981). *Oecologia* (Berlin) **51**: 284-288.
30. Itulya, F.M. (1987). *East African Agricultural and Forestry Journal* **52**: 33-36.
31. Jimenez-Osorino, J.J. and Gliessman, S.R. (1987). Allelopathic interactions in a wild mustard (*Brassica campestris* L.) and Broccoli (*Brassica oleracea* L. var. Italica) in intercrop agroecosystem. In *Allelochemicals: Role in Agriculture and Forestry* (Ed., G.R. Waller) **330**: 262-274. Washington DC: American Chemical Society..
32. Jimenez-Osorino, J.J. and Schultz, C.K. (1981). Relaciones cultivo-arvenses en una chinampa. *Teis delicenciatura*. Mexico: Facultad de Ciencias, UNAM.
33. Kalmykova, N.A. (1973). In *Physiological-Biochemical Basis of Plant Interactions in Phytocenosis* (Ed., A.M. Grodzinsky) **3**: 124-126. Kiev: Naukova Dumka.
34. Kao, M.M. (1989). Fungal metabolites as growth inhibitors of sugarcane and the mechanism of phytotoxicity. In *Phytochemical Ecology* (Eds., G.R., Waller and C.H. Chou). Monograph No. **9**: 329-341. Taipei, Taiwan: Institute of Botany, Academia Sinica.
35. Kaurov, I.A. (1970). Interaction of bird's foot and yellow lupine in pure and mixed cultures. *In Physiological-Biochemical Basis of Plant Interactions in Phytocenosis* (Ed., A.M. Grodzinsky) **1**: 66-77. Kiev: Naukova Dumka.
36. Kehr, W.R., Watkins, J.E. and Ogden, R.L. (1983). Alfalfa establishment and production with continuous alfalfa and following soybeans. *Agronomy Journal* **75**: 435-438.
37. Kitahara, Y., Yanagawa, H., Kato,T. and Takahashi, N. (1972). *Plant Cell Physiology* **13**: 923-925.
38. Klein, R.R. and Miller, D.A. (1980). Allelopathy and its role in agriculture. *Communication in Soil Science and Plant Analysis* **11**: 43-56.
39. Krasilnikova, N.A. and Garkina, N.I. (1946). *Microbiologiya* **15**: 109-141.
40. Lambert, R.G. (1959). *Plant Diseases Reporter* **43**: 1117-1119.
41. Lastuvka, Z. (1958). *Folia Biologia* **4**: 119-126.
42. Lastuvka, Z. (1970). Mutual effect of maize and pea in water cultivars with additional nutrition. In *Physiological-Biochemical Basis of Plant Interactions in Phytocenosis* (Ed., A.M. Grodzinsky) **1**: 55-59. Kiev: Naukova Dumka.
43. Lastuvka, Z. and Minarez, I. (1970). Mutual effect of maize and pea in water cultures with additional nutrition. In *Physiological-Biochemical Basis of Plant Interactions in Phytocenosis* (Ed., A.M. Grodzinsky) **1**: 56-59. Kiev: Naukova Dumka.
44. Lauffer, G.A. and Garrison, S.A. (1977). The effect of asparagus tissue on seed germination and asparagus seedling growth: possible allelopathic interactions. *Hort Science* **12**: 385.
45. Lodhi, M.A.K. (1981). Accelerated soil mineralization, nitrification and revegetation of abandoned fields due to the removal of crop-soil phytotoxicity. *Journal of Chemical Ecology* **7**: 685-694.
46. Lovett, J.V. and Jackson, H.F. (1980). Allelopathy a possible contributor to yield decline in sugarcane. *Plant Protection Quarterly* **7**: 180-182.
47. Lykhvar, D.F. and Nazarova, N.S. (1970). On importance of legume varieties in mixed cultures with maize. In *Physiological-Biochemical Basis of Plant Interactions in Phytocenosis* (Ed., A.M. Grodzinsky) **1**: 83-88. Kiev: Naukova Dumka.
48. Mitsui, S.H.O. and Nishiri, K. (1989). *Japanese Journal of Crop Science* **58**: 337-341.
49. McClure, P.R., Gross, H.D. and Jackson, W.A. (1978). Phosphate adsorption by soybean varieties. The influence of ferulic acid. *Canadian Journal of Botany* **56**: 764-767.
50. Miller, D.A. (1983). Allelopathic effects of alfalfa. *Journal of Chemical Ecology* **9**: 1059-1072.
51. Mishushtin, E.N. and Naumova, A.N. (1955). Izvestia Akademic Nauk USSR *Social Biology* **6**: 3-9.
52. Mosca, G., Merlo, D. and Toniolo, L. (1988). *Rivue Agronomie* **22**: 266-272.
53. Muller, Wilmes, V. and Zoschke, M. (1980). *Angewandte Botanik* **54**: 119-123.
54. Nadkernichnyi, S.P. (1974). On the problem of distribution of microscopic toxin forming in soddy medium podzolic soil under some farm crops. In *Physiological-Biochemical Basis of Plant Interactions in Phytocenosis* (Ed., A.M. Grodzinsky) **5**: 97-100. Kiev: Naukova Dumka.
55. Narwal, S.S. (1994). *Allelopathy in Crop Production*. Jodhpur, India: Scientific Publishers. pp. 288.

56. Narwal, S.S. (1996). Allelopathy related problems in crop production, agroforestry and horticulture. In *Allelopathy: Field Observations and Methodology* (Eds., S.S. Narwal and P. Tauro). Pp. 1-41, Jodhpur, India: Scientific Publishers.
57. Narwal, S.S., Singh, T., Hooda, J.S. and Kathuria, M.K. (1999a). Allelopathic effects of sunflower in succeeding summer crops. I. Field studies and bioassays. *Allelopathy Journal* **6**: 35-48.
58. Narwal, S.S., Yadav, S. and Gupta, S. (1999b). Allelopathic effects of sunflower on succeeding summer crops. II. Pot culture and biomass decomposition. *Allelopathy Journal* **6**: 209-226.
59. Neustrueva, S.N. and Terekhina, T.A. (1975). Effect of the residues in stubble of fallow-grown crops on germination and development of winter rye. In *Physiological-Biochemical Basis of Plant Interactions in Phytocenosis* (Ed., A.M. Grodzinsky) **6**: 49-54. Kiev: Naukova Dumka.
60. Nielsen, R.L., Stutnean, D.D. and Barnes, D.K. (1980). Interference between oats and alfalfa in mixed seedlings. *Agronomy Journal* **73**: 635-638.
61. Patrick, Z.A. (1955). *Canadian Journal of Botany* **33**: 461-486.
62. Patrick, Z.A. and Koch, L.W. (1963). The adverse influence phytotoxic substances from decomposing plant residues on resistance of tobacco to black root rot. *Canadian Journal of Botany* **41**: 747-758.
63. Periturin, F.T. (1913). *Izvestia* Moskva S-kh Institute. Kn. 4.
64. Petrova, A.G. (1977). In *Interactions of Plants and Microorganisms in Phytocenosis* (Ed., A.M. Grodzinsky) pp. 91-97. Kiev: Naukova Dumka.
65. Pronin, V.A.,, Voronkova, F.V. and Yakovlev,A.A. (1972). Plant relations in mixed stands as dependent on biological and biochemical processes in soil. In *Physiological-Biochemical Basis of Plant Interactions in Phytocenosis* (Ed., A.M. Grodzinsky) **3**: 121-124. Kiev: Naukova Dumka.
66. Rakhteenko, I.N. and Egrova, R.N. (1971). Physiological peculiarities of mutual influence of lupine and oat in mixed and pure cultures. In *Physiological-Biochemical Basis of Plant Interactions in Phytocenosis* (Ed., A.M. Grodzinsky) **2**: 79-84. Kiev: Naukova Dumka.
67. Rakhteenko, I.N., Kaurov, I.A. and Minko, I.E. (1973). Effect of water soluble metabolites of a series of crops on some physiological processes. In *Physiological-Biochemical Basis of Plant Interactions in Phytocenosis* (Ed., A.M. Grodzinsky) **4**: 23-26. Kiev: Naukova Dumka.
68. Read, J.J. and Jensen, E.H. (1989). *Journal of Chemical Ecology* **15**: 619-628.
69. Rice, E.L., Lin, C.Y. and Huang, C.Y. (1981). Effects of decomposing rice straw on growth and nitrogen fixation by Rhizobium. *Journal of Chemical Ecology* **7**: 333-344.
70. Rizvi, S.J.H. and Rizvi, V. (1987). Improving crop productivity in India: Role of allelochemicals. In *Allelochemicals: Role in Agriculture and Forestry* (Ed., G.R. Waller). ACS Symposium Series **330**: 69-75. Washington DC: American Chemical Society.
71. Schon, M.K. and Einhellig, F.A. (1982). Allelopathic effects of cultivated sunflower on grain sorghum. *Botanical Gazette* **143**: 505-510.
72. Schreiner, D. and Reed, H.S. (1907). *Certain Organic Constituents of Soil in Relation to Soil Fertility.* USDA Beaureaul of Soils, Bulletin No. 47.
73. Schreiner, O. and Shorey, E.D. (1909). *The Isolation of Harmful Substances from Soils.* USDA Beareau of Soils, Bulletin No. 53.
74. Shcreiner, O. and Sullivan, M.X. (1909). *Journal of Biological Chemistry* **6**: 39-50.
75. Seredyuk, L. (1982). On the role of organic substances in crop production. In *Role Allelopatii V Rastenie Vodstve. Collection of Scientific Papers.* pp. 155-159. Kiev: Naukova, Dumka.
76. Sigareva, D.D. (1982). Susceptibility of winter wheat, maize and sugarbeet to parasitic nematodes under continuous cropping and in crop rotation with organic and inorganic fertilizers. In *Role Allelopatii V Rastenie Vodstve. Collection of Scientific Papers.* pp. 129-137. Kiev: Naukova, Dumka.
77. Skinner, J.J. (1923). *Plant World* **16**: 342-348.
78. Sokolova, E.A. (1973). Effect of mustard root excretions on absorption of phosphorus and synthesis of nutrients in pea plants. In *Physiological-Biochemical Basis of Plant Interactions in Phytocenosis.* (Ed., A.M. Grodzinsky) **4**: 20-23. Kiev: Naukova Dumka.
79. Sokolova, E.A. and Mikryukov, G. (1970). On the influence of white mustard and barley on the morphology, anatomical structure and yield of pea. In *Physiological-Biochemical Basis of Plant Interactions in Phytocenosis* (Ed., A.M. Grodzinsky) **2**: 84-88. Kiev: Naukova Dumka.
80. Su, S.M. and Fang, X.L. (1981). *Acta Agronomica Sinica* **7**: 123-128.
81. T.A. (1845). *Gardening Chronicle* **5**: 159.
82. Takatori, F. and Souther, F. (1978). *Proceedings Workshop on Asparagus.* Riverside, California: Department of Plant Science, University of California.
83. Tomashevskaya, E.G. and Lugovskaya, E.Y. (1970). Mutual influence of plants on their assimilation of nutrient elements in mixed cultures of legumes and cereals in connection with different conditions of

nutrition. *In Physiological-Biochemical Basis of Plant Interactions in Phytocenosis* (Ed., A.M. Grodzinsky) **1**: 75-82. Kiev: Naukova Dumka.

84. Tsuzuki, E. (1980). Studies on the allelopathy in buckwheat plants. *Proceedings, Ist International Symposium on Buckwheat*. Pp. 13-23.
85. Ventura, W., Watanabe, I. and Castillo, M. (1984). *Soil. Sickness in Rice Monoculture*. IRRI Research Paper Series 99. Los Banos, Philippines: International Rice Research Institute.
86. Walker, D.W. and Jenkins, D.D. (1986). Influence of sweet potato plant residue on the growth of sweet potato vine cuttings and cowpea plants. *Horticultural Science* **21**: 426-428.
87. Waller, G.R.., Kenzer, F.G. and McPherson, J.K. (1987). Allelopathic compounds in soil from no tillage vs conventional tillage in wheat production. *Plant and Soil* **98**: 5-15.
88. Wang, T.S.C., Chang, S.Y. and Tung, H. (1967). Soil phenolics as plant growth inhibitors. *Soil Science* **104**: 138-144.
89. Wang, T.S.C., Cheng, S.Y. and Tung, H. (1967a). Extraction and analysis of soil organic acids. *Soil Science* **103**: 360-366.
90. Wang, T.S.C., Cheng, S.Y. and Tung, H. (1967b). Dynamics of soil organic acids. *Soil Science* **104**: 138-144.
91. Wang, T.S.C., Yang, T.K. and Chuang, T.T. (1967c). Soil phenolic acid as plant growth inhibitors. *Soil Science* **103**: 239-246.
92. Webster, G.R., Khan, S.D. and Moore, A.W. (1967). Poor growth of alfalfa (*Medicago sativa*) on some alberta soils. *Agronomy Journal* **59**: 37-41.
93. Wildermuth, G. and Rovira, A. (1977). Hyphal density as a measure of suppression of *Gaeumannomyces graminis* var. trici onwheat roots. *Soil Biology and Biochemistry* **9**: 203-206.
94. Wozniak, K.L., Sulivan, W.M., Flowerday, A.D. (1981). *Proceedings of 73rd* Annual Meeting, American Society of Agronomy. Pp. 116.
95. Wu, W.M.H., Liu, C.L. and Chao, C.C. (1976a). *Journal of Chinese Agricultural Chemistry Society* **96**: 16-37.
96. Wu, W.M.H., Liu, C.L., Chao, C.C. and Lin, M.S. (1976b). Journal of Agricultural Association of China, Taipei 96: 16-37.
97. Yamg, A. (1804). *The Farmers' Calendar*: London.
98. Zebyalyendzick, S.F. (1973). Vyestsi Akad Nauk, BSSR Syer Biyal Nauk **5**: 31-34. Byelorussia.

11

Prospects of allelopathic research in multi-storey cropping systems

JACOB JOHN* and A.M. NAIR
Department of Agronomy,
College of Agriculture, Vellayani 695 522, Thiruvananthapuram, Kerala, India

Abstract

The necessity to intensify agricultural land use has led to the development of multi-storey cropping and crop mixtures. The crop combinations in two most intensive multi-storey cropping systems practised World wide viz., plantation crop-based systems and homestead agroforestry system have been reviewed in this paper. In the past, much work on allelopathic interactions has been conducted in agriculture and forestry but studies in multi-storey cropping systems are limited. The study of allelopathic interactions in multi-storey cropping systems is highly complicated, owing to the species diversity and interferences arising from competition between different components of the system. This is further complicated by the unique microclimatic and edaphic factors of each system. Allelopathic research in these multi-storey cropping systems demands a multidisciplinary systems approach in which the different biological components of the system viz., microclimate, soil properties, microflora, spatial and temporal dimensions etc., should be investigated in totality. Considering the complexity of allelopathic interactions in multi-storey cropping systems, the thrust areas for allelopathic research are: separating allelopathic interference from competition, screening germplasm of crops/trees for suitability to these systems, monitoring the production of allelochemicals by the different components in relation to the microclimatic and edaphic factors of the system, exploitation of allelopathy for improving the productivity of the system and plant protection.

Key words: agroforestry, allelopathy, arecanut, cashew, coconut, homestead, multi-storey cropping, plantation crops, rubber.

1. Introduction

The too much pressure of the ever increasing human population on land for producing more food and wood have made it necessary to search for other alternatives to maximise the use of agricultural land. This has led to the concepts of multistoried cropping and crop mixtures. Multistoried cropping refers to multi-species combinations involving both annuals and perennials (60,68). Policy makers in tropical developing countries,

* Correspondence author, Present address: Division of Agronomy, College of Agriculture, Padannakkad, Padannakkad P.O., Kasaragod 671 328, Kerala, India.

S.S. Narwal et al. (eds.), Allelopathy in Ecological Agriculture and Forestry, 159–179.

have proposed integration of plantation crops, annual crops, livestock production and forestry. In multi-storey cropping systems, since many plant species grow together, plant-plant interactions involving allelopathy are presumed. Although, in the past much work has been conducted in agriculture and forestry, but the studies on the allelopathic interactions operating in multi-storey cropping systems are limited. This paper briefly reviews (a) the crop combinations, in two most intensive multi-storey cropping systems practiced World wide viz., Plantation crops-based systems and homestead agroforestry systems (b) discusses the nature of allelopathic interactions and (c) examines the research needs and prospects of allelopathy in these systems.

2. Plantation crops-based multi-storey cropping systems

Modern commercial plantations like coconut, rubber, arecanut, cashew, coffee, cardamom and tea represent a well managed, profitable and environmentally stable land-use systems in the tropics. Contrary to popular belief, a substantial proportion of tropical plantation crops are grown by small holders (62,63,108). Commercial tree crops are widely spaced and planted in rows and take a long time to produce yields, hence, raising of agricultural crops with them is widely practised system.

2.1. COCONUT-BASED

The coconuts are grown under diverse conditions, hence, large number of other economic species are interplanted (Table 1). The table shows that the intercrops range from staple food crops to cash and export crops. Medicinal plants like *Andrographis paniculata*, *Coleus vetiveroides*, *Kaemferia galanga*, *Maranta arundinaceae* and *Sida retusa* grow well as intercrops in coconut gardens (78).

Growing of cacao as a mixed crop with coconut was started in 1963 (69). Nelliat (67) reported the yield of a multistorey cropping experiment with coconut. Three crops viz., pepper, cacao or cinnamon and pineapple were interplanted with 20-year old coconut palms. All components were highly compatible, as the yield of the main crop, coconut was enhanced by 75 to 122 % over a period of eight years in cacao multistorey model and by 90 to 120 % in cinnamon model. A double row of cacao decreased the yield of both pepper and pineapple. In coconut-cacao crop mixture, the amount of litter added was 818 and 1785 kg/ha/year (oven dry) under single and double hedge systems of planting, respectively (103).

2.1. ARECANUT-BASED

Like coconut, due to its more height, the particular shape of its crown causing least shade under its canopy and wide spacing of planting; the arecanut palm permits the growing of annual, biennial and perennial crops under it. Banana, cacao, citrus, black pepper, pineapple, betel vine, elephant foot yam, tapioca, paddy, sorghum, cowpea, vegetables, turmeric, ginger and guinea grass are successfully grown under arecanut trees. Although the nature of intercrops grown vary from region to region, but banana is universally grown (45). The practice of mixed cropping perennials like citrus, jack fruit, cacao, coconut, clove, coffee has also been recorded (6,91). Cardamom-arecanut intercropping has been reported from certain parts of Karnataka (89).

Table 1. Crops commonly grown with coconut (excluding cover crops and fodder species)

Crop	Country	Reference
1. **Cereals**		
Oryza sativa L.	India	15, 21
Eleusine coracana Gaertn.	India, Sri Lanka	2, 15
Zea mays L.	Philippines	14
2. **Pulses**		
Vigna radiata (L.) Wilczek	India	2, 61
Vigna mungo L. Hepper		
Cajanus cajan (L.) Millsp	Sri Lanka	15
Vigna unguiculata (l.) Walp.	Philippines	74
Glycine max (L.) Merril		
Arachis hypogaea L.		
3. **Root crops**		
Manihot esculenta Crantz.	India, Sri Lanka,, Philippines	15, 61, 14
Ipomoea batatus (L.) Lamk.	India	61
Dioscorea spp.	India	61
Amorphophallus companulatus Roxb.ex Dene	India	61
Colocasia spp.	Philippines	31, 35
Xanthosoma spp.	Philippines, Fiji	31 , 35
4. **Spices and condiments**		
Zingiber officinale Rosc.	India	
Curcuma longa Koenig.	India	
Coriandrum sativum L.	India	3, 57, 61
Trigonella foenum-graecum L.	India	
Cinnamomum zeylanicum Blume.	India	
Capsicum annuum L.	Sri Lanka	15
Syzygium aromaticum (L.) Merr.	Tanzania	15
Piper nigrum L.	India, Philippines	15
5. **Fruits**		
Ananas comosus (L.) Merr.	India,	14, 15, 31
Mangifera indica L.	Sri Lanka	61
Musa spp.	Philippines	
Carica papaya L.	Malaysia	
Artocarpus altilis (Park.) Fosb.	Pacific Islands, Caribbean	
6. **Tree crops**		
Areca catechu L.	India	57
Theobroma cacao L.	India, Malaysia	15
Coffea canephora Pierre ex Froehner	Philippines	
7. **Other crops**		
Gossypim spp .	India, Sri Lanka	2, 15
Sesamum indicum L.		
Musa textilis Nees.	Philippines	92
Saccharum officinarum L.		

Source: 75

Studies in two high density multispecies cropping models, consisting of pepper, cocoa, coffee, mulberry and elephant foot yam (Model-1) and banana, betel vine, lemon, coffee and tapioca (Model-2) in 20-year old arecanut plantation showed that the arecanut yield increased by 7-20 %. Yield of the intercrops also increased as years passed (90). Cocoa mixed with arecanut, did not affect the performance of arecanut during the first ten years of plantation. In fact, a beneficial effect, as indicated by higher yield of arecanut in the mixed crop plot (50:50) was obtained (5).

2.3. RUBBER-BASED

In rubber plantation agroforestry systems based on the (a) growth stage of rubber trees, (b) amount of sunlight available, (c) growing habit of intercrops and (d) the growing season, the intercrops have been divided into three categories to minimise competition between the component plant species. Even this, does not eliminate the occurrence of competitions/interactions between the main crop and the intercrops.

I. China: In China, mixed cropping is being practiced in rubber plantations since the late 1950s. Intercrops in rubber plantations can be classified into three types, on the basis of their adaptability to the different rubber plantation development stages (Table 2).

Table 2. Intercrops in rubber plantations

Age of rubber plantation	Intercrops
	Crop tolerates less shade
1 - 6 years	Sugarcane (*Saccharum sinensis* Roxb) Pineapple (*Ananas comosus* L. Merr.) Upland rice (*Oryza sativa* L.) Groundnut (*Arachis hypogaea* L.) Sweet potato (*Ipomoea batatus* L. Lamk) Soybean (*Glycine max* L. Merr) Banana (*Musa* spp.) Mulberry (*Morus cathayana* Hemsh). Roselle (*Hibiscus sabdariffa* L.)
	Crop tolerates moderate to high shade
7 – 32 years	Rattan palm (*Calamus tetradactylus* Hance) Arrowroot (*Maranta arundinacea* L.) Edible canna (*Canna edulis* Ker.) Ginger (*Zingiber officinale* Rose) Yam (*Dioscorea alata* L.) Medicinal Indian mulberry (*Morinda officinalis* How.) Villous amomum (*Amomum villosum* Lour.) Sharp leaf galangal (*Alpinia oxyphylla* Miq.)
	Crop tolerates little to high shade
1 - 32 years	Tea (*Camillia sinensis* Kuntze) Pepper (*Piper nigrum* L.) Coffee (*Coffea robusta* Linden) Arecanut (*Areca catechu* L.) Coconut (*Cocos nucifera* L.) Cinnamon (*Cinnamomum zeylanicum* Blume)

Source: 52

The first type includes those intercrops which are suitable for immature rubber plantations, i.e. adequate solar radiation is available in the inter-row spaces; the second type includes those intercrops tolerant to limited or more shade in rubber plantations; the third type, includes those intercrops which are suitable for the whole rubber production period under the changing sunlight conditions. A large amount of litter is produced in some intercropping combinations (Table 3). For example, rubber/pepper (creeping) intercropping produces 7545-9205 kg litter/ha/year i.e. much higher even than that of rubber/legume cover (*Peuraria javanica*). The rubber trees exhibit faster growth and higher yield in intercropping than sole crop (52).

Table 3. Litter and crop residue in certain rubber-based intercropping systems

Intercropping combination	Litter (kg/ha/year)
Rubber + sugarcane	5382-8970*
Rubber + pineapple	9000*
Rubber + tea	5790
Rubber + pepper	7545-9205
Rubber + Peuraria	5352
Rubber + Amomum	6000*
Rubber + galangal	13891*
Rubber monoculture	2914-5854

* Including residues and litter Source: 52

II. Ivory coast: In Ivory coast, intercropping trials in rubber have been conducted with certain annual and perennial crops (44). The results showed the possibility of intercropping of food crops like yam, rice, maize, groundnut, plantain, banana and various vegetables with rubber in the first three to four years after rubber planting. Food crop rotations in rubber like Yam-rice/groundnut-maize/groundnut; Yam-rice/groundnut-plantain banana, Yam-Yam-Yam and Plantain-Plantain-Groundnut were also found feasible. Rubber tree growth was better when intercropped with food crops. In later stages of rubber tree growth, perennial intercrops like coffee, oil palm, pineapple, cocoa and lemon were found satisfactory.

III. Sri Lanka: In Sri Lanka, intercropping systems involving rubber/cinnamon and rubber/forage grasses/tree legumes were investigated (72). Cinnamon reached harvesting age 2 ½ years after establishment under immature rubber. Cinnamon also performed satisfactorily under mature rubber tree, where light availability to cinnamon was about 53 % of daylight. The yield of *Panicum maximum* grass was almost double than that of *Brachiaria brizantha* during the first two years after planting rubber tree but then yield declined with the growth of rubber trees. The growth and yield of trees legumes (*Leucaena leucocephala* and *Gliricidia sepium*) were poor, in rubber plantations.

IV. Indonesia: In South Sumatra, Indonesia, the most common food crops cropping pattern is rubber intercropping with corn+upland rice-soybean-cowpea. However, the development of roots and canopy of the rubber trees limit the productivity of the intercrops from the fourth year. Horticulture crops like pineapple, banana and chilli are also being tried as intercrops by some farmers (86).

V. India: In Kerala, India, small farmers usually grow banana, pineapple, ginger, turmeric and tuber crops as intercrops in immature rubber plantations (Figure 1). A multispecies cropping system, consisting of banana, pineapple, pepper (trailed on *Erythrina* stands) and cocoa as intercrops, and teak and fodder along the boundaries in immature rubber (paired rows) during the first 30 months was investigated in Kerala. The legume cover crop, *Peuraria phaseoloides* was established in the narrow interrow spaces. Banana residues and erythrina loppings were recycled in the system. Growth of rubber was significantly superior in the multi-storey cropping system compared to that in monoculture (42). Intercropping of rubber plantations with cardamom is also noticed in some parts of Kerala (93).

2.4. CASHEW-BASED

The large canopy of the grown up cashew trees protects the top-soil from erosion; but during its early growth period, a systematic inter- and mixed cropping system of compatible crops, with varying morphological frame and rooting habits, is desired. Such a cropping system will be ideal to maintain a balanced ecosystem. Cashew-based multi-storey cropping systems, especially in the first 5-6 years of orchard establishment, can enhance the return per unit area. Intercropping with annuals like pigeonpea and tapioca, biennials like pineapple and papaya and tree species like *Casuarina*, *Acacia*, *Leucaena* was found successful (53). A cover crop (*Mucuna* spp.) was also tried in the cashew orchard. More height, girth and spread of cashew was recorded in plots intercropped/mixed with red gram, pineapple, *Mucuna* spp., casuarina and acacia, than intercropping with tapioca (53). Multi-storey cropping systems of cashew + pepper + cardamom has also been reported (66).

2.5. OTHER MULTI-STOREY COMBINATIONS

I. Tea: Tea may be benefited from woody tree species having favourable shade and providing valuable organic matter. Silver oak (*Grevillea robusta*) is very useful shade tree for interplanting in tea. Its leaves cover the ground for about one year due to wide C/N ratio, hence, decomposes gradually (105). In south India, besides silver oak, *Acacia* spp., *Albizia* spp., *Erythrina lithosperma* and *Indigofera teysmanii* are also used as shade trees. In northeast India, *Albizia stipulata, A. odaratissima, A.lebbek, A. lucida, A. procera, Acacia lenticularis, Derris robusta* and *Dalbergia sericea* shade trees have been recommended for planting in mixed stands in tea plantations (24).

II. Coffee: Planting of shade trees in coffee plantations is common practice in India. The most popular shade trees are *Ficus glomerata, F. nervosa, Albizia stipulata, A. lebbek, A. moluccana,* A. *sumutrana, Dalbergia latifolia, Artocarpus integrifolius, Bischofia javanica* and *Grevillea robusta* (4). Intercropping of cardamom and coffee has been observed as a practice in some plantations in Karnataka (96).

III. Pepper: Pepper, being a vine of tropical forests, is shade loving and also needs a stand to climb. In pure plantations, live stand of *Erythrina indica*, *Garuga pinnata*,

Spondias mangifera, *Gliricidia maculata*, and *Grevillea robusta* are commonly used. However, *E. indica* is best. Besides, pepper is cultivated as a subsidiary crop in coffee, coconut, arecanut and orange gardens. It is also grown in home gardens with mangoes, jack fruit, coconut, arecanut and many other tree species (88).

3. Homestead Agroforestry systems

Homesteads or Home gardens have evolved over many centuries and is tradition in many tropical countries. Home garden refers to a system of production of subsistence crops, for the gardener and his family. It is a functional and self sustaining farm unit, consisting of many crops and multipurpose trees, with or without animals/poultry /apiculture and managed by the dwelling family to meet the basic family needs (food, fuel, timber) and producing marketable surplus for the purchase of non-producible items (40). With very high species diversity, complex structural arrangement of the components and strong ecological foundations, the home gardens simulates the natural tropical forest ecosystem. Homestead agroforestry practices have been reported from Java (94), Tanzania (29), India (59), Thailand (10), Pacific Islands (106), Indonesia (58), Nigeria (71), Malaysia (101), Sri Lanka (41), Bangladesh (51) and Mexico (82).

There are several types of home gardens in other geographical locations, each with its characteristic features. The home gardens existing in different continents exhibit various structural and functional characteristics (Table 4). In general, all home gardens consist of a herbaceous layer near the ground, a tree layer at upper levels and intermediate layers in between. The lower layer can usually be partitioned into two, with the lowermost (less than 1 m height) dominated by different vegetable and medicinal plants and the second layer (1-3 m height) being composed of food plants such as cassava, banana, papaya, yam etc. The upper tree layer can also be divided in two, consisting of emergent, fully grown timber and fruit trees occupying the uppermost layer of over 25 m height and medium-sized trees of 10-20 m occupying the next lower layer. The intermediate layer of 3-10 m height is dominated by various trees, some of which would continue to grow taller. This layered structure is never static (64).

Homestead agroforestry is most prevalent in Kerala, where, plants of different heights and architecture, occupy the available space both horizontally and vertically (Figure 2). The home gardens with a multitude of crops, presents a multi-tier canopy configuration. The major portion of the upper canopy (> 25 m) goes to coconut, arecanut, certain fruit and timber/fuel trees, followed by certain medium sized fruit, spice and timber/fuel trees (10-20 m). The third layer (3-10 m) comprises of crops like pepper, tree spices and certain fruit trees. The lower storey (1-3 m) of the harvesting plane is occupied by banana, cassava and other tuber crops. At the floor level, pineapple, vegetables and other herbaceous crops are grown (40).

4. Allelopathy and multi-storey cropping systems

4.1. CURRENT STATUS OF RESEARCH

Little work has been conducted on species compatibility in agroforestry and multi-storey cropping systems (109). Some species currently used in multi-storey cropping

Table 4. Major components of selected tropical home gardens

Region	Local name of system	Major crops		Usual number of vertical canopy strata	Reference
		Food	Cash		
Java	Pekarangan	Upland rice, maize, vegetable, coconut, fruit trees	Fruits and vegetables	5	94
Philippines	Homegardens	Sweet potatoes, coconut, banana	Tomatoes, brinjal, squash, peas, mango	4	95
Pacific	Homegardens	Coconut, colocasia, yams	Coconut	4	106
Sri Lanka	Kandyan gardens	-	Clove, pepper, tea, coconut	3	41
Southwest India	Compound gardens	Tuber crops, upland rice, banana, vegetables	Coconut, arecanut, cacao, cashew, spices	4	59
Southeast Nigeria	Compound farms	Yam, cocoyam, banana	Cola, oil palm	4	71
Northern Tanzania	Chagga homegardens	Banana, beans, colocasia, xanthosoma, yams	Coffee (arabica), cardamom	5	29
Burkina Fasso	Ka/Fuyo gardens	Maize and red sorghum	Tobacco	2	28
Southeast Mexico	Huertos Familiares	Maize, beans	Cacao	4	82
Grenada	Kitchen gardens	Colocasia, xanthosoma, yams, maize, pigeon pea	Banana, cocoa, nutmeg	4	13

Source: 28

Figure 1. Intercropping of pineapple and banana in two years old rubber plantation, Kerala, India (42)

Figure 2. A home garden of Kerala, India (40)

systems reportedly have allelopathic properties. For example, seed germination, root length and dry matter production of cowpea were depressed by aqueous leaf extracts of leucaena and when grown in leucaena top soil (98). Results of no significant effect of leucaena soil and decomposed leaf extracts on the germination of rice have also been reported (49). This finding contradicts the results of laboratory studies (50) where extracts of fresh leucaena leaves inhibited rice seed germination. The germination and growth of greengram, blackgram, cowpea, pigeonpea, cowpea, soybean, sorghum, sunflower, wheat, pea, maize and mustard were inhibited by litter extracts and top soil of *Casuarina equisetifolia* (43,97,98). Significant reduction in ground cover was noticed under *Albizzia lebbeck* (7). The litter extract of *Bombax ceiba* and *Albizzia lebbeck* proved stimulatory to germination of wheat, maize, pea and mustard (43). The inhibitory effects of the leaf extracts of *Azadirachta indica, Terminalia arjuna, Dalbergia sissoo, Albizzia lebbek, Sesbania grandiflora, Acacia auriculiformis* and *Leucaena leucocephala* on the germination of wheat, paddy and gram have been demonstrated (80). On increasing the concentration of leaf extract from 1 : 15 to 1 : 5 the increase in inhibition was higher in wheat and paddy. The inhibitory potential of *Coffea arabica* (18) has also been reported. The soil extracts and leachates of leaf, stem and debris of two clones of tea, China *(Camellia sinensis* (L.) O.Kuntze) and Cambodia (*C. assamica* ssp *lasiocalyx* (Planch.ex.Watt) Wight), applied to each other proved stimulatory. Hence, the nursery once used to raise cuttings could be reused without harmful effects on young plants (79). Leaf extracts of certain agroforestry tree species, *Acacia auriculiformis, Eucalyptus tereticornis, Casuarina equisettifolia, Albizzia lebbeck, Leucaena leucocephala, Artocarpus heterophyllus* (jack), *Mangifera indica* (mango), *Ailanthus triphysa* (matty), *Tamarindus indicus, Bombax malabaricum* (red silk cotton), *Myristica fragrans* (nutmeg), *Artocarpus hirsuta* (wild jack), *Thespesia populnea* (portia) and *Anarcadium occidentale* (cashew), commonly grown in Kerala were tested for allelopathic effects on rice and cowpea. Leaf extracts of all trees significantly inhibited germination and growth of rice and maximum inhibition was caused by *Ailanthus triphysa* and *Tamarindus indicus*. Growth of cowpea was suppressed by all trees, while germination was inhibited by *A. auriculiformis, E. tereticornis, C. equisettifolia, A. triphysa, T. indicus, T. populnea* and *A. occidentale* only. Inhibition of radicle and plumule growth was more pronounced (40).

4.2. MODE OF ALLELOPATHIC INTERFERENCE

In multi-storey cropping systems, allelopathic interference may result from natural products in intercrop foliage leachings, root exudates and volatiles. These chemicals are released into the environment through leaching, root exudation, volatilization and microbial decomposition of plant residues.

4.2.1. Leaching

Leaching is the removal of substances from plants by aqueous solutions such as rain, dew, mist and fog. Radioisotope labeling of plant tissue before leaching has shown that large quantities of both inorganic and many classes of organic natural products are leached from plant tissue (102). Both the quantity and quality of leachable natural products differ greatly with species, physiological age of tissue, stage of plant development, plant health, light, temperature, nutritional conditions, and the intensity and volume of the leaching solution (102). The secondary compounds released will be influenced by the type of crop being leached (32,33). To determine the presence of allelopathic activity, the protocol for leachate preparation should be similar as possible to that prevailing naturally during symptom development (39).

4.2.2. Root exudation

Root exudation is the release of substances into the surrounding medium by healthy, intact plant roots. A variety of natural products has been found in plant root exudates, though in comparison with leaves, the amounts of organic materials are much smaller (87). Many factors can affect the quantity and quality of natural products obtained e.g. plant species, plant age or stage of development, temperature, light, nutritional conditions, soil microorganisms, root supporting medium, soil moisture and root damage (87). Similarly, exudation in soils can be expected to vary with soil physical and chemical properties. Root exudation usually increases greatly during wilting and root damage (20). Soil microorganisms modify root-cell permeability and root metabolism and rhizosphere organisms may absorb or excrete qualitatively different natural products than plant roots. Studies of allelopathic activity of aggressor plant roots could not distinguish the natural products originating from root exudates or dead root tissue or microbial rhizosphere products (39).

4.2.3. Volatilization

Volatilization is the release of natural products into the atmosphere. A variety of plants either secrete or excrete metabolic products into special structures such as trichomes and glands, into intercellular spaces and canals, or onto leaf surfaces. In hot, dry weather, natural products with high vapour pressure are released into the atmosphere where they may be absorbed directly by plants or adsorbed onto soil surface (39).

4.2.4. Microbial decomposition of plant residues

Plant residue mulches commonly used in multi-storey cropping systems to protect soil from erosion, conserve moisture and supply nutrients may be the source of allelochemicals that interfere with crop productivity. To improve nitrogen nutrition of crop plants, plant-residue mulches, particularly of nitrogen-fixing species, are commonly used, which in fact may result in allelopathic interference. Mulching and conservation tillage, which leave plant residues on the soil surface or incorporate them into the soil, result in the liberation of large quantities of water soluble and partially water soluble products during residues decomposition. Large quantity of crop residues left annually in the fields results in soil sickness by allelopathic means (23).

The plantation crops, being mainly perennial crops produce a huge amount of waste biomass (Table 5). It has been estimated that a coconut garden with 175 trees per ha generates biomass of 7000 kg as dry leaves, sheathes, spadices, inflorescences and coconut husks. During monsoons, tannins oozing out of such heaps, creates problems of environmental pollution (8). The cut oil palm (*Elaeis guinensis* Jacq.) fronds constitute a major source of organic manure yielding 10 t dry matter per hectare (104). Cardamom is a shade loving crop, hence, grown underneath the trees in the forests, generally high in fertility status due to leaf fall and its decomposition (110). On an average, 5-8 t dry leaves fall annualy from shade trees in a hectare of cardamom (48). The wastes and surplus residues obtained from plantation crops, is recycled back to the soil by various methods such as mulching, in situ incorporation and composting. The wider C:N ratio coupled with low N content, presence of soluble tannins (8-12 %), low biodegradability are some of the problems associated with coir pith (27). Various measures to eliminate its phytotoxicity in the field to improve the crop productivity includes, removal of phytotoxins by flooding, crop rotation and detoxification through nutrient application (16).

Table 5. Availability of on farm wastes/byproducts in plantation sector in India

Crop	Waste available per year (million tonnes)
Coconut* (Excluding coir pith)	11.2
Areca leaves (dried)	0.13
Areca rachis	0.08
Areca husk	0.22
Cocoa shed leaves	0.0004
Cocoa prunings	0.012
Cocoa pod husk	0.033
Coir pith	7.50
Coffee husk	0.18
Tea waste	0.22

* includes spadices, bunch wastes, sheath, inflorescences and husks Source: 8

When plant tissues age and die, cell membrane integrity is lost. Allelochemicals that are compartmentalised in living cells are released into the surroundings and react with other natural products, resulting in qualitative changes in some of these products. Once natural products enter the soil as incorporated soil residues, additional qualitative changes occur as a result of physio-chemical action of the soil and the activities of soil microorganisms (22,47,54). The secondary compounds released from litter or formed will be influenced by microbial populations present in the soil (70).

Soil microorganisms can modify non-toxic materials to phytotoxic ones (37,73) or reduce phytotoxicity of crop residues (34). Microbial metabolism of organic compounds may increase or decrease the toxicity, due to release of organic carbon as CO_2 , fixation into microbial biomass, or transformation to other products. The soil microbial biomass has been studied in several multi-storey cropping systems. Coconut-cacao mixed cropping have shown greater microbiological activity than coconut monocropping system (65). The microbial biomass (bacteria, fungi, actinomycetes) was higher in arecanut-based high density multispecies cropping systems than in monocropping. The nature and activity of microorganisms associated with perennial monocrop are changed with introduction of other crops (11). In homesteads of Kerala, the soil microflora was found to vary with the cropping intensity, crop diversity, planting pattern of crops and the management practices adopted (40).

Initial experiments to determine the involvement of allelochemicals arising from residue decomposition should concentrate on simulating field conditions as closely as possible. For example, the same quantity, quality and age of residue documented during symptom description should be used; soil moisture and aeration conditions should also be similar (39). Experiments that use artificial media lacking active microbial populations may give results of little value in determining the cause of inhibition in field situations (54).

In multi-storey cropping , allelopathic interactions could most likely occur through leaching, root exudation and release of phytotoxins from decaying litter and roots in multi-storey cropping. The rain water passing through the foliage of the tree components leaches allelochemicals and transport them to under storey crop plants. Trees are integral part of multi-storey cropping systems, hence, adds large quantity of litter through dead and falling leaves, twigs, branches, fruits and prunings of trees as manure. For example, the rubber tree during its first five years of growth adds upto 5 t of leaf litter per hectare (52). Common trees of home gardens like jack (*Artocarpus heterophyllus*), wild jack (*Artocarpus hirsuta*), mango (*Mangifera indica*), mahogany (*Sweitania macrophylla*), bamblimass (*Citrus maxima*), nutmeg (*Myristica fragrans*), and coffee (*Coffea arabica*) annually contribute 3.51, 3.95, 2.43, 1.93, 1.73, 4.25 and 2.10 t litter per hectare, respectively (40). Leaf litter is potential source of phytotoxins and allelopathic interactions may occur through release of allelochemicals due to its decay or its leaching during rains or irrigation. This has several management implications in homestead agroforestry, where many intercrops are grown beneath the tree canopy. The increased amount of litter could lead to greater release of toxic chemicals into the soil, these remain active for a long time in low rainfall areas and may inhibit growth of subsequent intercrops. The crop or root residues remaining on soil surface after the harvest of intercrops or trees might serve as allelochemicals sources.

In multi-storey cropping systems, there is very high root density of component plant species (trees + crops), it seems that roots of component plant species intermingle with each other leading to allelopathic interactions through root exudates. Moreover, despite the deep rooting characteristics of trees, most of the fine feeder roots are found within the top 20 cm of the soil. For example, the rubber tree has its fine feeder roots concentrated in the top 15-30 cm soil layer and spreading up to several meters. These roots are continuously sloughed off. Substances exuding from the roots may affect adjacent species directly, or may influence them indirectly through decomposition of such biomass.

5. Research needs and prospects

5.1. SEPARATING COMPETITION FROM ALLELOPATHY

Interference occurs when one plant species fails to germinate, grows more slowly, shows symptoms of damage, or does not survive in the presence of another plant species. Interference can result from competition, allelopathy, or indirect influences. The challenge in plant interference work is identifying which of these factors causes the associated plant response. Allelopathy is an important mechanism of plant interference and is mediated through the addition of chemicals to the plant environment. In contrast, the other major mechanism of plant interference, competition or allelospoly, is a subtractive process that involves removal of resources e.g., water, minerals, or light energy. Allelochemicals originating in foliage leachings, root products, or mulches of crops or woody plants may result in reduced productivity or death of companion plants. Besides releasing phytotoxic natural products, one plant can influence another plant through competition for growth resources (sunlight, soil moisture, nutrients) (38); changing the organic matter content of the soil, soil pH or favouring or reducing pathogenic activity (36). Further, none of these factors are independent. Thus, the study of interference phenomena is difficult and to demonstrate a cause and effect relationship requires the study of not only allelopathy, but also of competition and other influences.

Ecologists and weed scientists often have ignored allelopathy as a possible mechanism in their competition studies. Systematic evaluation of crop and woody plant combinations and mulches for interference should provide useful information in designing new systems. Plant-plant interactions can occur over very extended periods of time and are complicated by various edaphic and climatic factors. Determining whether interference is competitive, allelopathic, or the result of other influences could affect the land management strategy that is chosen. It is a painstaking task to distinguish allelopathic effects from those related to competition. Five steps are required to demonstrate that allelopathy is the part of interference (39):

i. One or more phytotoxic natural products must be released from the aggressor.
ii. Phytotoxic natural products must be identified.
iii. The presumed allelochemicals must be transported and accumulated in the environment in sufficient quantities to account for the documented interference.
iv. Allelochemicals must be taken up by the susceptible plant.
v. Allelochemicals must reproduce the same symptoms (physiological action) of interference in the susceptible plant.

5.2. SCREENING GERMPLASM

Although, crops plants have the greatest potential for allelopathy, but research on allelopathic potential in germplasm of crops for agroforestry is limited. Allelopathic characteristics are more likely to occur in crop predecessors or wild types evolved in the presence of allelopathic and competitive influences from other species (77). Different cultivars of the same crop might release different quantities of toxins (33,46). The concept of mixtures to create diversity is appropriately extended, in home gardens, by growing woody and herbaceous perennials in association with seasonal annuals. These system have the real reservoirs of germplasm of cultivated plants, because the varieties with various environmental requirements, compatibility with various trees and resistance are raised. Attempts should be made to find allelopathic characteristics within the germplasm of crops and then incorporate these genetically into cultivars.

5.3. MONITORING PRODUCTION OF ALLELOCHEMICALS

The number and diversity of compounds involved in allelopathy are growing rapidly. Over 10,000 secondary plant products have been recorded, the total number may however exceed 4,00,000 (100). The diversity of these substances and our lack of knowledge of any synergistic effects add to the problems. It is also important to know whether allelopathic agents are produced by chance or specifically for an effect on other plants to ensure survival. The quantity and quality of the allelochemicals produced is regulated by the stage of the plant and is modified by environmental stresses like soil temperature, drought, flooding or poor drainage, ultraviolet light or sunlight, microorganisms soil salinity, diseases, herbicides, minerals and even growth regulators or hormones (16,26,56). However, research on the influence of these factors on allelopathic interactions in multi-storey cropping systems are lacking. For example, the tree/crop combinations in the home gardens acts as a buffer against drastic changes in ecoclimate. The tree canopy, helps to maintain the relative humidity in the home gardens at an optimum level. The relative humidity is prevented from exceeding a critical level during the rainy season and also from falling below a critical level during the dry months. The soil temperature and light intensity in home gardens is very low (40). In coconut-cacao crop mixture, soil temperature at 30 and 60 cm depth was 3-6 °C lower as compared to monoculture of coconut (103). Intercropping under rubber trees decreased the temperature by 1.2-7.0 ° C and increased the humidity by 2-10 % to create a situation similar to secondary tropical forests (52). The information on interactions between microclimate and allelopathy may be used to overcome the adverse effects of the allelochemicals on crops and to suppress the weeds in several other cropping systems.

5.4. PLANT PROTECTION

Many allelochemicals appear to be broad-spectrum inhibitors. The physiological modes of action of allelochemicals need to be determined. This information could offer land managers the opportunity to manipulate interference phenomena to control weeds with allelopathic ground covers or mulches (76). Utilization of mulches that suppress weed growth via allelopathy in perennial crop ecosystems should be probed into, because these ecosystems produce large quantity of plant residues. Phytotoxic substances released from many tree species retards the growth of associate weeds and crop species (17,19,99). The selective herbicidal activity of a chemical 1,3,7-trimethylxanthine isolated from seeds of *Coffea arabica* has been reported (83,84). This chemical inhibits the seed germination of the weed *Amaranthus spinosus* at 1200 ppm, but exerted no adverse effect on seed germination and growth of *Phaseolus mungo* plants (85). The use of allelochemicals to control both plant pathogens and weeds could outweigh the detrimental effects. For example, even though simple phenolic acids are implicated in allelopathic interactions, but the same compounds are associated with lipid layers at the plant surface to provide epidermal resistance to attack by plant pathogens (1).

Disease incidence in tree/crop components and weed growth under the tree canopy in home gardens of Kerala has been found to be extremely low. It is assumed that the occurrence of this low incidence may be due to the crop diversity. However, no definite reasons have been identified for this phenomenon. Allelopathic exclusion could be a possible reason.

5.5. CONCEPT OF THRESHOLD LEVEL

Once the secondary compounds are released into the immediate environment, they must accumulate in sufficient quantity to affect other plants, persist for some period of time, or be constantly released in order to have lasting effects (81). Few studies have been designed to prove that the plant produced a chemical in sufficient quantity over time or that it persisted long enough to affect other plants. Several investigators have noted that certain allelochemicals added to soil initially, could not be recovered later on, through laboratory extraction, in its same form or quantity (9,107). Researchers who initially reported that phenolic compounds were operative allelochemicals later concluded that their concentration in the soil was below the threshold required for biological activity (12,55). Allelochemicals arising either from the breakdown of lignin or by release from insoluble forms in the leaf litter are present in the soil (100) without causing adverse effects. Obviously, it must be the local concentration of such compounds which is important and very little information is available on such aspects. The presence of environmental stresses for growth factors (deficiencies of water, sunlight or nutrients) or the presence of pesticide residue, might modify the threshold for allelochemical activity (25). This makes the study of allelopathic phenomena more difficult. Moreover, the allelopathic action is impeded by the possible activity of other organisms. An examination of the washings from fresh plant leaves may reveal several potent phytotoxins, but each may be readily broken down in the soil or be released from the plant in an erratic manner and thus never pose a threat to other plants. Naturally the phytotoxins have to accumulate in enough quantity and have sufficient chemical stability to persist for sufficient time to exert their action. Maintenance of more residues as mulches on the soil surface might produce localized concentrations of allelochemicals. This aspect, attains more relevance in cases where we desire to control or eradicate weeds and could form part of the integrated weed management strategy.

Adsorbents have been effective in detoxification of contaminated soil. Soils containing abundant organic matter, probably have sufficient adsorption sites and microbial activity to inactivate the allelochemicals relatively rapidly (77). Adsorbents are particularly useful in perennial crops where toxins tend to accumulate. The use of companion plants that contribute organic matter or inoculation with microorganisms, which can readily metabolize the toxins might prove useful in perennial agroecosystems. Addition of a specific quantity of organic matter could eliminate the allelopathic effect and could be exploited for achieving selective action.

5.6. COMPATIBILITY BETWEEN COMPONENTS

Yield of certain crops are increased when grown in concert with other plant species. We should keep an open mind and not ignore certain advantages that may accrue with mixed cultures. We must obtain information about detoxification and other counter defenses and if possible, the biochemical basis for the many mutualistic associations which exist. Utilizing a companion plant that is selectively allelopathic, but does not appreciably interfere with crop growth has been reported (30). It is important that we expand our knowledge and capitalize on beneficial associations for long-term benefit.

6. Conclusions

The study of allelopathic interactions in multistoried cropping systems is highly complicated, owing to the species diversity and interferences arising out of competition between different components of the system. This is further complicated by the microclimatic and edaphic factors which are unique to each system. The results of bioassay and laboratory studies can certainly serve in ascertaining the allelopathic tendencies of the tested trees or crops or weeds. However, the findings cannot be used to judge its suitability for planting or inclusion as components in multistoried cropping systems. The crops though affected initially, might recover later. The microclimate and edaphic factors, unique to each system or combination, might nullify the adverse effects of allelopathy. In systems where organic matter is regularly recycled or leaf litter input is considerable, there are chances of build up of allelochemicals in the soil to toxic levels. Hence, the existing systems should be critically evaluated by taking into account both the spatial and temporal dimensions. Allelopathic research in multistoried cropping systems demands a systems approach in which the different biological components of the system, its microclimate, soil properties, microflora etc., should be considered in totality. Yet, certain common features like higher soil moisture content, microbial biomass, organic matter addition, soil nutrient status, lowered soil temperature, higher humidity, reduced light intensities may be studied in these high density multi-storey cropping systems and their effects on allelopathic interactions.

Considering the complexity of allelopathic interactions in these systems and based on the characteristics common to multi-storey cropping systems, the thrust areas for allelopathic research should include, (a) separating allelopathic interference from competition, (b) screening germplasm of crops/trees for genetical manipulations, (c) monitoring the production of allelochemicals by the different crop components in relation to the microclimatic and edaphic factors of the system, (d) exploitation of allelopathy for plant protection via mulches and plant derived chemicals, (e) determining the threshold levels/critical concentrations of allelochemicals needed to express its inhibitory/stimulatory influences for each crop/weed and (f) identify compatible and beneficial associations.

7. References

1. Agrios, G.N. (1969). *Plant Pathology*. London: Academic Press. pp. 629
2. Albuquerque, S.D.S. (1964). Coconut cultivation in Mysore. *Coconut Bulletin* **18**: 249-256.
3. Balasundaram, E.K. and Aiyadurai, S.G. (1963). Grow spices in the shade of coconut palms. *Coconut Bulletin* **17**: 182-184.
4. Bavappa, K.V.A. (1980). Plantation Crops. In: *Handbook of Agriculture*. New Delhi: Indian Council of Agricultural Research. pp. 908-920.
5. Bhat, S.K. (1978). Performance of cacao as mixed crop with arecanut. In *Proceedings of First Plantation Crops Symposium*.(Ed., E.V. Nelliat) Kottayam: Rubber Research Institute of India. pp 377-382.
6. Bhat, S.K. (1979). Mixed cropping in arecanut. *Technical Bulletin*, (Eds., E.V. Nelliat and K.S. Bhat) **3**: 35-38, Kasargod: Central Plantation Crops Research Institute..
7. Bhatt, B.P. and Todaria, N.P. (1990). Studies on the allelopathic effects of some agroforestry tree crops of Garhwal Himalaya. *Agroforestry Systems* **12**: 251-255.
8. Bidappa, C.C., Upadhyay, A.K., Hegde, M.R. and Palaniswami, C. (1996). Organic matter recycling in plantation crops. *Journal of Plantation Crops* **24**: 71-85.

9. Blum, U. and Rice, E.L. (1969). Inhibition of symbiotic nitrogen fixation by gallic and tannic acid and possible roles in old field succession. *Bulletin of the Torrey Botanical Club* **96**: 531-544.
10. Boonkird, S.A., Fernandes, E.C.M. and Nair, P.K.R. (1984). Forest villages : an agroforestry approach to rehabilitating forest land by shifting cultivation in Thailand. *Agroforestry Systems* **2** : 87 – 102.
11. Bopaiah, B.M. (1991). Soil microflora and VA-mycorrhizae in areca based high density multispecies cropping and areca monocropping system. *Journal of Plantation Crops* **18** (Supplement): 224-228.
12. Borner, H. (1971). German research on allelopathy. In *Biochemical Interactions among Plants.* Washington, DC: US National Committee for International Biology Programs, National Academy of Science. pp 52-56.
13. Brierley, J.S. (1985). The West Indian kitchen garden: a historical perspective with current insights from Grenada. *Food and Nutrition Bulletin* **7**: 52-60.
14. Celino, M.S. (1963). Increasing income from coconut lands. *Coconut Bulletin* **17**: 169-172.
15. Child, R. (1974). *Coconuts.* Second Edition, London: Longman.
16. Chou, C.H. (1986). The role of allelopathy in subtropical agroecosystems in Taiwan. In *The Science of Allelopathy* (Eds., A.R. Putnam and C.S. Tang), pp. 57-73, New York: John Wiley and Sons Inc.
17. Chou, C.H. and Kuo, Y.L. (1986). Allelopathic research of subtropical vegetation in Taiwan. III. Allelopathic exclusion of understorey by *Leucaena leucocephala* (Lam) de wit. *Journal of Chemical Ecology* **12**: 1431-1448.
18. Chou, C.H. and Waller, G.R. (1980). Possible allelopathic constituents of *Coffea arabica. Journal of Chemical Ecology* **6**: 643-653.
19. Chou, C.H. and Yang, C.M. (1982). Allelopathic research of subtropical vegetation in Taiwan. II. Comparative exclusion of understorey by *Phyllostachys edulis* and *Cryptomeria japonica. Journal of Chemical Ecology* **8**: 1489-1507.
20. Clayton, M.F. and Lamberton, J.A. (1964). A study of root exudates by the fog-box technique. *Australian Journal of Biological Science* **17**: 855-866.
21. CPCRI (1976). *Annual report for 1975.* Kasargod, Kerala, India: Central Plantation Crops Research Institute, pp. 24-27.
22. Dalton, B.R., Blum, U. and Weed, S.B. (1983). Allelopathic substances in ecosystems: effectiveness of sterile soil components in altering recovery of ferulic acid. *Journal of Chemical Ecology* **9**: 1185-1201.
23. Duke, S.O. (1985). Comparative ecophysiology of weeds and crops. In *Weed Physiology,* Vol. 1. *Reproduction and Ecophysiology.* (Ed., S.O. Duke). pp 101-129. Boca Raton: CRC Press.
24. Dutta, A.C. (1977). Shade trees, green crops and cover crop plants in the tea estates of northeast India. *Memorandum Tocklai Experiment Station,* Jorhat, India **30**: 128
25. Einhellig, F.A. (1987). Interactions among allelochemicals and other stress factors of the plant environment. In *Allelochemicals: Role in Agriculture and Forestry* (Ed., G.R. Waller). American Chemical Society Symposium Series. **330**: 343-357. Washington, DC: American Chemical Society.
26. Einhellig, F.A. (1989). Effect of allelopathic chemicals on crop produtivity. In *Bioregulators for Pest Control* (Ed., P.A. Hedin), American Chemical Society Symposium Series. **276**: 109-159.Washington, DC: American Chemical Society.
27. Fan, L.J., Young,-hyun, L. and Gharnay, M.M. (1982). The nature of lignocelluloses and their pre-treatment for enzymatic hydrolysis. *Advances in Biochemical Engineering* **23**: 157-187.
28. Fernandes, E.C.M. and Nair, P.K.R. (1986). An evaluation of the structure and function of tropical homegardens. *Agricultural Systems* **21** : 279 – 310.
29. Fernandes, E.C.M., O'ktingati, A. and Maghembe, J. (1984). The Chagga home gardens : A multistoreyed agroforestry cropping system on Mount Kilimanjaro (northern Tanzania). *Agroforestry Systems* **2** : 73 – 86.
30. Garb, S. (1961). Differential growth inhibitors produced by plants. *Botanical Review* **26**: 422-443.
31. Gomez, A.A. (1974). Intensification of cropping systems in Asia. In *Interactions of Agriculture with Food Science* (Ed., R. MacIntyre). p.93. Ottawa, Canada: International Development and Research Centre.
32. Guenzi, W.D. and McCalla, T.M. (1962). Inhibition of germination and seedling development by crop residues. *Soil Science Society of America Proceedings* **26**: 456-458.
33. Guenzi, W.D., McCalla, T.M. and Norstadt, F.A. (1967). Presence and persistence of phytotoxic substances in wheat, oat, corn and sorghum residues. *Agronomy Journal* **59**:163-165.
34. Haider, K. and Martin, J.P. (1975). Decomposition of specifically carbon 14 labelled benzoic and cinnamic acid derivatives in soil. *Soil Science Society of America Proceedings* **39**: 657-662.
35. Hampton, R.E. (1972). Vegetables. In *Record of a Coconut Industry Workshop,* Fiji: Department of Agriculture. pp. 72-74.

36. Harper, J.L. (1977). Mechanisms of interactions between species. In *Population Biology of Plants*. New York: Academic Press. pp. 347-381.
37. Hassan, M.S., Alsaadawi, I.S. and Behaldi, A.E. (1989). Citrus replant problem in Iraq. II. Possible role of allelopathy. *Plant and Soil* **116**: 157-160.
38. Horsley, S.B. (1988). Nitrogen species preference of *Prunus serotina* EhrH. and *Betula alleghaniensis* Britt. seedlings. *American Journal of Botany* **75**: 75-81.
39. Horsley, S.B. (1991). Allelopathy. In: *Biophysical Research for Asian Agroforestry* (Eds., M.E. Avery, M.R.G. Cannell and C. Ong). pp. 167-183. USA: Winrock International.
40. Jacob, J. (1997). *Structure Analysis and System Dynamics of Agroforestry Home Gardens of Southern Kerala*. Ph. D. Thesis, Vellanikkara, Kerala, India: Kerala Agricultural University pp.213
41. Jacob, V.J. and Alles, W.S. (1987). Kandyan garden of Sri Lanka. *Agroforestry Systems* **5** : 123-137.
42. Jessy, M.D., Varghese, P., Punnoose, K.I. and Sethuraj, M.R. (1996). Multispecies cropping system with rubber: a preliminary report. In *Proceedings of the Symposium on Farming System Aspects of the Cultivation of Natural Rubber (Hevea brasiliensis)*. Hertford, United Kingdom: International Rubber Research and Development Board. pp. 81-89.
43. Joshi, P.C. and Prakash, O. (1992). Allelopathic effects of litter extract of some tree species on germination and seedling growth of agricultural crops. In *Proceedings of First National Symposium on Allelopathy in Agroecosystems*. (Eds., P. Tauro and S.S. Narwal), pp. 127-128. Hisar, India: Indian Society of Allelopathy.
44. Keli, J.Z., Omont, H., Dea, G., Kouadia, C. and Obouayeba, S. (1996). Rubber-based farming systems with annual and perennial intercrops in the Ivory coast: Overview of 20 years of trials. *See Reference No.* 42, pp. 46-55.
45. Khader, K.B.A. (1982). Inter and mixed cropping in areca gardens. *Indian Farming* **32** (9): 21-23.
46. Kimber, R.W.L. (1973a). Phytotoxicities from plant residues. II. The effect of time of rotting of straw from grasses and legumes on the growth of wheat seedlings. *Plant and Soil* **38**: 347-361.
47. Kimber, R.W.L. (1973b). Phytotoxicity from plant residues. III. The relative effect of toxins and nitrogen immobilization on the germination and growth of wheat. *Plant and Soil* **38**: 543-555.
48. Korikanthimath, V.S. (1994). Nutrition of cardamom. In *Advances in Horticulture* (Eds., K.L. Chadha and P. Rethinam), Vol. 9 (1): 467-476. New Delhi: Malhotra Publishing.
49. Koul, V.K. (1990). The effect of soil beneath *Leucaena leucocephala* and its decomposed leaves on germination of rice. *Leucaena Research Report* **10**: 54-55.
50. Koul, V.K., Raina, A., Khanna, Y.P., Tickoo, M.L. and Singh, H. (1991). Evaluation of certain farm grown tree species on rice. *Indian Journal of Forestry* **14**: 54-57.
51. Leuschner, W.A. and Khalique, K. (1987). Homestead Agroforestry in Bangladesh. *Agroforestry Systems* **5** : 139 – 151.
52. Lin, W., Chen, Q., Zhou, Z. and Huang, S. (1996). Mixed farming in China's rubber plantations. *See Reference No.* 42, pp. 110-112..
53. Mandal, R.C., Yadukumar, N. and Mohan, E. (1990). Cashew-based farming system. *Journal of Plantation Crops* **18** (Supplement): 156-160.
54. Martin, J.P., Haider, K. and Wolf, D. (1972). Synthesis of phenols and phenolic polymers by *Hendersonula toruloidea* in relation to humic acid formation. *Soil Science Society of America Proceedings* **36**: 311-315.
55. McCalla, T.M. (1971). Studies on phytotoxic substances from soil microorganisms and crop residues at Lincoln, Nebraska. In *Biochemical Interactions Among Plants*. Washington, DC: US National Committee for International Biology Programs, National Academy of Science. pp 39-43.
56. Menges, R.M. (1987). Allelopathic effects of palmer amaranth and other plant residues in soil. *Weed Science* **35**: 1059-1072.
57. Menon, K.P.V. and Pandalai, K.M. (1958). *The Coconut Palm-A Monograph*. Ernakulam, India: Indian Central Coconut Committee. pp. 43.
58. Michon, M.G., Mary, F. and Bompard, J. (1986). Multistoreyed agroforestry system in West Sumatra, Indonesia. *Agroforestry Systems* **4** : 315-338.
59. Nair, M.A. and Sreedharan, C. (1986). Agroforestry Farming Systems in the homesteads of Kerala, Southern India. *Agroforestry Systems* **4**: 339-363.
60. Nair, P.K.R. (1977). Multispecies tree combinations with tree crops for increased productivity in the tropics. *Gartenbauwissenschaft* **42**: 145-150.
61. Nair, P.K.R. (1979). *Intensive Multiple Cropping with Coconuts in India: Principles, Programme and Prospects*. Berlin, Germany: Verlag Paul Parey. pp. 147.

62. Nair, P.K.R. (1983). Agroforestry with coconuts and other tropical plantation crops. In: *Plant Research and Agroforestry* (Ed., P.A. Huxley), pp.79-102. Nairobi, Kenya: International Centre for Research on Agroforestry.
63. Nair, P.K.R. (1989). *Agroforestry Systems in the Tropics*. Dordrecht, The Netherlands: Kluwer Academic Publishers. pp. 39-84.
64. Nair, P.K.R. (1993). *An Introduction to Agroforestry*. Dordrecht, The Netherlands: Kluwer Academic Publishers. pp. 499.
65. Nair, S.K. and Rao, S.N.S. (1977). Microbiology of root region of arecanut and cacao under mixed cropping. *Plant and Soil* **46**: 511-519.
66. Nambiar, M.C. (1976). Cashew, pepper and cardamom in the new system. *Indian Farming* **25** (11): 22-23.
67. Nelliat, E.V. (1979). Prospects of multiple cropping. *Technical Bulletin*, (Eds., E.V. Nelliat and K.S. Bhat).Vol. **3**: 3-5. Kasargod, Kerala: Central Plantation Crops Research Institute.
68. Nelliat, E.V., Bavappa, K.V.A. and Nair, P.K.R. (1974). Multi-storied cropping-a new dimension of multiple cropping in coconut plantations. *World Crops* **26**: 262-266.
69. Nelliat, E.V., Gopalasundaram, P., Varghese, P.T. and Sivaraman, K.S. (1979). Mixed cropping in coconut. *Technical Bulletin*, (Eds., E.V. Nelliat and K.S. Bhat) Vol. **3**: 28-35. Kasargod, Kerala: Central Plantation Crops Research Institute.
70. Norstadt, F.A. and McCalla, T.M. (1963). Phytotoxic substance from a species of *Penicillium*. *Science* **140**: 410-411.
71. Okafor, J.C. and Fernandes, E.C.M. (1987). The compound farms of south-eastern Nigeria : a predominant agroforestry homegarden system with crops and small livestock. *Agroforestry Systems* **5**: 153-168.
72. Pathiratna, L.S.S. and Perera, M.K.P. (1996). Performance of cinnamon and forage grass+tree legume mixtures under rubber (*Hevea*). *See Reference No.* 42, pp. 41-45.
73. Patrick, Z.A., Toussoun, T.A. and Koch, L.W. (1964). Effect of crop residue decomposition products on plant roots. *Annual Review of Phytopathology* **2**: 267-292.
74. PHILCOA (1974). *Annual Report 1973-74*. Quezon City, Philippines: Agricultural Research Department, Philippine Coconut Authority.
75. Plucknett, D.L. (1979). *Managing Pasture and Cattle Under Coconuts*. Boulder, Colarada, USA: Westview Press.
76. Putnam, A.R., DeFrank, J. and Barnes, J.P. (1983). Exploitation of allelopathy for weed control in annual and perennial cropping systems. *Journal of Chemical Ecology* **9**: 1001-1010.
77. Putnam, A.R. and Duke, W.B. (1978). Allelopathy in agroecosystems. *Annual Review of Phytopathology* **16**: 431-451.
78. Rajagopalan, A., Viswanathan, T.V. and Nirmala, D. (1992). Medicinal plants as intercrops in coconut gardens-a preliminary study. *Journal of Plantation Crops* **20** (Supplement): 50-51.
79. Rajasekhar, R. and Sharma, V.S. (1991). Allelopathy between tea (*Camellia* L. spp.) cultivars. *Journal of Plantation Crops* **18** (Supplement): 122-125.
80. Rao, O.P., Saxena, A.K. and Singh, B.P. (1994). Allelopathic effects of certain agroforestry tree species on the germination of wheat, paddy and gram. *Annals of Forestry* **2**: 60-64.
81. Rice, E.L. (1974). *Allelopathy*. New York: Academic Press pp. 353.
82. Rico-Gray, V., Jose, G.G.F., Alexandra, C., Armando, P. and Paulino, S. (1990). Species composition, similarity and structure of Mayan homegardens in Tixpeual and Tixcacaltuyub, Yucatan; Mexico. *Economic Botany* **44**: 470-482.
83. Rizvi, S.J.H., Jaiswal, V., Mukerji, D. and Mathur, S.N. (1980). Antifungal activity of 1,3,7-trimethylxanthine isolated from seeds of *Coffea arabica*. *Naturwissensschaften* **67**: 459-460.
84. Rizvi, S,J.H., Mukerji, D. and Mathur, S.N. (1981). Selective phytotoxiity of 1,3,7-trimethylxanthine between *Phaseolus mungo* and some weeds. *Agricultural Biology and Chemistry* **54**: 1255-1256
85. Rizvi, S.J.H. and Rizvi, V. (1985). Possible role of allelochemicals in India in improving crop productivity. In: *Allelochemicals: Role in Agriculture and Forestry* (Ed., G.R. Waller). American Chemical Society Symposium Series 330. Washington, DC: American Chemical Society.
86. Rosyid, M.J., Wibawa, G. and Gunawan, A. (1996). Rubber based farming systems development for increasing small holder income in Indonesia. In *Proceedings of the IRRDB Symposium* 1996, Sri Lanka. pp. 17-24.
87. Rovira, A.D. (1969). Plant root exudates. *Botanical Review* **35**: 35-57.
88. Salam, A.M., Mohankumaran, N., Jayachandran, B.K. (1991). Thirty one tree species support black pepper vine. *Agroforestry Today* **3** (4): 16.

89. Saleem, C.P. (1981). Cardamom-areca intercropping. *Cardamom* **13** (4): 3-7.
90. Sannamrappa, M. (1993). Arecanut based cropping system in maidan parts of Karnataka. *Journal of Plantation Crops* **21** (Supplement): 3-6.
91. Sannamarappa, M. and Muralidharan, A. (1982). Multiple cropping in arecanut. In: *Monograph Series No.2* (Eds., K.V.A. Bavappa, M.K. Nair and T. Premkumar). pp. 33-49. Kasargod, Kerala, India: Central Plantation Crops Research Institute..
92. Seshadri, C.A. and Sayeed, P.M. (1953). Profitable subsidiary crops. *Coconut Bulletin* **7**: 19-21.
93. Sivadasan, C.R. and Nair, C.K. (1988). Rubber-cardamom intercropping-a case study. *Spices India* **1** (6): 28-29.
94. Soemarwoto, O. (1987). Homegardens: a traditional agroforestry system with a promising future. In *Agroforestry - A Decade of Development*, (Eds., H.A. Steppler and P.K.R. Nair). pp. 157-170. Nairobi, Kenya: International Centre for Research on Agroforestry.
95. Sommers, P. 1978. *Traditional Home Gardens of Selected Philippines Households and their Potential for Improving Human Nutrition*. M.Sc. Thesis. Los Banos, The Philippines: University of Philippines.
96. Srinivasan, C.R., Sudarshan, M.R., Nair, C.K. and Naidu, R. (1992). Study on feasibility of cardamom coffee intercropping. *Journal of Plantation Crops* **20** (Supplement): 60-63.
97. Srinivasan, K., Ramasan, M. and Shanta, R. (1990). Tolerance of pulse crops to allelochemicals of tree species. *Indian Journal of Pulses Research* **3**: 40-44.
98. Suresh, K.K. and Rai, R.S.V. (1987). Studies on the allelopathic effects of some agroforestry tree crops. *International Tree Crops Journal* **4** : 109 - 115
99. Suresh, K.K. and Rai, R.S.V. (1988). Allelopathic exclusion of understorey by a few multipurpose trees. *International Tree Crops Journal* **5** : 143 - 151
100. Swain, T. (1977). Secondary compounds as protective agents. *Annual Review of Plant Physiology* **28**: 479-501.
101. Tajuddin, I. (1986). Integration of animals in rubber plantations. *Agroforestry Systems* **4**: 55-66.
102. Tukey, H.B.Jr. (1970). The leaching of substances from plants. *Annual Review of Plant Physiology* **21**: 305-324.
103. Varghese, P.T., Nelliat, E.V. and Balakrishnan, T.K. (1978). Beneficial interactions of coconut-cacao crop combination. In *Proceedings of First Plantation Crops Symposium*, (Ed., E.V. Nelliat). pp 383-392. Kasargod Kerala: Central Plantation Crops Research Institute.
104. Varghese, P.T. and Rethinam, P. (1994). Nutrition of oilpalm. *Advances in Horticulture* **9**: 407-422
105. Venkataram, C.S. (1978). Mixed cropping in tea. *See Reference No.* 5, pp. 438-440.
106. Vergara, N.T. and Nair, P.K.R. (1985). Agroforestry in the South Pacific region: an overview. *Agroforestry Systems* **3**: 363-379.
107. Wang, T.S.C., Cheng, S.Y. and Tung, H. (1967). Dynamics of soil organic acids. *Soil Science* **104**: 138-144.
108. Watson, G.A. (1983). Development of mixed tree and food crop systems in the humid tropics: a response to population pressure and deforestation. *Experimental Agriculture* **19**: 311-332.
109. Wood, P.J. (1988). Agroforestry and decision making in rural development. *Forest Ecology Management* **24**: 191-201.
110. Zachariah, P.K. (1978). Fertiliser management for cardamom. *See Reference No.* 5, pp. 148-156.

SECTION IV

AGROFORESTRY SYSTEMS

12

Allelopathy in forest ecosystems

M.S. REIGOSA*, L. GONZALEZY, X.C. SOUTO[1] AND J.E. PASTORIZA

Department of Plant Biology and Soil Science
Faculty of Science, University of Vigo, Apdo 874, Vigo-E 36200, Spain.

Abstract

In forest plantations, many types of plant spp. grow together for long periods and hence, are ideal location for allelopathic research. The high yielding exotic spp. introduced to increase productivity of forests are most allelopathic than indigenous spp. These reduce the vegetation diversity and understorey flora due to inhibitory allelopathic efforts. The information about transformation and degradation processes of allelochemicals in soil is little known, hence, needs further studies. Various types of stresses (biotic/abiotic) greatly influence the production and release of allelochemicals in environment. *Eucalyptus globulus* is most allelopathic tree spp. in forest plantations and its leachates contain large number of allelochemicals. Allelopathic interactions are mediated by a mixture of many allelochemicals. Under field conditions, the concentrations of individual allelochemicals is too low to cause allelopathic effects.

Key words: *Acacia dealbata, Acacia melanoxylon*, allelochemicals, allelopathy, *Eucalyptus globulus*, microorganisms, *Pinus radiata, Quercus robur*

1. Introduction

In the living plants, allelochemicals are present in cells, hence, do not exert any influence on the growth and development of its own plant or other plants. The allelochemicals influence other organisms after these are released into the environment. In forest ecosystems, many types of tree spp. and understory vegetation grow together hence, these plant spp. influence each other for quite long period of time. There may be

---

*Correspondence author
Department of Natural Resources, Plant Biology, School of Forestry Industries, University of Gvigo Pontevedra, Spain.

S.S. Narwal et al. (eds.), Allelopathy in Ecological Agriculture and Forestry, 183–193.

various types of above and below ground interactions. This paper reviews various types of allelopathic interactions which occurs in forest ecosystems.

2. Allelopathic interactions

The trees release allelochemicals through leaching, root exudates, decomposition of leaf litter and volatilization (Fig. 1). The allelopathic interactions play significant role in forest ecosystems (14,19) due to following reasons (a) trees release

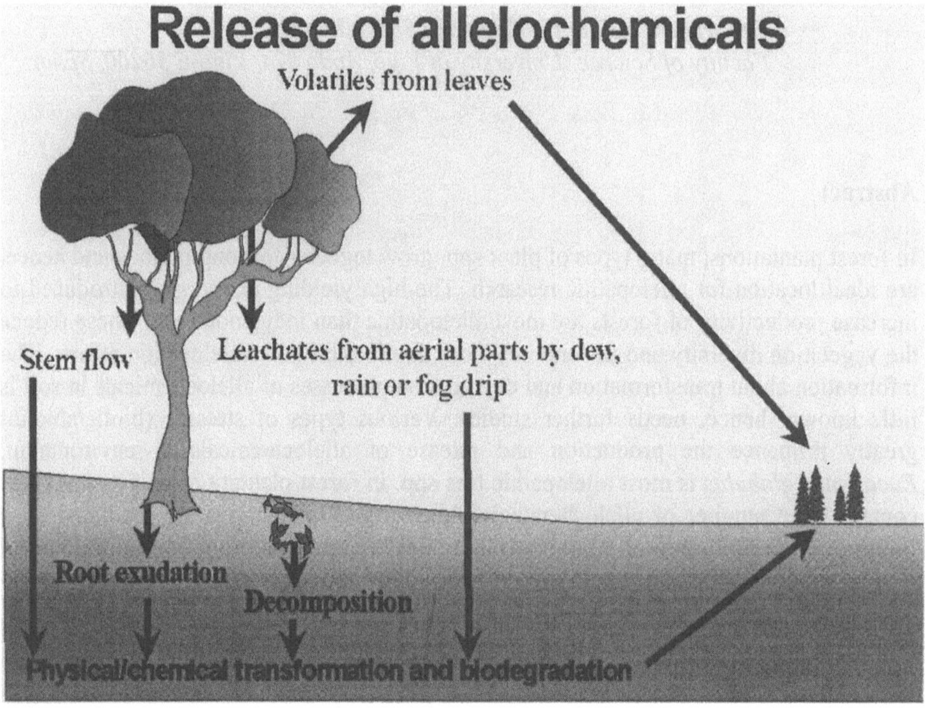

Figure 1. Possible ways of allelochemicals release

allelochemicals for long periods, which with passage of time may accumulate in soil to toxic levels. The accumulation may also occur due to reduction in microbial decomposition under certain conditions. (b) In forest plantations, generally there are one or few dominant tree spp, which could lead to accumulation of allelochemicals of these spp. (c). The allelochemicals released from forest trees affect the understory spp., with least affect on donor spp., however, problems of natural regeneration also occur in some tree spp. in forests (7, 8). (d) The introduction of high yielding exotic tree spp. in forest plantation may also increase accumulation of allelochemicals in soil due to two reasons: (I) The very high requirements of such tree spp. for water and nutrients may cause their deficit in soil, leading to increased production of allelochemicals. (II) The soil microflora, which plays an important part in the allelopathy (16) may not be adapted to such allelochemicals, hence, the allelochemicals may accumulate in soil to toxic levels. The allelochemicals released from these spp. may not affect the understory flora in the region, where the tree spp., has co-evolved but may produce allelopathic phenomenon in regions of its new introduction.

2.1. FATE OF ALLELOCHEMICALS

The allelochemicals journey between the release from donor plant and to the production of physiological effects in receiver plant, passes through many important stages and processes (Fig. 2). In general, allelochemicals released from the plants to the

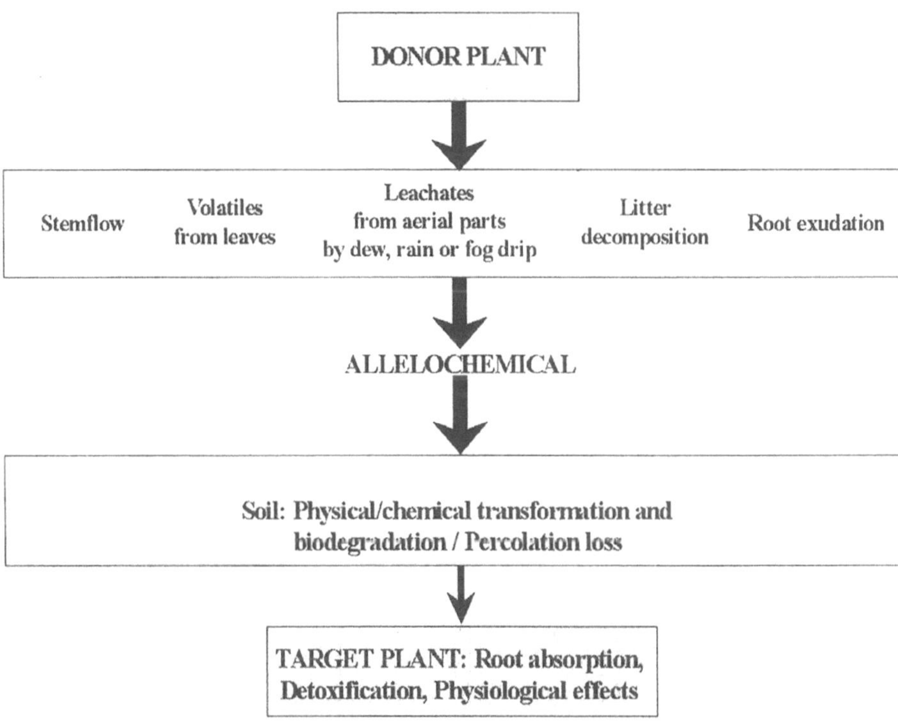

Figure 2. Journey route of allelochemicals from the release from donor plant to the production of allelopathic effect in donor plant

environment through any mode of release, ultimately reaches the soil. In soil, these compounds undergo various transformations (physical, chemical), biodegradation or percolation losses with or without mediation of soil microbes, but full details are unknown (15). A theoretical scheme of what happens to allelochemicals in soil has been proposed (Fig.3). In allelopathy studies, we call the soil `black box' because very little information is available about the soil complex and various processes occurring in soil.

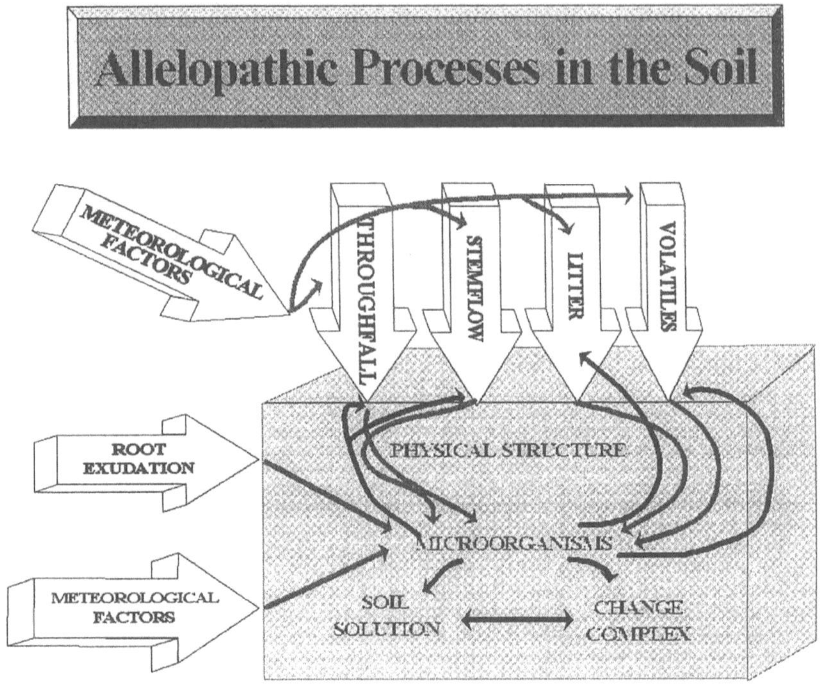

Figure 3. Allelopathic processes in the soil

2.2. EFFECTS OF ALLELOCHEMICALS

Little information is available about the effects of allelochemicals on various physiological and biochemical processes in the plants. This is because (a) there are large number of allelochemicals, (b) each allelochemical usually have more than two physiological effects, (c) allelopathic effects occur as a result of many compounds acting together, perhaps, synergistically (4) and (d) allelopathy effects occur in nature alongwith stress factors. Allelochemicals can simultaneously affect many different physiological processes. It is evident from the studies of Devi (3) with ferulic acid (Fig. 4). Einhellig (4) also did many such studies and have summarized the effect of allelochemicals on various physiological processes in plants (Fig 5).

2.3. STRESS AND ALLELOPATHIC INTERACTIONS

In nature, plants undergo various types of stresses. It is true in natural forests, where, one or other type of stress (biotic or abiotic) affects the trees or understorey spp.

most of the time. The various types of stresses which affect the growth and development of plant spp. are depicted in Fig. 6. It is known that allelopathy is the defense mechanism of plants to overcome the stresses.

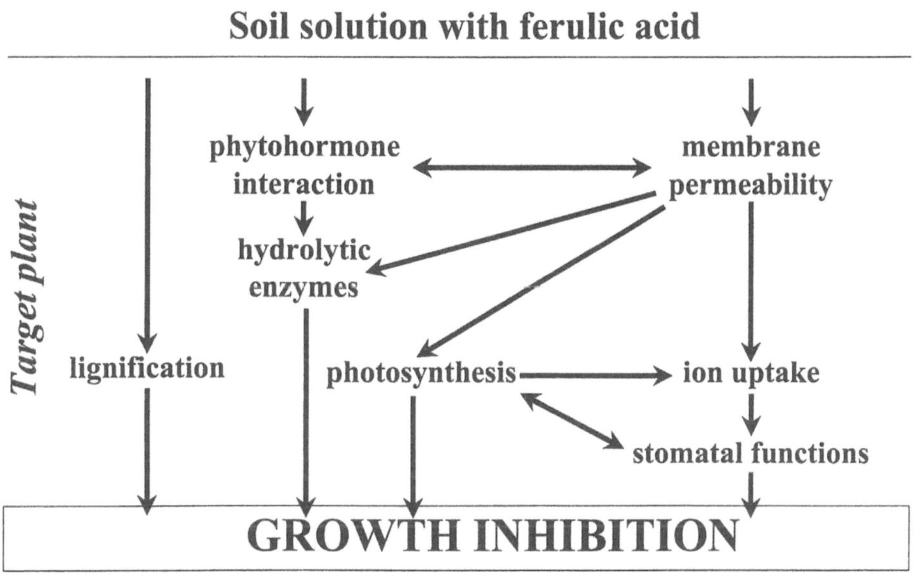

Figure 4. Effect of ferulic acid application on physiological processes in maize plants (3).

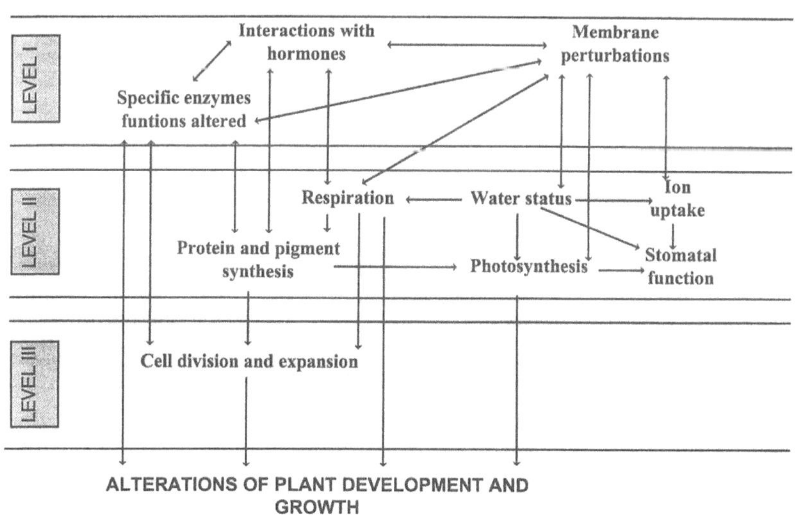

Figure 5. Summary of modes of action of allelochemicals on various physiological and biochemical processes in recipient plant.

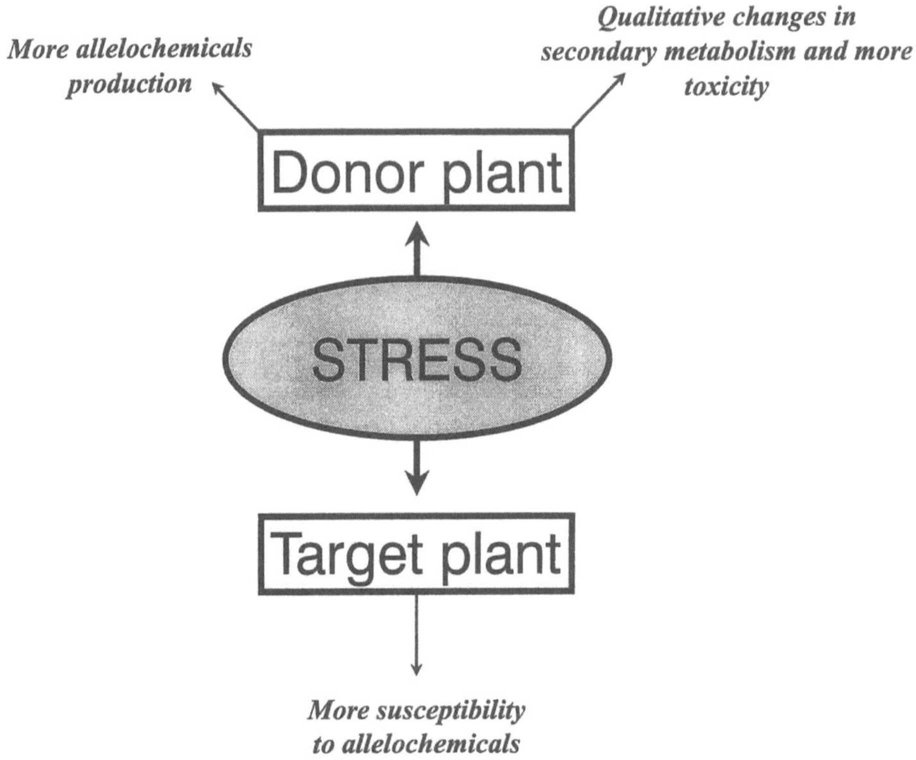

Figure 6. Various types of stresses which influence the allelochemicals production in plants

The stress exert variable effects on donor and receiver plants and some times plants undergo multiple stresses simultaneously. Under stress, donor plants, may produce (a) many types of allelochemicals, (b) large quantity of allelochemicals and (c) more phytotoxic compounds. While on the other hand, receptor plants under stress become more susceptible to allelochemicals (15,20) (Fig.7). Therefore, we should be very careful in designing experiments under stress conditions and for interpretation of their results. Because minimising a stress would artificially reduce the allelopathic potential of donor plant.

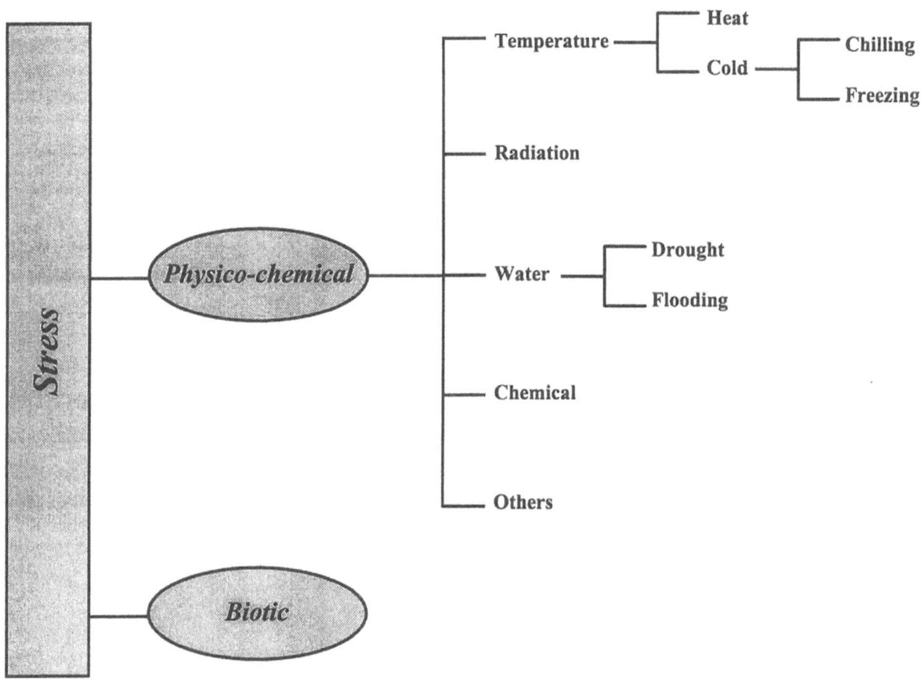

Figure 7. Interactions between the allelopathy and other stresses

2.4. TRESS SPP

Forest plantations are the ideal sites to study allelopathic interactions (11), hence, we studied the allelopathic potential of four major exotic tree spp. viz., *Acacia dealbata, A. melanoxylon, Eucalyptus globulus*, pine (*Pinus radiata*) vis-a-vis the indigenous oak tree (*Quercus robur*) of Galician forests in forest plantation.

I. ***Acacia dealbata :*** It was introduced in Galician region as an ornamental tree but it spread rapidly due to its strong invasive capacity. We have found that its trees release many allelochemicals, mainly phenolic compounds (14). The allelopathic effects of its aqueous rain leachates vary with climatic conditions, phenological phase of tree (1,14) and time (2,13).

II. ***Ecalyptus globulus :*** Owing to its very high productivity, it is the main tree in forest plantations of Galicia. It has drastically reduced the plant species richness and understorey flora and consumes too much water. We have studied the allelopathic effects of its rain leachates and decomposing litter (9, 10).

III. Exotic versus indigenous trees : We have compared the allelopathic potential of *Acacia melanoxylon, Eucalyptus globulus, Pinus radiata* and *Quercus robur* for one year under field conditions. The *Q. robur* being indigenous tree was used as control. To minimise variability, these spp. were planted at the same place, in same soil, with same slope, orientation and altitude. Using the litter bag technique, the plant materials of these four spp. was buried in four plantations stands to determine the effect of microflora on the decomposition process and the release of allelochemicals. (14). The phytotoxicity of exotic spp. followed the order : *E. globulus* > *A. melanoxylon* > *P. radiata*. Perhaps, because *P. radiata* was introduced in Galicia region long ago. These studies showed that exotic spp. were more phytotoxic than indigenous spp. It was also found that pH, relative humidity and allelopathy greatly influence the population of important soil microbes viz., proteolitic, dinitrogen fixers, nitrosomonas, nitrosobacter, ammonifiers, general Bacteria, cellulose hydroyzers and starch hydrolyzers.

3. Eucalyptus allelopathy

Among the *Eucalyptus* spp., *E. globulus* is most allelopathic tree spp. in forest plnatation of Galicia. We have identified many allelochemicals in the shoot leachates collected from its trees in forest plantations (Fig. 8). These include phenolics and

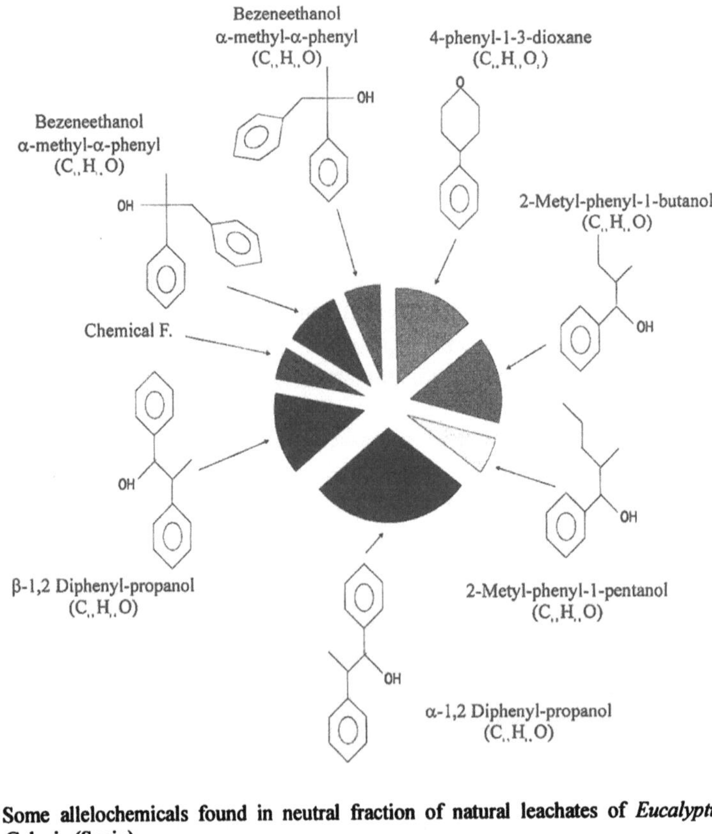

Figure 8. Some allelochemicals found in neutral fraction of natural leachates of *Eucalyptus globulus* in Galacia (Spain)

terpenoids and some samples also contained volatile compounds. This show the great allelopathic potential of shoot leachates. It also confirmed that allelopathy is mediated by a mixture of several allelochemicals and under field conditions, the concentrations of none of individual allelochemicals is too high to cause strong allelopathic effect.

Laboratory studies to determine the persistence of allelochemicals in soil have shown that the rate of degradation of allelochemicals increased with passage of time and was maximum at 20 days (Fig. 9).

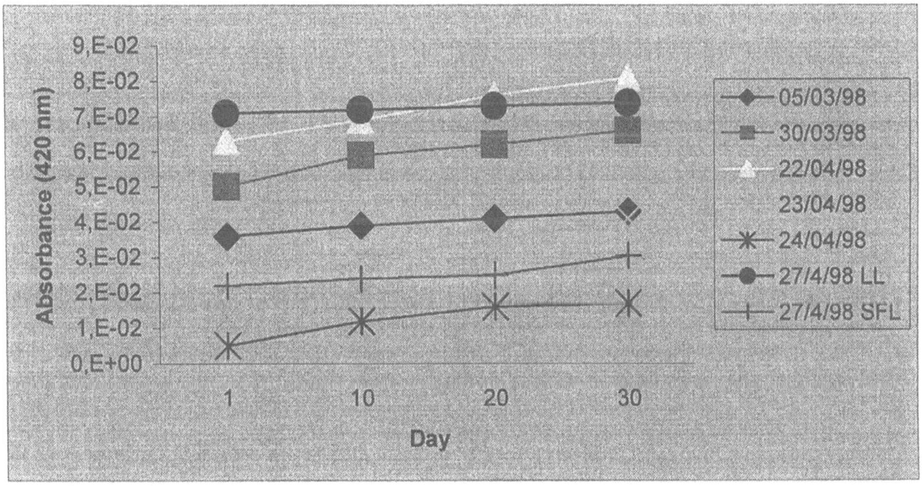

Figure 9. Degradation of allelochemicals in soil samples with time (days)

4. Conclusions and future lines of research

The forest plantations are ideal locations for conducting allelopathic research. The tree leaves flowers, bark, roots and decomposing parts release to the environment, large number of compounds, some with allelopathic potential. The production and release of these compounds is influenced by environmental changes, climatic conditions and phenological phases of plant spp. We have developed methods to study allelopathy in forest plantations.

4.1. Future lines of research

To fully understand the allelopathic interactions in forest plantations the research may be conducted in following areas.

I. To understand the transformations (physical and chemicals) and decomposition of allelochemicals in soil (Black Box in allelopathy).

II. To identify the allelochemicals produced during the above processes as per S.No.1.

III. Effects of these allelochemicals on the activity and composition of soil microbes.

IV. Effects of various types of stresses on different physiological and biochemicals processes in soil and on production of various allelochemicals.

References

1. Carbelleria, A. and Reigosa, M.J. (1999). Effects of natural leachates *of Acacia dealbata* Link in Galicia (NW Spain). *Botanical Bulletin of Academia Sinica* **40**: 87-92.
2. Casal, F.J., Reigosa, M.J. and Carballeria, A. (1985). Potential allelopathiue de *Acacia dealbata* Link. *Revue Ecologie Biologie Sol* **22**: 1-12.
3. Devi, S.R. (1994). *Physiological Responses of Maize (Zea mays L.) to Exogenous Ferulic Acid.* Ph.D. thesis, Hyderabad, India : University of Hybderabad.
4. Einhellig, F.A. (1986). Mechanisms and modes of action of allelochemicals. In. *The Science of Allelopathy* (Eds., A.R. Putnam, and Ch-Sh.Tang) pp. 184-198. New York : Wiley Interscience.
5. Gonzalez, L., Souto, X.C. and Reigosa, M.J. (1992). Effectos alelopáticos producidos por la especie *Pinus radiata* D. Don durante el proceso de decomposicion cuatro suelos naturales en Galicia. *Nova Acta Cientifica Compostelana* 3: 93-102.
6. Gonzalez, L., Souto, X.C. and Reigosa, M.J. (1995). Allelopathic effects of *Acacia melanoxylon* R. Br. phyllodes during their decompsoición. *Forest Ecology and Management* **77**: 53-63.
7. Kohli, R.K. (1999). Allelopathic interactions in Forestry System. In *Environmental Forest Scicnce*. pp. 269-283. Dordrecht, Netherlands : Kluwer Academic Publishers.
8. Malik, A.U. (1999). Allelopathy and competition in coniferous forests. In *Enviornmental Forest Science* pp. 309-315. Dordrecht Netherlands : Kluwer Academic Publishers.
9. Molina, A., Reigosa, M.J. and Carballeira, A. (1984). Efectos alelopaticos durante la descomposicion de residuos de *Eucalyptus globulus* Labill. *Cuadernos Ciencias Agraria Publicacions Seminario Estudios Galegos* **5**: 117-131.
10. Molina, A., Reigosa, M.J. and Carballeria, A. (1991). Release of allelochemic agents from litter, through fall and topsoil in plantations of *Eucalyptus globulus* Labill in Spain, *Journal of Chemical Ecology* **17**: 147-160.
11. Rabotnov, T.A. (1974). On Allelopathy in the phytocenoses. *Izvestia Akad Nauk SSSR Ser. Biologia* **6**: 811-820.
12. Reigosa, M.J. (1998). Forest ecosystem. *McGraw-Hill Yearbook of Science and Technology*, pp. 143-146. New York: McGraw-Hill.
13. Reigosa, M.J., Casal, J.F. and Carballeira, A. (1984). Efectos alelopaticos de *Acacia dealbata* Link durante su Floración. *Studia Oecologica* **5**: 135-150.
14. Reigosa, M.J., Souto, X.C. and Gonzalez, L. (1996). Allelopathic research : methodological, ecological and evolutionary aspects In : *Allelopathy : Field Observations and Methodology* (Eds., S.S. Narwal and P. Tauro), pp. 213-231. Jodhpur, India: Scientific Publishers.

15. Reigosa, M.J., Sanchez, A.M. and Gonzalez, L. (1999a). Ecophysiological approach to allelopathy. *Critical reviews in Plant Sciences*. In Press.
16. Reigosa, M.J., Souto, X.C., Gonzalez, L. and Bolano, J.C. (1999b). Allelopathic effects of exotic tree species on microorganisms and plants in Galicia (Spain). In *Environmental Forest Science*, pp. 293-300. Dordrecht, Netherlands : Kluwer Academic Publishers.
17. Souto, X.C., Gonzalez, L. and Reigosa, M.J. (1994). Comparative analysis of the allelopathic effects produced by four forestry species during the decomposition process in their soils in Galicia (NW Spain). *Journal of Chemical Ecology* **20** : 3005-3015.
18. Souto, X.C., Gonzalez, L. and Reigosa, M.J. (1995). Allelopathy in forest environemnt in Galicia (NW Spain). *Allelopathy Journal* **2**: 67-68.
19. Souto, X.C., Gonzalez, L., Pedro, M.N. and Reigosa, M.J. (1992). Allelochemical effects produced by *Eucalyptus globulus* Labill during the decomposition process in four natural soils in Galicia (NW Spain). Comparative analysis. *Comparative Physiology and Ecology* **18** : 114-123.
20. Tang, CH.-SH.,Cai, W-F., Kohl, K. and Nishimoto, R.K. (1995). Plant stress and allelopathy. American Chemical Society, pp. 142-157.

13

Allelopathic interactions in agroforestry systems

A.S.GILL* and J.V.N.S.PRASAD
Division of Crop Production,
Indian Grassland and Fodder Research Institute, Jhansi-284003,India

Abstract

Agroforestry systems, growing of multipurpose tree species (MPTS) with arable crops has the potential to increase the total biomass production per unit area. It is an ideal technology to meet the food, fuel, fruit and fodder requirements for the future generations. MPTS not only fulfil the above requirements but are also capable of improving soil fertility, controlling erosion and ensures sustainability of the system in which they exist. In recent years, lot of interest has been generated regarding the allelopatic effects of MPTS in various agroforestry systems. The tree which is the dominant component in the agroforestry system affects the associated crops plants through the leaf fall, exudates and by chemicals which are released at various stages of decomposition of leaf litter. Large number of allelopathy experiments conducted on Eucalyptus, the main species in social forestry programmes and industrial plantations in India, conclusively established its allelopathic effects. In addition to Eucalyptus, the allelopathic proclivities of some commonly grown MPTS have also been discussed in this chapter.

Key words : Acacia, allelopathy, agroforestry, crops, eucalyptus, growth attributes, leucaena.

Introduction

Cultivating agricultural crops and trees together is an ancient practice world-wide. Agroforestry, is defined as a land use that involves deliberate retention, introduction of mixture of trees or other woody perennials in crop/animal production fields to benefit from the resultant ecological and economic interactions. In India, it has been a traditional practice and has received greater emphasis in recent years as a sustainable land use option of high potential. Studies conducted on agroforestry systems in arid and semi arid regions of India (20,21) revealed that the association of some trees with crops improved the crop yields. Trees constitute an integral part of the

* Correspondence author

S.S. Narwal et al. (eds.), Allelopathy in Ecological Agriculture and Forestry, 195–207.

agroforestry systems and were deliberately retained on farm lands. The contribution of tree towards the total farm out put is often unaccounted in traditional agroforestry practices. To increase the economic importance of the tree component, the multipurpose tree species (MPTS) have been introduced in agroforestry systems. The multipurpose trees are those trees and shrubs having more than one preferred use, product and/or service.

The success of agroforestry lies in the exploitation of positive interactions and limiting the negative interactions. Numerous metabolic chemicals involved in plant-plant interactions are released from the plants, primarily through leaching from above ground parts and play a significant role in plant interactions (78). The effects of these chemicals on other plants depends on the concentration of the chemical and the combination in which one or more of these substances are released into the environment (52). Thus, trees build a characteristic profile of their own and the magnitude of substances supplied to the soil layer depends upon the rate of leaching, litter fall and chemical disintegration of compounds in the environment (40). The dominant plant species exert influence on the floor conditions and understorey vegetation through indirect means such as shade, moisture status of the soil and nutrient availability. However, recent researches have shown that the allelopathic interactions between the component crops determines the habitat characteristics in an agroforestry system. The amount of metabolites released and their accumulation beneath the tree over a period of time influence the development of other plants in the community.

Allelopathy has been accepted as a factor having ecological significance in plant dominance, patterning of vegetation succession, crop productivity and agroforestry systems (40). This chapter describes the role of allelopathy in agroforestry systems containing trees which are common under Indian conditions.

2. Major tree species

2.1. EUCALYPTUS

Many species of Eucalyptus are grown for timber and essential oils in tropical and subtropical countries. Of these, *E. citriodora* is commonly grown in India on borders of crop fields. Owing to its fast growth, wide adaptability to soil and climatic conditions and least after care, the government in recent years, has recommended its plantations in large areas. However, the relative density, dominance and frequency of understorey vegetation under *Eucalyptus* sp. was significantly poor compared to *A.nilotica* and *D.sissoo* (64). Although, Learner and Evenari (36) reported the existence of germination inhibitors in the leaves of *E. rostrata*, but Del Moral and Muller (17) were the first to study the allelopathic effect of *E. camaldulensis* and *E. globulus* in California. They demonstrated that light, nutrients, moisture etc. were not responsible for the formation of bare areas in the plantation but the inhibitory effect on vegetation was due to the leaf leachates. Similarly, allelopathic action of Eucalyptus was found responsible for suppression of heath species beneath the

Eucalyptus canopy in Australia (10, 38). Now it is known that allelopathic properties of Eucalyptus was the main cause for understorey suppression especially in drier climates (10).

2.1.1. *Field Studies*

Igboanugo (23) reported the deleterious effects of Eucalyptus on germination of beans and maize. Shivanna *et al* (60) observed a significant reduction in the germination of fingermillet and cowpea in Eucalyptus plantations. In Eucalyptus plantation spaced at 3x3m, the seeds of *Phaseolus vulgaris* germinated but seedlings did not survive (51), however, at 5x5m spacing the seedlings survived for 30-35 days. A clear circular suppression zone close to the stem (<1M circle) was observed around the stem in most crop plants wherein seedlings died and dried within 20-25 days (5). In a 3-5 year old Eucalyptus plantations, reduction of 57.9% and 50.2% in plant population/m^2 was observed in irrigated wheat over the normal crop up to a distance of 5.0m and 6.6m, respectively (16). Narwal and Sarmah (42) reported maximum reduction in plant population and plant height of test crops up to a distance of 6.7m from the Eucalyptus tree. The reduction in plant population in sunhemp, sesbania, sorghum, maize, pigeonpea, castor and pearlmillet was 50.5, 27.8, 25.8, 25.0, 18.2, 12.5 and 6.67%, respectively, over the control. Similarly, the reduction in plant height in sesbania, sunhemp, pigeonpea, castor, maize, sorghum and pearlmillet was 80.9, 67.8, 65.4, 65.3, 58.6, 53.3 and 50.4%, respectively, over the control. The negative effect of Eucalyptus tree was lessened with an increase in the distance from the tree species.

Crops grown under the Eucalyptus tree produces less number of shoots per plant, less biomass of root and shoots/m^2 as compared to the open area (31). Likewise the biomass of redgram and greengram was also drastically reduced (66). Thus a decrease either in the germination percentage or plant population/unit area leads to reduction in biomass accumulation/unit area which adversely influences the yields. The yields of rice, wheat, barley, potato and cymbopogon decreased considerably when grown with Eucalyptus trees compared with their sole crops, (77). Wheat, rice, potato, pigeonpea, castor and maize yields were affected considerably when planted at a distance of 5.0 m and not beyond 6.7m from the Eucalyptus trees (42). The yields of chickpea, lentils, cauliflowers and berseem and toria were reduced up to 50% in a 12 m wide strip adjacent to trees (63). Thus the extent of reduction in crop yield in agroforestry systems depends upon the distance from the tree.

Bund plantation as wind breaks or shelterbelts proved harmful to jute, groundnut and rice up to a distance of 15-20m (15). Similarly, chickpea, linseed, wheat, rape and Egyptian clover were affected up to a distance of 12m (31), potato and wheat unto a distance of 5-6m from the trees (5,16). The most susceptible crops were *Cicer arietinum, Lens esculentum, Triticum aestivum, Brassica oleracea, B. campestris* and *Trifolium alexandrium* (63). The harmful effect of Eucalyptus on jute, groundnut and rice could not be ameliorated even with the application of fertilisers (15).

The direction of crop sowing also had great affect, the crop sown on southern side of the tree did better than the crop sown on the northern side (16). Both the rhizosphere and atmosphere rich in allelochemicals results in poor performance of crops. The cell survival and chlorophyll content were reduced drastically with increasing distance from the shelter belts in case of *Triticum aestivum* (28). Moreover the lack of vegetation around the trees of *E.citriodora* might be due to the lixiviation of allelochemicals from the leaves by the rain and also due to the decomposition of leaves in the soil. Allelochemicals thus released during decomposition are retained by the soil and concentrated on the top layer with time. Perhaps due to this, the germination and plant population of the crop plants was reduced resulting in reduced yield. Most of the times, the closed canopy of the plantation limits the amount of light intercepted which might have further enhanced the accumulation of allelochemicals (60).

2.1.2. *Bioassay Studies*

2.1.2.1. Leaf extracts. The results of bioassay studies with leaf of various Eucalyptus species have been depicted in Table No. 1. Fresh intact leaves caused little growth suppression, in contrast to coarsely chopped leaves and oils extracted from leaf. Fresh litter extracts of Eucalyptus hybrid (*E. tereticornis*) reduced the germination of wheat, maize, pea and mustard (27). Compared to control, the process of germination was slow with the Eucalyptus leaf extracts. Similar reduction due to Eucalyptus leaf extracts was reported in groundnut, castor, sunflower, wheat, maize, pea, mustard, barley, lentil, chickpea, mustard and in finger millet (8,27,46,61). The inhibitory action was found to be proportional to the content of phenolic compounds (27). Puri (51) reported the decreasing order of toxicity for germination in lab studies, brown leaves > green leaves > dried leaves > bark. Generally, the germination percentage decreased with an increase in the aqueous extract concentration. There may be slight stimulation at lower concentration of the leachates. Rao and Reddy (53) observed stimulatory effect of Eucalyptus leaf extracts on the germination of greengram, horsegram and cowpea and inhibitory effect on wheat, pearlmillet and coriander. Inhibition at higher concentration might be due to the presence of more allelochemicals in the extract (8).

The radicle growth of four Ethiopean crops viz., chickpea, maize, beans and teff were significantly reduced by the aqueous leaf extracts of *Eucalyptus globulus, E. camaldulensis and E.saligna*. Treatment with leaf extracts for a 10 week period significantly reduced the root and shoot dry weight of various crops (37). Of the various crops studied chickpea was the most susceptible in germination while, teff was most susceptible in germination and growth. The treated seedlings were deformed in shape and their performance was poor. Puri (51) observed a curved blunt extension region devoid of roots at root-shoot transition region. Rao and Reddy (53) found a gradual recovery in field grown seedlings, treated earlier with Eucalyptus extracts indicating that once the inhibitory substances are leached out from the soil, harmful effects will be minimised.

Table 1. Allelopathic effect of Eucalyptus leaves on various crops

Eucalyptus leaf	Test crops	Nature of effect	Reference
E. tereticornis	Greengram, horse ram, cowpea	Stimulatory effect of germination	53
	Wheat , pearlmillet	Inhibitory to germination	53
	Wheat, Sweet sorghum, lentil, pea, mustard, maize, finger millet, green gram, horse gram, coriander	Inhibitory to germination and growth	1, 51, 50, 59
	Sorghum, wheat, rice, pearlmillet	Inhibitory to germination and growth	6, 7, 27
	Wheat, mustard, lentil, pea, chickpea, cowpea, triticale, sorghum	Inhibitory to germination and growth	2, 3, 30, 57
	Radish, fenugreek, brinzal, onion	Inhibitory to germination and growth	41, 8
	Wheat, chickpea	No effect on wheat and inhibitory to germination in chickpea	39
E.globulus	Chickpea, pea, maize	Inhibitory to germination and growth	37
	Green gram, cowpea	Inhibitory to germination and growth	17
	Greengram, chickpea, sorghum, lentil, cowpea, clover	Inhibitory to germination and growth	30
E.citriodora	Pulses, forages, cereals, vegetables and oilseeds	Inhibitory to germination and growth	34
	wheat mustard	Inhibitory to germination and growth	68
	wheat rice, maize, triticale, mustard, vetch	Inhibitory to germination and growth	4
E.macrorhyncha	Wheat and radish	Inhibited root and shoot growth	4

Volatile terpenes present in Eucalyptus contribute maximum to allelopathic potential of the tree. Among various Eucalyptus species, *E. citriodora* contain maximum content of terpenes, while minimum was in *E. tereticornis* (29). *E. citriodora* volatile oils were rich in citronellol, whereas *E. globulus* and *E. tereticornis* volatile oils were rich in cineole, α-pinene and β- pinene. The effect of crude volatile oils from the leaves of *E. globulus* and *E. citriodora* and also pure terpenes, cineole and limonene were studied on lentils, barley, oats and *Phaseolus aureus*. Germination, seed vigour and seedling water content decreased as volatile oil concentration increased (28). Exposure of *V. radiata* and lentil leaves to 13.5 ml/cm^{-3} vapour from *E. globulus* or *E. citriodora* oil reduced cell survival and total chlorophyll content. The leaf leachates were also responsible for reducing the germination, shoot and root growth and lateral root development (46).

The shoot dry weight of maize crop was reduced drastically when the tree biomass was applied as mulch or incorporated into the soil (56). A significant reduction in plant height and number of leaves was observed (7) due to the addition of powdered leaf litter of Eucalyptus. Suresh and Rai (73) found maximum reduction in shoot length, root length and dry matter production in sorghum, when the dry leaves of Eucalyptus were applied as mulch (50). Sorghum was most susceptible to the Eucalyptus mulch among all the test crops (73).

The bulk of the allelochemicals of Eucalyptus are added in the rhizosphere through the death and decay of fallen plant parts, mainly the leaves. These allelochemicals released through the process of decay inhibits photosynthesis and respiratory metabolism in crops. Kohli (29) extracted 624 ± 204 µl of Eucalyptus oil per kg of soil from various locations of Eucalyptus monoculture plantations. Since the rate of inhibition of crop growth is dependant on the concentration of oil vapour, hence, the inhibitory activity of Eucalyptus leaves is more, because they contain large amounts of volatile oils (29).

2.1.3. *Pot Culture*

Blaise and Tyagi (9) reported that different plant parts of Eucalyptus had variable influence on crop plants and the percentage emerged seedlings were lower with fresh leaves. Studies conducted with top and rhizosphere soil from Eucalyptus plantations exhibited harmful effect on the germination and growth of crops. The top soil taken from *E. camaldulensis*, *E. grandis* and *E. tereticornis* inhibited the germination of cowpea, pigeonpea, horesgram and Indianbean (60). Eucalyptus hybrid proved most inhibitory and legumes were less affected than cereals. Cowpea and sorghum proved most tolerant crops, while Italian millet was found most sensitive. The seedlings of the test crops in the top soil did not survive for more than a month. Similarly, Suresh and Rai (73) observed inhibitory effect of both top soil and rhizosphere soil from Eucalyptus plantation on the germination, root growth and dry matter production in sorghum, cowpea and sunflower. In similar studies with the top soil, Srinivasan *et al* (70) found cowpea as the most tolerant and soybean as most sensitive crop among blackgram, greengram, cowpea, pigeonpea and soybean. The top soil from Eucalyptus plantation had a toxic effect on the germination and hypocotyl growth of lettuce, gardencress, foxtailmillet and barnyard grass and had no effect on rice (44). It also adversely affects the above ground and root dry matter, number of nodules and their weight in *Glycine max*, *Desmodium intortum* and *Panicum maximum*. (49). In pot trials with groundnut and maize, irrigation with 5, 10, 15 or 20% water extracts of abscised leaves of *E.globulus* decreased plant height, leaf area and chlorophyll content compared with plants irrigated with water. Inhibitory effects were more pronounced at higher concentration of the extract. Sidhu and Hans (62) reported a decrease in plant height, tillering and shoot and root dry matter production in wheat with application of heaped stem litter. Most of the allelochemicals leach out of the plant in their water soluble form (33,43). These are called glycosides, which after their release might have been adsorbed by the soil in the plantation. In nature, Eucalyptus oil vapour also travel downwards and get adsorbed on the soil surface (64). They not only adversely affect the germination and root growth of surface feeders but also the above ground vegetative growth (29). In addition to this, the soil in the Eucalyptus plantation is found to contain gallic acid, gentisic acid, syringic acid, vanillic acid, caffeic acid, *p*-coumaric acid, ferulic acid and cinnamic acids besides a few unidentified phenolics (29).

2.2. LEUCAENA

Studies on the floristic composition and biomass production under leucaena plantations revealed that it harboured only 5 species as compared to 29 species in the open (74). Studies conducted in Taiwan revealed that the biomass of ground cover beneath the leucaena plantation was relatively low as compared to adjacent open grassland and there was about 80% bare ground in the leucaena plantation. The decomposing leaves might have suppressed the growth of test plants but not that of leucaena (22).

2.2.1. *Field Studies*

The presence of leucaena tree in the agroforestry system, significantly reduced the germination of sorghum, cowpea and sunflower. Other seedling attributes such as root length, dry matter production were also restrained due to the presence of leucaena (73). The top soil collected from 3-year old wood lots of the tree, the rhizosphere soil collected from the wood lots and field soil mulched with dry leaves of leucaena, depressed the germination of the test crops and the magnitude of reduction was maximum with the top soil (73). The phytotoxic affects of leucaena were attributed to mimosine, a toxic non-proteinaceous amino acid present in the seed and leaves (35) which might have gradually accumulated in the top soil of the plantation over a period of time thus, resulting in maximum reduction in germination. Mimosine is reported to have caused germination inhibition in moongbean (69) and reduced the hypocotyl growth of lettuce (79). However, Koul (32) found no significant effect of leucaena soil and decomposed leaf extracts on the germination of rice. This might have been due to rapid leaf decomposition and consequent leaching of toxins from the soils by rain which reduces the phytotoxicity (32). Among *L. leucocephala, E. tereticornis, P. juliflora, A. auriculifomis, P. dulce* and *C. siamea*, leucaena had the least allelopathic influence on the drymatter production of the associated grasses viz. *Sehima nervosum, Heteropogon contortus, Apluda mutica, Cenchrus ciliaris* and *C.setigerus*, etc. (14). Contrarily, Singh (65) observed a marked increase in productivity of pigeonpea, sesame, castor and sorghum under leucaena trees. The stimulatory effect of leaf extracts of *L. leucocephala* on rice radicle growth has also been reported (12).

2.2.2 *Bioassay Studies*

Even though toxic effects of mimosine have been reported (26,54) yet little information is available about its allelopathic properties. Mimosine inhibited the germination of greengram, but the growth of understorey leucaena seedlings was not affected under mature trees (69). This might be due to the detoxification mechanism possessed by leucaena where the toxic mimosine has been degraded to 3-4 hydroxy pyridine and other non-toxic metabolites (69). Kuo *et al* (35) also observed a significant reduction in the radicle growth in lettuce and rice at mimosine concentrations of 10 and 20 ppm, respectively. The germination of rye grass has been strongly inhibited by 2.5% aqueous litter extracts of K-8, K-72 and S, cultivars of leucaena. Tawata and Hongo (76) tested mimosine at various concentrations (1, 10, 25, 50 and 100 ppm) for its allelopathic activities against rice, radish, turnip, kidneybeans, carrot and *Bidens pilosa.* Radicle growth of all species was inhibited at 10 ppm concentration, while the growth of rice, radish, turnip and kidneybean was promoted at 1 ppm. Carrot, Kidneybean and *Bidens pilosa* were less sensitive to mimosine than the other species. Similar germination inhibition has been reported in lettuce, rice, rye (13, 35), wheat, maize, pea and mustard (13,27,35). A reduced cotyledon weight in mimosine treated seeds suggested the possibility of inhibited mobilisation of stored food from cotyledon to embryo as the food mobilisation efficiency was found to be significantly low (55). This might have resulted in poor seedling growth resulting in poor plant stand.

2.3. *ACACIA* SPECIES

A.albida being leguminous deciduous and deep rooted tree, recycles the plant nutrients by natural leaf drop and thus may reduce the competition for nutrients to some extent. However, tree species like *A. tortilis* and *A. nilotica* even though recommended for introduction into different habitats, exhibit very poor understorey vegetation and found to have allelopathic proclivities (16,71).

2.3.1. *Field studies*

In Acacia plantations, considerable portion of cultivated land around the tree remains bare due to the inhibition in growth of various agricultural crops. Even a single row of *A.nilotica* tree plantation inhibited the plant height, shoot number, ear length, grain number and grain yield of associated wheat crop unto a distance of 4 m from the tree under irrigated conditions in Haryana (58). However, under rainfed conditions at Hisar, the mustard crop was significantly influenced upto a distance of 26m from the tree. The plant stand was reduced to 71% at 28-30m and 57% at 30-32m over control (16). With a canopy cover of 5.0m the inhibition zone was observed upto a distance of 15m. When the canopy cover was 8.0m complete suppression effect of Acacia was observed upto 26m. However, normal growth was recorded at 35m away from the tree (16).

Duhan and Laxminarayana (18) reported similar suppression effect of *A .nilotica* on pearlmillet and clusterbean upto a distance of 2m from the tree. A similar reduction in chickpea crop was reported by Jakhar and Rana (25). In majority of the cases the adverse effect of tree species decreased gradually on all the parameters as the distance from the tree species is increased.

2.3.2. *Bioassay studies*

The aqueous extract of bark and leaves of *A. nilotica* trees significantly inhibited the seed germination, radicle and plumule growth of various crops. Crops vary in their response to the bark and leaf leachates. Swaminathan *et al* (75) found that the extracts of leaf and bark of *A.nilotica* had inhibitory effects on tomato, sunflower, sorghum and cotton. Duhan and Laxminarayana (18) observed similar decrease in germination, plumule and radicle length in pearlmillet and clusterbean crops due to the bark extracts of *A.nilotica*. A similar effect in chillies, greengram, maize, horsegram and groundnut with 1:10 (w/v) bark and leaf leachates of *A. magium* was observed by Palani *et al* (48). The inhibition caused by the bark extracts was more compared to the leaves. It was presumed that tannins were responsible for the above allelopathic properties (75).

The leaf leachate of *A. nilotica* was found to reduce the root and shoot length and vigour index of mustard, lentil, fieldpea and wheat (59). Singh and Nandal (68) reported the inhibitory effect of *A.nilotica* leaf extract on clusterbean, pearlmillet, cowpea and sorghum. When cowpea was grown in 10% and 20% combined plant extract of *A.nilotica*, the nodulation, nodule fresh weight, shoot dry weight, was significantly reduced, thus effecting the total biomass production (19). Sundaramoorthy *et al* (72) found that the soluble proteins and free amino acid content in cotyledons and shoot axis decreased with an increase in the concentration of aqueous leaf extract. However, the sugar content increased with the increase the concentration of the applied extract. They isolated the allelopathic principle responsible for above effects and were found to be organic acids with Rf values of 0.986, 0.969 and 0.185 and 0.176 and three phenolic compounds. The phenolic compounds were coumaric, vanillic and phloretic acids. Similar phenolics were identified in the underneath soils of *A. tortilis*.

2.3.3. *Plantation soil*

Sundaramoorthy and Kalra (71) studied the allelopathic properties of *A. tortilis* plantation at Jodphur and found that the properties of the soil under *A.tortilis* and *Prosopis cineraria* did not differ significantly. The soil extract of *A.tortilis* exhibited inhibitory effect on germination and seedling growth of pearlmillet, clusterbean and sesame. The growth of sesame was most severely inhibited. Of the various crops tested, clusterbean exhibited better tolerance but the rate of inhibition per unit increase in concentration was similar in all the test crops (71). Duhan and Laxminarayana (18) observed a significant decrease in the germination, plumule length and radicle length in pearlmillet and clusterbean due to the extracts of *A.nilotica* plantation soil. Even when the soil beneath the *A.nilotica* and acid washed soil was mixed in 1:1 proportion, the seed germination was inhibited by 25%, plumule length by 49.3% and radicle length by 30.5% in pearlmillet crop compared to control (18). The nodule number in cowpea crop was significantly reduced when grown in the plantation soil. Similarly, the shoot dry weight, nodule fresh weight, nitrogenase activity was also reduced in the three strains of *Rhizobia* when compared to the non-allelopathic soil (19). Compared to the ordinary soil, the other three soil media i.e. top soil, rhizosphere soil and soil mulched with dry leaves from 3-year old *A.nilotica* plantation significantly depressed the growth and yield of test crops viz., cowpea, sesame, horsegram and sorghum (47).

Casal *et al* (11) reported that the allelopathic properties of *A.dealbata* were responsible for the poor understorey vegetation. The leaves, flowers and soil collected beneath the plants of *A.dealbata* significantly influenced the germination, root and shoot length of *Trifolium repens, T.pratense* and *Lolium perenne*. The inhibition might be due to *p*-hydrobenzoic, protocatechuic, vanillic, genstisic, *p*-coumaric, caffeic, ferulic and sinapic acids which were found in the extracts. The leaf leachates of *A.auriculiformis* decreased the percent germination, plumule and radicle length in rice and cowpea. Length and dry matter accumulation were more severely affected in rice than in cowpea (24).

3. Future areas of research

Most of the experiments related to the allelopathic effects of tree crops were conducted in laboratory conditions mainly with the leachates of various tree parts. So little information is available regarding the allelopathic effects of various parts of tree under field conditions.

i. Efforts should be made to extract the chemicals responsible for allelopathic action and they should be tested for the replicability of similar effects as in Koch's postulates.
ii. Separation of allelopathic affects from competition due to light, moisture and nutrients for precise quantification.
iii. Development of suitable methodologies for determining the allelopathic effects under various systems such as intercropping, agroforestry and natural grasslands.
iv. Information on the nature and quantity of allelochemicals released. and their mode of action is lacking.
v. Isolation, identification and characterisation of allelochemicals.
vi. Ability of various crops to tolerate the allelopathic effect of tree and vice-versa.
vii. Screening of crops for their ability to tolerate the allelopathic effect.
viii. Development of technologies capable of reducing/minimising the allelopathic activity of trees in the agroforestry system.

References

1. Anwar, C. (1991). Study of the allelopathic effect of Eucalyptus species on the growth of corn seedlings. *Bulletin Penelitian Hutan* No. **543** : 9-17.
2. Ballester, A., Arias, A.M., Cobian, B., Lopez, C.E. and Vietez. (1982). Study of the allelopathic potentials caused by *Eucalyptus globulus, Pinus pinaster* and *Pinus radiata. Don pastos* **12** : 239-254.
3. Bansal, G.L. (1988). Allelopathic effects of aqueous extracts of stem and leaves of three tree species on the germination of some crops and weeds. In *Trends in Tree science*. pp. 118-23. Solan, India: Indian Society of Tree Scientists.

4. Bansal, G.L., Nayyar, H. and Bedi, Y.S. (1992). Allelopathic effect of *Eucalyptus macrorrhyncha* and *Eucalyptus youmanii* on seedling growth of wheat and radish. *Indian Journal of Agricultural Science* **62** : 771-772.
5. Basu, P.K., Kapoor, K.S., Nath, S. and Banerjee, S.K.(1987) Allelopathic influence : An assessment of the agricultural crops growing near *Eucalyptus tereticornis*. *Indian Journal of Forestry* **10**: 267-71.
6. Bedi, S. and Kohli, R.K. (1989). Comparison of allelopathic potential of root and shoot system of *Eucalyptus globulus*. In *Proceedings of the Forest Biology in the Service of Mankind*. pp. 70 Madurai, India: Indian Society of Tree Scientists.
7. Bhaskar, V., Arali, A. and Shankaralingappa, B.C.(1992). Alleviation of allelopathic effects of *Eucalyptus tereticornis* through litter burning. In *Proceedings First National Symposium. Allelopathy in Agroecosystems (Agriculture and Forestry)*, (Eds., P. Tauro, S.S. Narwal) pp.118-119. Hisar, India: Indian Society of Allelopathy.
8. Bisla, S.S., Nandal, D.P.S. and Narwal, S.S.(1992). Influence of aqueous leaf extracts of Eucalyptus and Poplar on the germination and seedling growth of winter crops. *See Reference No.* **7** pp. 95-97.
9. Blaise, D. and Tyagi, P.C.(1996). Allelopathic effects of Eucalyptus on crops. In *Abstracts. First World Congress on Allelopathy*. pp.16-20. Cadiz, Spain.
10. Bowman, D.M. and Kirkpatrick, J.B. (1986). Establishment, suppression and growth of *Eucalyptus delagatensis* in multi aged forest.III Interspecific allelopathy, competition between adults and juveniles for moisture and nutrients and frost damage to seedlings. *Australian Journal of Botany* **34**: 81-94.
11. Casal, J.F., Reigosa, M.J., Carbelleira, A. (1985). Allelopathic potential of *Acacia dealbata*. *Rev. Ecol. Biol. Sol.* **22** : 1-12.
12. Chaturvedi, O.P. and Jha, A.N.(1992). Studies on the allelopathic potential of important agroforestry species. *Forest Ecology and Management* **53**: 91-98.
13. Chou, C.H. and Kuo, Y.L. (1986). Allelopathic exclusion of understorey by *Leucaena leucocephala*. *Journal of Chemical Ecology* **12**:1431-48.
14. Chouhan, G.S., Mathur, A. N., Bhandari, M.M.C. and Jat, P.K. (1992). Allelopathic effect of some tree species on associated grasses under silvi pastoral system. *See Reference No.*7.pp.130-131.
15. Craig, I.A. and Saenalo, M.(1988). Effect of paddy bund planted *Eucalyptus* trees on the performance of field crops. *Working Paper No. 1*, The Phar, Thailand : Nerad project.
16. Dalal, M.R., Dahiya, D.S., Sarmah, M.K. and Narwal, S.S. (1992). Supression effects of arid-zone trees on plant stand and growth of crops. *See reference No* .7. pp. 132-135.
17. DelMoral, R. and Muller, C.H. (1970). The allelopathic effect of *Eucalyptus camaldulensis*. *American Midland Naturalist* **83**: 254-282.
18. Duhan, J.S. and Laxminarayana, K.(1995). Allelopathic effect of *Acacia nilotica* on cereal and legume crops grown in fields. *Allelopathy Journal* **2**: 93-98.
19. Duhan, J.S., Sharma, P.K. and Laxminarayana, K. (1994). Allelopathic effect of *Acacia nilotica* on nodulation and nitrogen fixation by *Rhizobium* (cowpea). *Allelopathy Journal* **1**: 47-52.
20. Gill, A.S. (1992). Allelopathy in agroforesty. *See Reference No.*7.pp. 7
21. Gill, A.S. (1994). Allelopathic effect of MPTs in agroforestry. In *Abstract, International Symposium on Allelopathy in Sustainable Agriculture, Forestry and Environment*, (Eds., S.S. Narwal, P. Tauro, G.S. Dhaliwal and Jai Prakash), pp. 6. New Delhi , India : Indian Society of Allelopathy.
22. Gill, A.S. and Parihar, S.S. (1996). Allelopathy in agroforestry systems. In *Allelopathy. Field Observations and Methodology* (Eds., S.S. Narwal and P. Tauro.) pp. 121-126.Jodhpur, India: Scientific publishers.
23. Igboanugo, I.B. (1988). Effects of some Eucalypts on yields of *Vigna unguiculata, Zea mays* and *Sorghum bicolor*. *Agriculture Ecosystem and Environment* **24**:453-8.
24. Jadhav, B.B. and Gaynar, D.G. (1992). Allelopathic effects of *Acacia auriculiformis* on germination of rice and cowpea. *Indian Journal of Plant Physiology* **35**:86-89.
25. Jhakhar, S.S. and Rana, V.S. (1994). Synergistic and antagonistic effect of *Prosopis cineraria* and *Acacia nilotica* on pearlmillet and chickpea. *SeeReference No.* **8**. pp. 167.
26. Jones, R.K. (1981). Does ruminal metabolism of mimosine explain the absence of leucaena toxicity in Hawai. *Australian Vetarinary Journal* **57**:55-59.
27. Joshi, P.C. and Prakash, O.(1992). Allelopathic effects of litter extract of some tree species on germination and seedling growth of agricultural crops. *See Reference No.* 7. pp. 128-8.
28. Kohli, R.K. (1990). Allelopathic properties of Eucalyptus. *Project Report* MAB. Do En..Project, pp. 199.

29. Kohli, R.K. (1994). Allelopathic implications of Eucalyptus in Agro ecosystems. In *Allelopathy in Agriculture and Forestry* (Eds., S.S. Narwal and P. Tauro). pp. 75-91. Jodhpur,India: Scientific Publishers
30. Kohli, R.K., Kaur, K., Chaudhari, P., Kumari, A. and Saxena, D.B. (1987). Negative aspects of Eucalyptus farming. In *Agroforestry for Rural Needs.* (Eds., P.K. Khosla and D.K. Khurana) pp. 225-41, Solan, India: Indian Society of Tree Scientists.
31. Kohli, R.K., Singh, D. and Verma, R.C.(1990). Influence of shelterbelts on winter season agro ecosystems. *Agriculture Ecosystems and Environment* **33**: 23-30.
32. Koul, V.K. (1990). The effect of soil beneath *Leucaena leucocephala* and its decomposed leaves on germination of rice. *Leucaena Research Report* **11**:54-55.
33. Kumari, A. and Kohli, R.K. (1987). Autotoxicity of Ragweed. (*Parthenium hyterophorus*). *Weed Science* **35**: 629-632.
34. Kumari, A., Singh, D., Verma, R. C. and Kohli, R.K. (1989).Correlation response of germination and vigour on different plant seeds to the *Citronella* oil from Eucalyptus. *See Reference No.***6**. pp.107-19.
35. Kuo, Y.L., Chou, C.H., Itu, T.W. (1983). Allelopathic potential of *Leucaena leucocephala.* In *Allelochemicals and Pheromones*, (Eds., C.H. Chou and G.R. Waller), pp. 107-19. Taipei, Taiwan: Academia Sinica.
36. Learner, R.H. and Evenari, M. (1961). The nature of germination inhibitors present in the leaves of *Eucalyptus rostrata* . *Physiologia Plantarum* **16**:221-229.
37. Lisanework, N. and Michelsen, A. (1993). Allelopathy in agroforestry systems. The effects of leaf extracts of *Cupressis lucitanica* and three *Eucalyptus sps.* on four Ethiopian crops. *Agroforestry Systems* **21**:63-74.
38. Lovett, J.V. (1986). Allelopathy, The Australian experience. In *The Science of Allelopathy* (Eds., A.R. Putnam and C.S. Tang). pp. 75-100. New York. Wiley Interscience.
39. Melkania, N. P.(1984). Influence of leaf leachates of certain woody species on agricultural crops. *Indian Journal of Ecology* **11** :82-86.
40. Melkania, N.P. (1992). Allelopathy in fodder and agroecosystem in the Himalayan region. In *Allelopathy :Basic and Applied Aspects.* (Eds., S.J.H. Rizvi and V. Rizvi), NewYork: Champan and Hall.
41. Nandal, D.P.S., Bisla, S.S. and Narwal, S.S.(1992).Allelopathic influence of Eucalyptus and Poplar leaf extracts on the germination and seedling growth of winter vegetables. *See Reference No.* **7**: 82-86.
42. Narwal, S.S. and Sarmah, M.K. (1992). Supression effect of *Eucalyptus tereticornis* on .the field crops. *See Reference No.* 7 pp. 111-13.
43. Neiman (1952). Estimation of water soluble matter, gums and starches and crude glycyrrhizin in *Gleyrrhiza glabra. Chein Wealblad* **48**:213.
44. Nishamura, H., Kaku, K., Nakamura, T., Fukazawa, T. and Mizutani, J. (1982). Allelopathic substances, *p*-methane-3, 8-diols isolated from from *Eucalyptus citriodora* . *Agriculture Biology and Chemistry* **46**:319-20.
45. Nishamura, H., Nakamura, T. and Mizutani, J. (1988). Allelopathic effects of methane-3, 8-diols isolated from *Eucalyptus citriodora. Phytochemistry* **23**: 277-79.
46. Padhy, B., Khan, P.A., Acharya, B. and Buxipatra, N.P. (1992) Allelopathic effects of Eucalyptus leaves on seed germination and seedling growth of finger millet. *See Reference No.* 7. pp. 102-104.
47. Palani, M., Dastagir, M.G. Thiyageswari, S. and Balaji, S. (1998a). Effect of bark and leaf leachates of *Acacia mangium* on germination and seedling growth of agricultural crops. In *Abstracts National Seminar on Integration of Livestock and Agroforestry Systems in Wasteland Development.* pp. 26. Kattupakkam, Tamil Nadu : India.
48. Palani, M. Dastagir, M.G., Thiyagaswari, S. and Balaji, S. (1998 b). A study on the allelopathic effect of *Acacia nilotica* on agricultureal crops. *See Reference No.* 47, pp 29.
49. Paulino, V.T., Sanchez, M.J.F., Werner, J.C. and Consla, V. (1987). Allelopathic effects of Eucalyptus on forage growth. *Res. Agric.* **62**: 17-35.
50. Prasad, J.V.N.S (1998). *Biomass and Ethanol Production Potential of Sweet Sorghum Varieties as Influenced by Plant Densities, Nitrogen and Allelogenic Tree Biomass application.* Ph.D Thesis, New Delhi : Division of Agronomy, Indian Agricultural Research Institute.
51. Puri, S. (1992).The allelopathic effects of *Eucalyptus tereticornis* in an agroforestry system. *See Reference No* .7. pp.101.
52. Putnam, A.R. and Tang, C.S.(1986). *The Science of Allelopathy.*, New York, USA :Wiley Inter Science.

53. Rao, N.S. and Reddy, D.C. (1984). Studies on the inhibitory effects of Eucalyptus leaf extracts on the germination of certain food crops. *Indian Forester* **110**: 218-22.
54. Reis, P.J., Tunki, D.A. and Chapman, R.E.(1975). Effects of mimosine. A potential chemical defleecing agent on wool growth and the skin of the sheep. *Australian Journal of Biological Sciences* **28**: 69-84.
55. Rizvi, V., Sinha, R.C. and Rizvi, S.J.H. (1994). Allelopathic effects of mimosine and its role in agroforestry. *See Reference No.***21**. pp. 50.
56. Sanginga, S. and Swift, M.J. (1992). Nutritional affects of Eucalyptus litter on the growth of Maize. *Agriculture Ecocystems and Environment* **41**: 55-56.
57. Saxena, S. and Singh, J.S.(1978).Influence of leaf leachate from *Eucalyptus globulus* and *Aescules indica* on the growth of *Vigna radiata* and *Lolium perenne*. *Indian Journal of Ecology* **5**:148-59.
58. Sharma, K.K. (1992). Wheat cultivation in association with *Acacia nilotica* (L.). Field bund plantation - A case study. *Agroforestry Systems* **17**: 43-51.
59. Sharma, K.M.S., Dhillon, M.S. and Dhingra, K.K. (1967). Presence of germination inhibitors in the leaf leachates of some farm grown trees. *Indian Forester* **113**: 816-20.
60. Shivanna, L.R., Prasanna, K.T. and Mumtaz, J. (1992). Allelopathic effects of *Eucalyptus*: An assessment on the response of agricultural crops. *See Reference No.* **8**. pp. 108-110.
61. Shridevi, B., Nimbole, N. N. and Rao, P.S. (1994). Influence of Eucalyptus leaf extract on germination of crop seeds. *See Reference No* .**21**.pp.53.
62. Sidhu, D.S. and Hans, A.S. (1988). Preliminary studies on the effect of *Eucalyptus* leaf litter on acumulation of biomass in wheat. *Journal of Tropical Forestry* **4**: 318-23.
63. Singh, D. and Kohli, R.K.(1992). Impact of *Eucalyptus tereticornis* shelterbelts on crops. *Agroforestry Systems* **20**: 253-66.
64. Singh, D., Kohli, R.K. and Saxena, D.B. (1991). Effect of *Eucalyptus* oil on germination and growth of *Phaseolus aureus*. *Plant and Soil* **137** : 223-227.
65. Singh, G.B. (1993). Role of agroforestry in improving the environment. *Indian Farming* **33**: 15-19.
66. Singh, P.N., Gupta, S., Azmi, S. and Singh, G. (1992). Allelopathic effects of *Eucalyptus citriodora* leaf litter leachate on germination and seedling growth of wheat, chickpea and *Toria*. *See Reference No.* **7**. pp. 105-7.
67. Singh, R. and Bawa, R. (1982). Effect of leaf leachates from *Eucalyptus globulus* and *Aesculus indica* on germination of *Glacium flavum*. *Indian Journal of Ecology* **9**:21-28.
68. Singh, R.P. and Nandal, D.P.S. (1993). Allelopathic effect of aqueous leaf extract of important agroforestry tree species on some fodder crops. *Forage research* **19**: 55-61.
69. Smith, I.K. and Fowden, C.J. (1966). A study of mimosine toxicity in plants *Journal of Experimental Botany* **17**: 750-761.
70. Srinivasan, K., Ramasan, M. and Shanta, R. (1990). Tolerence of pulse crops to allelochemicals of tree specics. *Indian Journal of Pulses research* **3**:40-44.
71. Sundaramoorthy, S. and Kalra, A. (1991).Allelopathy and vegetation in Acacia tortilis plantations in Indian desert. *Annals of Arid Zone* **30** : 259-66.
72. Sundaramoorthy, S., Kalra, N. and Sen, D.N. (1992). Allelopathic potentials of *Acacia tortilis* on seed germination and seedling growth of some legumes. *See Reference No.* **8**. pp. 123-124.
73. Suresh, K.K. and Rai, R.S.V. (1987). Studies on the allelopathy affects of some agroforestry tree crops. *International Tree Crops Journal* **4**:109-15.
74. Suresh, K.K. and Rai, R.S.V. (1988). Allelopathic exclusion of understorey by a few MPT. *International Tree Crops Journal* **5**:143-51.
75. Swaminathan, C., Rai, R.V.S. and Suresh, K.K. (1989). Allelopathic activities of *Acacia nilotica* in India. *Journal of Tropical Forest Science* **2**: 56-60.
76. Tawata, S. and Hongo, I. (1987). Mimosine allelopathy of Leucaena. *Leucana Research Reports* **8**: 40-41.
77. Tomar, G.S. and Srivastava, S.K. (1986). Preliminary studies of rice cultivation in association with trees. In *Agroforestry Systems a New Challenge* (Eds., P.K. Khosla, S. Puri) pp. 207-211. Solan, India : Indian Society of Tree Scientists.
78. Tukey, H.B. Jr. (1970). The leaching of substances from plants. *Annual Review of Plant Physiology* **21** : 305-24.
79. Wilson, W.F. and Bell, E.A. (1979). Amino acids and related compounds as inhibitors of lettuce growth. *Phytochemistry* **18**: 1883-84.

14

Allelopathy research in agroforestry systems of South India

C.S. HUNSHAL*, H.T. CHANNAL, A.R. ALAGAWADI and ND R.H. PATIL
Department of Agronomy
University of Agricultural Sciences, Dharwad – 580 005, Karnataka, India

Abstract

In agroforestry, trees produce lot of litter the and the allelochemicals released from it influence the companion/understorey crops, however, its influence varies with the tree - crop combinations. Pigeonpea was found least susceptible and sesame the most sensitive to the bark leachates of acacia, casuarina, dalbergia, eucalyptus and tectona. *Acacia nilotica* reduces shoot biomass of wheat, berseem, chickpea and cotton. Pearlmillet grown with *Acacia tortilis* showed only 12% germination at 5 m distance from the tree and germination was enhanced to 78 and 92% at 10 and 15 m distance. *A.nilotica* had a greater depression effect on pearlmillet than clusterbean. Plant and soil extracts of *A.nilotica* and *Cassia siamea* inhibited the nodulation and N fixation of clusterbean (*Cyamopsis tetragonoloba*). Contrarily dried leaves of glyricidia promoted the growth of tomato seedlings and the productivity of pigeonpea, sesame, castor and sorghum was increased under leuceana trees and that of sorghum under eucalyptus. Leaf extracts of tectona, *Albizia procera* and *Acacia nilotica* showed stimulatory effects on germination, growth, chlorophyll, protein, carbohydrate and proline content of soybean.

The trees in association with rice, fingermillet (*Eleusine coracana*) and *Echinochloa frumantacea* increased organic carbon by three times, available P and K status significantly in soils. Eucalyptus in association with legumes enhanced the organic carbon, total N, mineral N and exchangeable cation and available P in soils. The dehydrogenase activity in soil under tree species was in the order of *Leuceana leucocephala* < acacia < eucalyptus. The population of bacteria, fungi and azotobacter increased under *Bambusa bambos*, *Casurina equisetifolia*, *Ceiba pentandra*, *Eucalyptus tereticornis*, *Leucaena leucocephala*, *Tectona grandis*.

Therefore, to improve the productivity of agroforestry systems, it is essential to determine the allelopathic compatibility of crops with trees.

Key words: Agroforestry, allelochemicals, allelopathy, inhibition, leachates, metabolites, phenols, phytotoxins, rhizosphere and tannins.

---

*Corresponding author

S.S. Narwal et al. (eds.), Allelopathy in Ecological Agriculture and Forestry, 209–227.

1. Introduction

The burgeoning human and animal population is exerting great pressure on the forests for fuel, timber and fodder and on agricultural land for food grains, thus to meet these demands agroforestry had been introduced. It combines the production of crops and forest / fruit trees simultaneously on the same unit of land and applies management practices compatible with the local cultural practices (27). Since in agroforestry, trees are grown in association with the crop and fodder plants, its productivity depends on the growth resources (sunlight, soil moisture, nutrients) and allelopathic interactions. Therefore, it is essential that the allelopathic compatability of crops with trees should be determined before planting in the agroforestry systems (49). Allelopathy is defined as "any direct or indirect harmful or beneficial effect of one plant (including microorganisms) on another through production of chemical compounds that escape into the environment" (33, 48). Numerous metabolic chemicals potentially involved in plant – plant chemical interactions are released from plants primarily through leaching from above ground parts and thus play a significant role in plant interactions (72).

It is often asserted uncritically that dominant plant species exert influence on the floor conditions and understorey vegetation through competition for growth resources, besides allelopathic interactions which may also influence habitat characteristics and interactions with other factors (20, 34, 49). The wide variations in quality and quantity of metabolites released and their accumulation beneath a tree determines the development of species and community underneath the canopy. Leachates also influence the number, and behaviour of microorganisms present in underneath soil. This paper reviews the allelopathic research carried out in the agroforestry species of South India. Most of the studies in Agroforestry (AF) have concentrated on the testing of aqueous extracts of parts of tree species on the associated field crops under laboratory conditions (53). Since AF aswell as allelopathy are younger areas of research, the field studies on allelopathic effect of tree species on associated crops are limited. In India, most of the allelopathic investigations are based on bioassay studies (13) and very few attempts have been made to analyse the allelochemicals and to confirm allelopathy in field conditions (13). Further, only a limited number of species have been screened against the allelopathic potential of over storey species and other herbs.

2. Major multipurpose tree species

In south India, some of the major multipurpose tree species (MPTS) grown in agroforestry systems are : *Eucalyptus tereticornis, Eucalyptus hybrid, Acacia nilotica, Acacia auriculiformis (Australian babul), Ailanthus excelsa, Ceiba penton, Melia azedarach, Tectona grandis, Casuarina equisetifolia, Terminalia alata, Glyricidia maculata, Azadirachta indica, Bauhinia purpurea, Bauhinia recemosa, Dalbergia latifolia, Dalbergia sissoo, Leucaena leucocephala, Prosopis cineraria* and *Tamarindus indica.*

2.1. *EUCALYPTUS* SPP.

2.1.1. *Pot experiments*

Allelopathic effect of *Eucalyptus tereticornis* fresh leaves, litter and root material on the performance of wheat, maize and cowpea indicated that initially the litter had more adverse effect on the seedling emergence, while later the root material proved most phytotoxic. Perhaps, the slow decomposition of roots released the phenolics slowly, there by affecting plant growth at later phases. The root growth was affected more than the shoot, probably the roots remain in direct contact with allelochemicals (5).Testing of cereals and pulses for germination in soils from plantation of eucalyptus revealed that sorghum and rice were sensitive, whereas *Setaria italica* and *Eleucina coracana* and pulses like pigeonpea, horsegram, cowpea and *Lablab purpureus* were tolerant (55).The crude oil of *E. globulus* and *E. citriodora* contains, cineole and limonene. Its vapours decreased the germination, seedling growth, relative growth rate, water content, plant height and number of leaves in *Phaseolus aureus* var. ML-267 (57) (Table 1). This is an important observation which partly explains, that is why densely grown eucalyptus trees are devoid of other vegetation under them.

Table 1. Effect of eucalyptus oil (Cineole rich) and citriodora oil (Citrionella rich) in concentration of 13.5 nl ml^{-1} on growth of twenty days old *P. aureus* var. M-267 plants

Treatment	RGR w.r.t. dry wt.	Mean No. of leaves per plant	Plant height (mm)	Water content (%)
Control	0.122^a	6.10^a	192.8^a	85.0^a
Eucalyptus oil	0.079^b	4.20^b	152.6^b	63.4^b
Citriodora oil	0.068^c	3.70^c	142.6^c	53.0^c

Source : 57

2.1.2. ***Bioassay studies***

Aqueous extracts of fresh leaves, dry leaves, fresh bark, dry bark, fruits and floral buds of *E. camaldulensis* decreased the germination, tillering and height of wheat (44). Aqueous leaf extracts of eucalyptus sp. were more effective than stem extracts in reducing germination and seedling growth (50). Among pearlmillet, sorghum, sunflower and soybean, pearlmillet was most tolerant and sunflower was most susceptible to eucalyptus extracts.

Mature leaf leachates of *E. globulus* (10%) caused maximum inhibition in the total biomass of *Costus speciosus* (its rhizomes contain diosgenin, a steroid precursor) followed by senescent and juvenile leaves and the inhibition decreased with dilution. Leachates of mature leaves of *E.globulus* indicated that concentration of phytotoxic compounds like phenolic acids in leaves increased with age (29). The young leaves of eucalyptus are hydrophobic due to waxy layer on their surface, hence, had less leaching (18).

The fresh leaves, litter, root, bark and seeds of eight year old plantation of *E. citriodora, E. globulus* and *E. tereticornis* contained caffeic, coumaric, gallic, gentisic, hydroxybenzoic, syringic and vanillic acids and catechol.

Phenolic content (μg/ml) in leachates from various parts of eucalyptus species

Eucalyptus spp.	Seeds	Fresh leaves	Litter
E.citriodora	550	455	290
E.globulus	205	255	240
E.tereticornis	210	150	475

Higher total phenol content in fresh leaves may be attributed to defense mechanism of plants against pathogens, insects and browsers (73) and the release of secondary metabolites during degradation of litter (60).

In bioassays catechol, coumaric, gallic, hydroxybenzoic and syringic acids, inhibited the germination and vigour index in blackgram, greengram, rice and tomato. Syringic acid at 1mM concentration was inhibitory to blackgram and greengram, whereas, gallic and hydroxybenzoic acids and catechol were inhibitory to paddy at 2 mM concentration. At the same concentration, mixture of phenolic compounds showed greater inhibition than the individual phenolic compound (61). Similarly, eucalyptus leaf extract (100 g/l) decreased the germination of sorghum, soybean and maize, but not that of groundnut and cowpea (15). Aqueous leachate of eucalyptus leaves had an adverse effect on the rate of seed germination, seedling growth, chlorophyll, protein and nucleic acid synthesis and certain metabolic enzymes like amylase, catalase, peroxidase, DNAse, RNAse in rice (37). The degree of toxicity of leachates on germination, shoot and root growth of seedlings followed the order: brown leaves > green leaves >decayed leaves > bark. Similarly, the fresh leaf leachates of *E. globulus* inhibited the germination, shoot and root growth in fingermillet (38). Leaf litter leachate of *E. citriodora* at 3 ml was most detrimental to germination of wheat > mustard > chickpea (59).

2.1.3. *Field studies*

Experiment conducted with five year old plantation of *Eucalyptus camaldulensis* in rainfed areas, under acid lateritic soils of South Karnataka, indicated that germination of cowpea and sesame was reduced in 4m x 4m plantation and seedlings started dying 20-25 days after sowing and the death was enhanced by prolonged drought. Whereas in 5m x 5m plantation the seed germination was not affected and the seedlings survived upto 30-35 days and there was a clear zone of suppression near tree stems (enhanced by drought) where seedlings died more rapidly (55). Although moisture was a limiting factor but it could be due to the allelopathic effect as well. The effects are more pronounced during the drier periods, as the drought conditions intensified the allelopathic action (74).Various crops like chickpea, lentil, wheat, cauliflower, berseem (*Trifolium alexandrium*) and *Brassica compestris* were grown inbetween *E.tereticornis* 12 m wide strip. The yield of chickpea was greatly reduced. In analysing the phytotoxins, maximum content was found at 1m distance from tree line for all depths. Therefore, the poor performance of crops in the sheltered area was related to allelopathic effect of the shelterbelts (56). An area of 1m radius around *E. citriodora* becomes barren due to accumulation of allelochemicals from leaves by the rain and leaf decomposition in the soil (2). Allelochemicals thus released are adsorbed by the soil in the top layer, where their effects are accomplished. The tree row had maximum effect on sunnhemp, moderate in pigeonpea, castor, maize and least in sorghum. To overcome the allelopathic effect of eucalyptus leaves, burning has been suggested (4). *Phaseolus vulgaris* seeds germinated but failed to survive when sown between 3m x 3m spaced *Eucalyptus tereticornis* (43). Studies on seven year old plantations of several tree species, revealed that *Grevilea robusta* had an understorey of 54 plant species with a density of 2400 plants/m^2, followed by *Casarina equisetifolia* with 30 species, *Acacia auriculiformis* with 27 species and *E. tereticornis* with only 14 species and 9 plants/m^2 (3). Hence, the soil surface under eucalyptus plantation was very sparse compared to other trees.

Adverse effects of 20 year old plantations of *Eucalyptus citriodora, E.camaldulensis* and *E. grandiflora* were seen on growth of okra, wheat, cowpeas and maize sown in the alleys (22). The reduction ranged from 8 to 73.4 %. Okra was the most sensitive crop and its flowering was reduced. The grain yield of maize and cowpeas was declined by 11.2 and 60.8 %, respectively, while that of sorghum improved slightly (1.5%). The growth resources (light, moisture and temperature) were not limiting, therefore, the inhibition may be attributed to an allelopathic effect of eucalyptus. Bund plantations of eucalyptus proved harmful to jute, groundnut and rice upto a distance of 15 – 20 m (11), chickpea, linseed, wheat, rape and egyptian clover upto a distance of 12 m (28) and potato and wheat upto a distance of 5-6 m (5). The deleterious effect on jute, groundnut and rice could not be ameliorated even with the addition of fertiliser (11). The direction of sowing had marked effect, the crops grown on the southern side of eucalyptus did better than the crops on the northern side (12).

2.2. *ACACIA* SPECIES

2.2.1. *Bioassay studies*

Aqueous extracts of roots, leaves and inflorosence of *A. tortilis* and dry leaf leachates of *Acacia auriculiformis* reduced germination, plumule and radicle length and dry matter of cowpea and rice and the decrement increased with increasing leachate concentration. Root length and dry matter accumulation were severely affected in rice than in cowpea (24) (Table 2).

Similar was the response with the soil extracts. The inhibition was greater with increasing concentration of aqueous extracts suggesting that phytotoxic effects could be more pronounced in low or erratic rainfall areas, where phytotoxic substances may accumulate over the time and moisture is insufficient to dilute them by runoff or leaching from the top soil (52).

The aqueous extracts of bark and leaves of six year old *A.nilotica* trees significantly inhibited the seed germination, radicle and plumule growth of sorghum, cotton, okra, chillies, tomatoes and sunflowers (68). The bark extract was more inhibitory than leaf extract. It was assumed that the phytotoxin present in bark was tannin.The extracts of leaf and soil collected beneath the stand of *A. tortilis* exhibited inhibitory effects on germination and growth of pearlmillet (52).

2.2.2. *Field studies*

Growth of pearlmillet and clusterbean was severely inhibited at distances of 1-2 m from the tree of *Acacia nilotica* to the extent of 88.2 and 64.9% whereas at 7.5 m it was 80.8 and 43.6% respectively compared to their normally grown plants at 30 m distance.This reduction in growth was confirmed due to allelopathic effect when soil and bark extracts inhibited the seed germination of pearlmillet by 40 and 46% and clusterbean by 26 and

Table 2. Effect of *Acacia auriculiformis* A. Cunn. leaf leachates on germination, length and dry matter of plumule and radicle of rice (var. Palghar-1) and cowpea (variety CM-8) (Expressed as percentage)

Leachate soaking Period (h)	Germination (%)		Plumule (cm)		Radicle (cm)		Dry matter (mg)	
	3 DAS	11 DAS	3 DAS	11 DAS	3 DAS	11 DAS	Plumule	Radicle
Rice								
Control	77.66 (100.0)	94.66 (100.0)	0.55 (100.0)	6.64 (100.0)	1.56 (100.0)	9.24 (100.0)	18.73 (100.0)	30.57 (100.0)
4	66.96	71.47	85.45	69.54	28.20	33.51	48.40	16.92
8	62.38	71.13	78.18	45.63	25.00	32.86	48.40	15.94
12	54.58	70.77	58.18	44.42	14.74	27.92	48.39	15.22
16	40.82	60.56	54.54	41.26	8.97	26.74	45.62	14.22
20	33.03	58.80	40.00	41.21	3.20	25.34	47.70	12.91
24	20.64	58.80	27.28	40.21	1.92	23.63	38.37	10.33
SE	0.68	0.79	0.04	0.08	0.04	0.15	1.43	0.38
CD 5%	2.08	2.44	0.16	0.23	0.12	0.47	4.40	1.19
Cow pea								
Control	95.33 (100.0)	99.33 (100.0)	0.82 (100.0)	16.06 (100.0)	1.73 (100.0)	5.95 (100.0)	307.53 (100.0)	81.76 (100.0)
4	46.15	96.33	98.78	29.32	69.36	95.97	52.54	45.46
8	32.16	76.51	73.17	27.95	40.46	86.21	46.08	44.53
12	20.97	53.68	71.95	27.89	39.30	75.46	44.53	26.18
16	15.37	45.63	59.76	25.90	38.72	70.25	38.21	26.10
20	13.98	44.96	54.87	25.40	23.70	65.88	35.94	21.18
24	11.88	42.94	36.58	25.15	23.12	57.31	32.60	18.41
SE	3.55	3.47	0.09	0.40	0.09	0.46	17.93	4.83
CD 5%	10.936	10.70	0.28	1.24	0.28	1.42	55.27	14.89

DAS : Days after soaking, Source : 24

40% respectively (16) . The grain yields of castor, sunflower and pigeonpea intercropped in *Acacia albida* alley widths were not decreased thus indicating no competition between the component crops. Sharma (54) reported that the plant height, shoot number, ear length, grain number and grain yield of irrigated wheat were decreased only upto a distance of 4 m from the tree line of *Acacia nilotica.* Whereas, under rainfed conditions it showed suppression of mustard upto a distance of 26 m from the tree line (12). As the distance from the tree is increased, the adverse effect is gradually decreased.

Kumar *et al.* (31) reported various changes in the physical and chemical characteristics of the soil under *A. nilotica* due to accumulation of plant litter. Plant stands of pearlmillet and clusterbean were markedly inhibited upto 2.5 m and were not affected beyond 30 m from the trees (16). Sundarmoorthy and Kalra (63) found no differences between the properties of soil beneath *A. tortilis* and *Prosopis cineraria.* This indicates that reduction in crop growth was due to chemical inhibition.

2.3. CASUARINA

The leachates of *Casuarina equisetifolia* significantly decreased the germination, plumule and radicle growth and dry matter of rice and cowpea and the inhibition increased with an increase in the soaking period of leachates (26) (Table 3).

Table 3. Effect of *Casuarina equisetifolia* (C E) and *Tectona grandis* (TG) leaf leachates on germination, length and dry matter of plumule and radicle of rice and cowpea at 11 days after sowing

Leachate soaking period (h)	Germination (%)		Plumule length (cm)		Radicle length (cm)		Dry matter / plant (mg)			
							Plumule		Radicle	
	CE	TG	CE	TG	CE	TG	CE	TG	CE	TG
					Rice					
Control	99.6	98.33	6.68	6.60	8.92	9.06	18.85	19.72	28.99	31.04
4	96.3	98.65	75.8	81.51	19.5	72.84	83.1	95.08	38.9	83.28
8	96.3	89.15	75.0	81.01	16.3	70.86	74.0	93.81	38.3	75.80
12	93.9	88.48	68.1	79.86	13.5	62.03	68.3	84.12	34.2	53.02
16	92.9	86.78	65.1	72.42	13.1	57.83	62.7	54.25	30.3	46.68
20	84.9	84.06	62.8	72.42	11.8	56.71	59.2	53.24	19.6	38.90
24	82.9	78.98	58.2	68.34	8.63	53.75	43.4	50.05	12.7	36.75
SE	1.04	00.98	0.05	0.14	0.03	0.14	0.04	0.83	0.04	1.46
CD at 5%	3.19	3.04	0.16	0.44	0.05	0.44	0.12	2.57	0.10	4.31
					Cow pea					
Control	99.3	99.33	16.64	16.06	6.81	5.95	310.6	296.40	85.1	81.23
4	95.6	100.0	41.4	34.86	55.5	128.23	84.8	67.96	89.7	108.42
8	94.6	97.31	35.8	32.00	50.0	116.97	77.2	56.08	88.9	96.07
12	93.2	95.97	35.4	31.75	51.9	105.11	72.0	52.77	86.2	90.47
16	93.2	59.10	34.4	31.69	51.1	82.35	70.8	42.04	70.8	80.00
20	89.9	40.91	34.1	29.45	49.1	67.39	69.9	35.09	66.4	73.10
24	84.56	36.21	33.6	24.71	46.6	58.65	64.9	28.70	52.1	71.17
SE	2.32	2.20	0.14	0.76	0.04	0.33	0.38	9.02	0.27	2.47
CD at 5%	7.23	6.25	0.44	2.34	0.12	1.03	1.16	27.81	0.83	7.62

Source : 25, 26

The reduction was more pronounced in radicle length than plumule length in rice but reverse was true in cowpea. Similar report was made in sorghum and sunflower. In another study, cowpea was found to be less susceptible than sorghum to the phytotoxins in leaf extracts of casuarina (58). Similar effect was observed with cowpea radicle development by casuarina leaf leachates (66). The plumule dry matter of rice and cowpea was reduced by same magnitude by the leaf leachates, whereas, the dry matter of radicle of rice reduced more than cowpea. The variability in inhibition by leaf leachates may be due to differential response of cells to allelochemicals. Five allelochemicals *viz.*, Kaempferol – 3 – alpha – rhamnoside, quercetin – 3 alphaarboside, luteolin – 3.4-dimethoxy-7 beta-rhamnoside, kaempferol-3-beta dirhamnoside and quercetin-3-beta-glucoside have been isolated from casuarina of which the first three have greater auto-toxicity(14).

2.4. GLYRICIDIA

It is a fast growing shrub used for green manuring in rice culture. Glyricidia aqueous leaf extracts of 4% concentration were more inhibitory to germination of fingermillet, sorghum and sunflower than 2%. However, blackgram, rice and horsegram were least affected. Besides, the radicle and plumule length of all test crops (cowpea, greengram, blackgram, sunflower, horsegram, rice, sorghum, fingermillet, mustard, lablab bean) except groundnut were retarded significantly at both concentrations. The maximum inhibition in radicle growth occurred in blackgram, greengram, fingermillet, sorghum and rice, whereas inhibition in plumule growth was noticed in greengram, sunflower, rice, sorghum, fingermillet and horsegram. On the contrary, the leaf extracts promoted the radicle and plumule growth in groundnut (39) (Table 4).

Table 4. Effect of *Glyricidia maculata* leaf extract on the germination and seedling growth of test crops

Test crop	Germination (%)			Radicle length (cm)			Plumule length (cm)		
	Distilled water (Control)	Leaf extract 2%	Leaf extract 4%	Distilled water (Control)	Leaf extract 2%	Leaf extract 4%	Distilled water (Control)	Leaf extract 2%	Leaf extract 4%
Cowpea	97	-18.0	-30.0	2.6	-30.8	-38.8	3.8	-44.7	-60.5
Greengram	97	-14.0	-25.8	6.2	-62.9	-77.4	4.9	-61.2	-77.6
Blackgram	97	-5.2	-12.4	5.2	-51.9	-75.0	3.7	-51.3	-56.8
Sunflower	84	-42.8	-47.6	1.6	-37.5	-37.5	3.0	-66.6	-66.6
Horsegram	93	-7.5	-10.8	3.1	-22.9	-14.3	4.4	-61.3	-36.4
Rice	97	-6.2	-12.4	3.2	-50.0	-65.6	4.1	-68.3	-60.9
Sorghum	56	-50.0	-64.2	3.3	-48.5	-69.6	3.7	-43.2	-72.9
Fingermillet	85	-67.0	-84.7	1.8	-55.6	-61.1	2.3	-69.6	-39.1
Mustard	96	-21.9	-64.4	4.0	-47.5	-25.0	4.7	-31.9	-74.4
Groundnut	75	-21.3	-45.3	2.7	-3.7	-25.9	2.8	+46.4	+10.7
Lablab bean	97	-36.0	-39.2	4.2	-33.3	-33.3	4.8	-35.4	-52.1
Mean	85			3.4			3.8		

Positive (+) or negative (-) indicate effect over distilled water control, Source : 39

In the laboratory studies, leachates of *G. sepium* prunings significantly reduced the growth of maize seedlings but in the field conditions mulching with *G. sepium* prunings caused leaf chlorosis in maize and cowpea seedlings and number of chloratic leaves increased with increase in mulch rate. However, this phytotoxic effect did not reduce growth of maize and cowpea seedlings in the field and addition of the mulch one week before planting eliminated the phytotoxic effects on maize (70). Toxic compounds from leaves of *G. sepium* including gallic acid, protocatechuic acid, *P*-hydroxybenzoic acid, gentisic acid, *ß*-resorcyclic acid, vanillic acid, syringic acid, 3-coumaric acid, ferulic acid, sinapinic acids, coumarin and myricetin were identified (45).

2.5. *LEUCAENA LEUCOCEPHALA*

Out of leaf extract made by 3 successive extractions, the first extract had an inhibitory effect and second stimulatory effect on the radicle growth of rice seedlings (6). The leachates from *Leucaena leucocephala* growing medium (soil) did not effect the growth of cocoa seedlings, unlike that of *Cassia siamea* and *Adenanthera microsperma* trees (42). The rice germination was not affected by extracts of soil as well as decomposed leaves below the canopy of *Leucaena leucocephala* (30).

Contrary to above, aqueous extracts of *Leucaena leucocephala* seeds, litter and soil showed significant phytotoxic effects on rice and lettuce (10). Ten phytotoxins namely mimosine, quercetin, gallic, protocatechuic, *p*-hydroxybenzoic, *p*-hydroxyphenylacetic, vanillic, ferulic, caffeic and *p*- coumaric acids were identified. The mature leaves of *L.*

leucocephala contains about 5 % mimosine on dry weight basis and its amount varies with the varieties.

2.6. *PROSOPIS* SPECIES

The fruit and seed extracts of *P. juliflora* considerably reduces the germination and seedling growth of maize and wheat as compared to root, leaf and flower extracts. The extract of soil from underneath the canopy inhibits the germination but did not influence the seedling growth of test crops (35). Aqueous leachates of leaves, stems, litter and soil from *P. juliflora* inhibited the germination of pearlmillet and sesame but leaf leachates were most inhibitory (64). The litter and soil from below *P. juliflora* inhibited the seed germination and seedling growth of *Crotolaria medicaginea* and *Indigofera linnaei* while that from below *P. cineraria* was stimulatory (21). Fresh leaf leachates of both the tree species were not inhibitory. Competition for growth resources was not responsible for the bare areas observed around *P. juliflora* trees.

Soil incorporation of 2 % litter (w/w) significantly reduced the germination of blackgram, sorghum and not sunflower, but the extracts of fresh leaves and root did not affect germination and dry matter production. The allelopathic effect was due to phenolic compounds present in leaf litter and persisted in soil for two weeks. Hence, germination and dry matter production was not inhibited when blackgram was sown one week after leaf incorporation, whereas, the inhibition continued for two weeks in sorghum (7) (Table-5).

There was no effect of *Prosopis cineraria* on chickpea germination, growth and plant population, hence, the crop grew near to the trunk of the tree (12). On the contrary *P. juliflora* contains germination and growth inhibitors in its litter which restrict the establishment and growth of ground cover (51).

2.7. *TECTONA GRANDIS*

The effect of leaf leachates of teak on germination, plumule, radicle length and drymatter in rice and cowpea indicated that increase in the soaking period (till 24 h) significantly decreased the germination and plumule length more in cowpea (minimum 36 and 24% of control) than in rice (minimum 79 and 68% of control). The radicle length and dry matter of cowpea was stimulated more than control at lower concentrations but not in rice (25).

2.8. *TERMINALIA* SPECIES

The aqueous extracts of dried leaves and fruit pulp of *Terminalia belerica, T. chebula* and *T. tomentosa* inhibited the germination and radicle elongation of *Eleusine coracana, Brassica compestris, Hordeum vulgare* and *Glycine max* (46). Aqueous leachates of *Terminalia tomentosa* leaves had no adverse effect on germination of rice and cowpea but growth of rice plumules and radicle and cowpea plumules was decreased with increasing leachate concentration, while cowpea radicle growth increased by high concentration of leachates (19). Fruit leachates of *Terminalia pallida* at 100 % reduced germination (upto 92%) and seedling growth (8).

Table 5. Effect of different times of sowing to mitigate the adverse influence of allelopathy of *Prosopis juliflora* leaf litter on the germination and dry matter of test crops (%)

Treatments	Sorghum		Blackgram	
	1% LI	2% LI	1% LI	2% LI
Seed germination (%) at 10 DAS				
Sowing at LI	-48.4	-73.0	-63.9	-71.4
Sowing at 1 WALI	-51.6	-71.9	+0.7	+5.5
Sowing at 2 WALI	+4.6	+4.8	-4.3	-2.9
Sowing at 3 WALI	-4.3	0.0	+1.4	+0.5
Sowing at 4 WALI	-10.5	-4.5	+1.3	+3.2
Sowing at 5 WALI	-1.49	+2.6	0	-4.7
Sowing at 6 WALI	-3.1	+2.1	+2.6	+0.4
SEd	4.5	3.8	3.7	2.6
CD (P=0.05)	9.6	8.3	7.8	5.5
Dry matter (g / pot) at 30 DAS				
Sowing at LI	+76.60	-40.4	-48.9	-87.4
Sowing at 1 WALI	+38.60	-44.3	+94.8	+124.2
Sowing at 2 WALI	+119.2	+157.7	+10.3	+2.78
Sowing at 3 WALI	+109.7	+146.5	+36.1	+107.6
Sowing at 4 WALI	+112.0	+156.5	+29.4	+33.9
Sowing at 5 WALI	+115.8	+150.9	+55.2	+56.7
Sowing at 6 WALI	+88.70	+110.5	+69.3	+55.3
SEd	1.91	4.39	2.33	3.88
CD (P=0.05)	3.98	9.16	4.86	8.09

Source : 7
LI :Leaf incorporation, WALI : Week after leaf incorporation
(-) = Inhibition (+) = Stimulation; DAS = Days after sowing.

2.9 MISCELLANEOUS TREE SPECIES

2.9.1. *Bark leachates*

The bark leachates from six year old stands of *Acacia nilotica, Acacia leucophloes, Ailanthus excelsa, Bauhinia sp, Casuarina equisetifolia, Dalbergia sissoo, Eucalyptus tereticornis* and *Tectona grandis* inhibited germination of maize, pigeonpea and sesame. However, *Ailanthus excelsa* was most inhibitory to germination, radicle growth and plumule elongation followed by *Acacia nilotica* and *Dalbergia sissoo*. The leachates inhibited the sesame the most and the pigeonpea the least (67) (Table- 6).

2.9.2. *Leaf leachates*

The leaf leachates of 4 tree species inhibited wheat seed germination and were ranked in the order : *L . leucocephala* > *P. deltoides* > *E. tereticornis* > *B. variegata.* Ten per cent leaf leachates of *Leucaena leucocephala* and *P. deltoides* inhibited shoot and root growth but 5% leachates of all species (except leucaena) increased shoot growth (70). Glucosides and polyphenols were the major allelochemicals in these species except *Eucalyptus tereticornis* (62). Similar reports were made in wheat (70). In another

Table 6. Response Index Values for three receiver crops exposed to bark leachates of multipurpose trees

Bark leachates	Maize			Pigeonpea			Sesame		
	Germi-nation	Shoot length	Root length	Germi-nation	Shoot length	Root length	Germi-nation	Shoot length	Root length
A. nilotica	-0.43	-0.56	-0.60	-0.30	-0.33	-0.57	-0.59	-0.60	-0.88
A. leucophloea	-0.19	-0.55	-0.14	-0.41	-0.53	-0.60	-0.28	-0.36	-0.51
Ailanthus	-0.61	-0.71	-0.82	-0.57	-0.82	-0.63	-0.57	-0.66	-0.78
Bauhinia	-0.10	-0.50	-0.10	-0.09	-0.28	+0.003	-0.11	-0.11	-0.42
Casuarina	-0.13	-0.51	-0.06	-0.10	-0.16	-0.05	-0.42	-0.38	-0.76
Eucalyptus	-0.14	-0.55	-0.04	-0.24	-0.48	-0.33	-0.12	-0.05	-0.49
Dalbergia	-0.41	-0.75	-0.41	-0.35	-0.46	-0.54	-0.57	-0.70	-0.75
Tectona	-0.26	-0.62	-0.21	-0.04	-0.07	-0.02	-0.25	-0.04	-0.76
CD (P=0.05)	0.19	0.12	0.12	0.18	0.20	0.12	0.17	0.14	0.21

Source : 67

experiment the aqueous leaf extracts of *Azadirachta indica, Terminalia arjuna, Dalbergia sissoo, Albizzia lebbek, Sesbania grandiflora, Acacia auriculiformis and Leucaena leucocephala* inhibited the germination in wheat, paddy and chickpea but maximum inhibition was caused by *Sesbania grandiflora* and *Acacia auriculiformis.* The magnitude of reduction was least in wheat and most in chickpea (46).

The leaf leachates of *Eucalyptus camaldulensis, Acacia nilotica, Derris indica, Cassia siamea* and *Sesbania grandiflora* had no inhibitory effect on germination of rice, except for 16 h extract of pongamia, whereas, groundnut germination was inhibited by all tree species at all concentrations. For *Eleusine coracana* all leachates inhibited germination except 8 h extract of *Acacia nilotica* (41).The leaf extracts of *Tectona grandis, Albizia procera* and *Acacia nilotica* showed stimulatory effect on germination, growth, chlorophyll, protein, carbohydrate and proline content of soybean (71).

2.10. CHEMICAL COMPOUNDS PRESENT IN MAJOR TREE SPECIES

The tree species known to produce chemical compounds (67) and some of them are listed in Table-7.

Table 7: Chemical compounds present in major tree species

Tree species	Plant part	Chemical compound
***ACACIA* (Mimosaceae)**		
A. auriculiformis	heart wood	3- isimeric flavan-3, 4-diols-(-) tetracidin, isotetracacidin and mollisacacidin
A. chundra	gum autohydrolysis	Rhamnose and arabinose.
A. farnesiana	seeds	N-acetyl-L-djenkolic acid.
A. melanoxylon	flowers	Quercetin-3-galactoside.
A. mollissima	bark	Catechin and alacatechin,
	heart wood	Fustin and fisetin.
A. nilotica	gum	Arabinobiose-2-beta-L-arabinofuranosyl- L-arabinose.
	bark	Quercetin, gallic acid, catechin & epicatechin.
	degraded gum	Galactose residues and glucuronic

		crop residues.
A. planifrons	bark & heart wood	d-catechin.
A. torta	bark	Aliphatic hydroxy compound triterpenoid and stigmasterol acacinoid acid and lupeon on hydrolysis.
***ACALYPHA* (Euphorbiaceae)**		
A. indica		Sitosterol.
***AILANTHUS* (Simaroubaceae)**		
A. altissima	seeds	2,6-dimethoxyquinone ailantone, chaparrinone and ailanthone.
	bark	Fatty acid- C22 acids.
A. excelsa	bark	Malanthine.
A. triphysa	trunk exudate	New triterpenoids baricol, epoxymala baricol and malabaricanediol.
***ATROPA* (Solanaceae)**		
A. belladonna	leaves	Hyoscymine, hyoscine flavone glycosides, kaempferol-7-monoglucoside and quercetin-7-monoglucoside.
***AZADIRACHTA* (Meliaceae)**		
A. indica	trunk bark	Nimbin, nimbidin and nimbidol paraffinalcohol-sugiol, oxophenol-nimbiol nimbosterol.
	seed oil	Tetranortri terpenoids, azadiradione, azadirone, meliantriol, meldenin and vipinin.
	leaves	New meliacin-nimbolide, quercetin beta-sitosterol.
	seed, & bark	Deacetylnimbin.
	wood oil	Cyclolucalenol, 24-methylenecycloartanol and β-sitosterol.
	trunk wood	Nimbolin and nimbolin-B.
***BAUHINIA* (Caesalpiniaceae)**		
B. purpurea	fresh flowers	Isoquercetin, astragalin and quercetin.
B. tomentosa	flowers	Isoquercitin and quercitin.
CASSIA		
C. absus	seed	β-sitosterol, b-glucoside, D-galacto-D-mannan having galactose and mannose.
C. autifolia	leaves	New glycoside.
	pods	Aglycone-sennidin, gluco-aloeemodin and rhein-8-monoglycoside.
C. angustifolia	bark	Fistucacidin.
C. auriculata	pod	Rheingluciside.
C. fistula	sap wood	Lupeol, beta-itosterol, hexa cosanol and leucoanthocyanidin-5.
C. obtusifolia	seeds	Obtusin, chrysoobtusin and aurantioobtusin.

C. siamea	root bark	Lupeol, chrysophanol and cassiamin-A.
	trunk bark	Dianthraquinone pigments-cassianin, siameanin, siameanin and lupenone.
	heart wood	Tetrahydrostilbene.
C. tora	heart wood	Sterol and myricyl alcohol.
***DALBERGIA* (Papilionaceae)**		
D. latifolia	heart wood & bark	Latifolin, dalbergin, dalberginone and red quinone-dalatinone.
	bark	Hentriacantone, beta-sitosterol, sucrose and tannins.
D. sissoo	flowers	Biochanin-A, tetrarigenin, new isoflavone-5, 4-dihydroxy-6, 7-dimethoxy isoflavone.
	immature green pods	7-O-methyl tectorigenin. biochanin-A-7-glucoside, isoflavone-O-glycoside.
***EUCALYPTUS* (Myrtaceae)**		
E. globulus	leaves	Cis- and trans- pinocarvel, alpha-pinene and unsaturated alpha-ketone.
***GLYRICIDIA* (Papilionaceae)**		
G. maculata	leaves	Kaempfterol-3-O-rhamno-glucoside.
G. sepium		
***MELIA* (Meliaceae)**		
M. azadirachta	leaves	Kaempfterol-3-L-rhamno-D-glucoside and rutin.
M. azedarach	fruits	Melianone and melianol.
	bark	Meliantriol and kuline.
	tree bark	Tetracyclin triterpenoids-kulactone, kulolactone and kulinone.
***PROSOPIS* (Mimosaceae)**		
P. cineraria	flowers	Patulitrin glucoside.
P. glandulosa	aerial parts	Prosopol, prosopenol and beta-sitosterol.
***TAMARINDUS* (Caesalpiniaceae)**		
T. indica	leaves, bark & flowers	Hordenine.
***TECTONA* (Verbenaceae)**		
T. grandis	heart wood	Anthraquinone-2-carboxylic acid and anthraquinone-2-
carboxaldehyde.		
		essential oil tectol and dehydrotectol tectoquinone.

***TEPHROSIA* (Papilionaceae)**		
T. maxima	roots	Maximum substance A with 3,4,6,7-tetra methoxy isoflavone.
***TERMINALIA* (Combretaceae)**		
T. alata *T. tomentosa*		β-sitosterol, oleanolic acid, Arjunolic acid, triterpene acid and tomentosic acid.

Source : 47

3. Physico-chemical properties of soils

Litter of agroforestry tree species improve soil physical properties. Study of effect of different tree species on water holding capacity (WHC) of soil below the tree indicated that *Acacia nilotica* recorded the highest WHC (56.6%) followed by *Casuarina equisetifolia* (52.1%), *Eucalyptus tereticornis* (50.5%) whereas the barren site had the least (35.1%)(65) (Table 8).

Table 8 :Effect of a few tree species on physical and chemical properties of soil

Tree species	Water holding capacity (%)	Organic carbon (%)	Available nutrients (kg / ha)		
			N	P	K
Acacia leucophloea	46.2	0.16	191	6.02	240
Acacia nilotica	56.6	0.16	197	5.32	385
Albizzia lebbeck	48.1	0.29	190	5.36	327
Azadirachta indica	40.8	0.15	189	6.29	274
Casuarina equisetifolia	52.1	0.44	194	6.37	219
Eucalyptus camaldulensis	46.6	0.21	184	5.26	273
Eucalyptus tereticornis	50.5	0.42	190	4.51	281
Hardwickia binata	52.5	0.38	188	5.00	287
Leucaena leucocephala	48.2	0.19	188	4.29	237
Prosopis juliflora	41.5	0.25	190	4.37	224
Barren site	35.1	0.15	183	3.88	284
CD at 0.05	4.4	0.19	6.6	0.92	NS

Source : 65

Agroforestry trees showed considerable reduction in runoff (fallow 25.3%, cassava 18.9% and eucalyptus plus cassava 7.6%) and soil loss (fallow 11.9 t/ha, cassava 7.3 t/ha and eucalyptus plus cassava 2.9 t/ha) (40). The litter of various trees is known to contribute various nutrients. The litter fall of *Prosopis juliflora* was 3187 kg/ha/yr. which contributed 16.2, 6.1, 34.6, 72.8 and 29.5 kg/ha N, P, K, Ca and Mg, respectively, and thus improved the soil fertility (65) (Table- 9).

Understorey crops also influence the soil fertility. Eucalyptus forest soils without understorey recorded less organic carbon, total nitrogen, mineral nitrogen (NH_4^+ NO_3^-), exchangeable cations (Ca^{2+}, Mg^{2+}, K^+, Na^+) and available phosphorus compared to eucalyptus forest soils with legume (*A. pulchella*) understorey (36).

Table 9 :Nutrients input through litterfall in different tree species

Tree species	Litter fall (kg / ha / yr)	Nutrient return (kg / ha)				
		N	P	K	Ca	Mg
A. nilotica	720	3.3	1.0	4.6	8.4	4.1
A. leucophloea	702	3.1	0.5	2.9	7.8	3.4
A. planifrons	1085	5.5	0.7	2.5	18.6	6.1
C. equisetifolia	2840	23.6	2.2	11.7	74.0	26.3
E. tereticornis	1620	15.3	2.8	14.0	29.2	10.4
H. binata	1792	8.3	1.1	7.7	31.1	10.4
L. leucocephala	3905	20.8	13.1	16.7	67.2	30.7
P. juliflora	3187	16.2	6.1	34.6	72.8	29.5
CD ($P < 0.5$)	619	3.3	1.1	4.5	11.9	4.8

Source : 65

The dehrydrogenase, urease and phosphatase activity increased with increase in soil organic carbon and nitrogen status. The dehydrogenase activity depends on the chemical composition of the litter (32). The dehydrogenase activity followed the order leucaena < silver oak < acacia < eucalyptus litter incorporated treatments. The variation was due to native carbon source in these spp. Leucaena with lowest C/N ratio, phenols and lignins recorded the highest activity showing faster decomposition. Whereas, eucalyptus with higher C/N ratio, phenols and lignins recorded the least activity showing slower decomposition.

Analysis of soil samples under the leaf litter of 11 tree species for the population of total bacteria, fungi, actinomycetes and free-living N_2 – fixers reported both stimulatory and inhibitory effects on different groups of soil microorganisms (1). While bamboo and *Acacia auriculiformis* showed inhibitory effect on total bacteria, *Eucalyptus sp* was inhibitory to actinomycetes and bamboo, *Acacia catechu* and *Dalbergia sissoo* were inhibitory to soil fungi. However, *Casuarina equisetifolia*, *Pinus fatula*, *Eucalyptus sp.* and *Tectona grandis* exhibited maximum stimulatory effect on bacteria, actinomycetes, fungi and free-living N_2 – fixers respectively. All the tree species stimulated the azotobacter in soil. Nodulation of clusterbean by three rhizobium strains was reported to be reduced by 10 to 15 % when rhizosphere soil of *Acacia nilotica* was used for experimentation compared to non-rhizosphere control soil (17) (Table 10).

Table 10 :Nodulation and nitrogen fixation in clusterbean as influenced by Allelopathic effect of *Acacia nilotica* (pot trial)

Rhizobium	No. of nodules per plant		Nodule fresh weight per plant (mg)		ARA* µmoles C_2H_2 reduced g / hr	
	Non allelopathic soil	Allelopathic soil	Non allelopathic soil	Allelopathic soil	Non allelopathic soil	Allelopathic soil
G-3	42	36	445	386	2.096	1.653
IC-3100	30	27	394	323	1.433	1.292
PS-1	31	27	265	221	1.204	0.999
Control	29	16	139	129	0.706	0.571
CD 5%	8.36	5.97	28.34	49.82	0.246	0.477

*ARA = Acetylene reduction assay, Source : 17

In Chillum jar experiments with 10 and 20 % allelopathic soil extracts, the nodulation of clusterbean was inhibited by 41 and 54 % whereas that with 10 and 20 % combined plant extract, the inhibition of nodulation was 46 and 65 % respectively. Inhibition of nitrogenase activity varied from 12-39 % with both types of extracts (17) (Table- 11).

Table 11: Effect of combined plant extract of *Acacia nilotica* and allelopathic soil extract on nodulation and nitrogen fixation of clusterbean in Chillum jar

Treatment	Number of nodules per plant		Nodule fresh weight per plant (mg)		ARA* µmoles C_2H_2 reduced g / h	
	G-3	G-23	G-3	G-23	G-3	G-23
Control	49	46	408	465	2.963	2.270
Allelopathic soil extract (10%)	29	28	286	237	2.318	1.746
Allelopathic soil extract (20%)	23	22	224	183	2.080	1.610
Combined plant extract (10%)	21	25	262	213	2.249	1.689
Combined plant extract (20%)	17	19	208	168	2.006	1.461
CD at 5%	7.88	9.59	67.63	45.76	0.504	0.443

*ARA = Acetylene reduction assay, Source : 17

4. Problems and future areas of research

Trees produce large quantity of litter which accumulates on the soil surface and allelochemicals released from it influence the companion/understorey crops in agroforestry. This influence is very much selective and varies with the tree – crop combinations. Therefore, it is essential to determine the allelopathic compatibility of crops with trees which may influence the total productivity of an agroforestry system.

4.1. PROBLEMS

I. Isolation, identification and characterisation of allelochemicals released by various components in Agroforestry system is most difficult and time consuming.
II. Partitioning of the allelopathic effect from resource competition is essentially impossible in natural systems. Any experimental design to separate allelopathy and resource competition would create conditions that will never occur in nature (23).
III. The *in-vitro* results cannot be perfectly true (validity) under *in-vivo* conditions.

4.2. FUTURE LINES OF RESEARCH

I. To understand the existing allelopathic interactions in the traditional tree-crop associations.
II. To determine the allelopathic potential of local and exotic tree spp. and their influence on crops in agroforestry.

III. Design techniques to separate the allelopathic effects from resource competition, to understand the mechanisms involved in tree -crop associations which may help in formulating new agroforestry models suitable for different agroclimatic conditions.

IV. To conduct more *in-vivo* studies than *in-vitro* studies.

V. To monitor the production of allelochemicals in different plant parts in relation to environmental changes, age of crop, persistence of chemicals in the soil and other related interactions.

References

1. Alagawadi, A.R. and Patil, V.C. (1991). Soil microflora under forest tree species. *National Seminar on Ecology of Western Ghats*. Dharwad , Karnataka, India: Karnataka University.
2. Alves, P.L.C.A., Toledo, R.E.B., Pavani, M.C. M.D. and Pitelli, R.A. (1996). The allelopathic phenomenon in *Eucalyptus citriodora* Hook. In *Abstracts I World Congress on Allelopathy, The Science of Future*,Abstract No A-11 pp 66,September 16-20,Cadiz, Spain.
3. Bhaskar, V. and Dasappa (1986). Ground flora in eucalyptus plantation of different ages. In *Eucalyptus in India - Past,Present and Future*. (Eds., J.K. Sharma, C.T.S. Nair, S. Kedarnath and S. Konda) pp 213-224. Peechi, Kerala, India: Kerala Forestry Institute India.
4. Bhaskar, V, Arali, A. and Shankaralingappa, B.C. (1992). Alleviation of allelopathic effects of eucalyptus hybrid through litter burning. *In proceedings of First National Symposium Allelopathy in Agroecosystems* (Eds., P. Tauro and S.S. Narwal). pp 118-119. Hisar, India: Indian Society of Allelopathy, Haryana Agricultural University.
5. Blaise, D., Tyagi, P.C., Khola, O.P.S. and Ahlawat, S.P. (1997). Effects of eucalyptus on wheat, maize and cowpea. *Allelopathy Journal* **4** : 344-344.
6. Chaturvedi, O.P. and Jha, A.N. (1992). Studies on allelopathic potential of important agroforestry species. *Forestry Ecology and Management* **53** : 91-98.
7. Chellamuthu, V, Balasubramanian, T.N., Rajarajan, A. and Palaniappan, S.P. (1997). Allelopathic influence of *Prosopis juliflora* (Swartz) DC. in field crops. *Allelopathy Journal* **4** : 291-302.
8. Chetty, M.C. and Rao, K.N. (1994). Effect of *Terminalia pallida* fruit leachates on the biochemical composition of seedlings in *Vigna radiata*. *Allelopathy Journal* **1** : 58-65.
9. Chou, C.H. (1986). The role of allelopathy in subtropical agroecosystem in Taiwan. In *The Science of Allelopathy* (Eds., A.R. Putnam and C.S. Tang), pp. 57-73. New York : Wiley Interscience.
10. Chou, C.H. and Kuo, Y.L. (1986). Allelopathic exclusion of understorey by *Leucaena leucocephala* (Lam.) de wit. *Journal of Chemical Ecology* **12** : 1431-1448.
11. Craig, J.A. and Sacnalo, M. (1988). Effect of paddy bund planted eucalyptus trees on the performance of fields. *Working paper* No. 1. The Phar, Khan Kaen, Thailand: Nerad Project.
12. Dalal, M.R., Dahiya, D.S. Sarmah, M.K. and Narwal, S.S. (1992). Suppression effects of arid -zone trees on plant stand and growth of crops. *See Reference No*.4. pp 132 - 135.
13. Datta, S.C. and Chatterjee, A.K. (1980). Pollution by plants. In *proceedings of the Symposium on Environmental Pollution and Toxicology*. pp. 195-214. New Delhi: Today and Tommorrow Publishers.
14. Deng - Langbui, Chuittua, K. , Shiming, L., Deng, L.G., Kong, L.H. and Luo, S.M. (1996). Isolation and identification of extract from *Casuarina equisetifolia* branchlets and its allelopathic action on seedling growth of casuarina. *Chinese Journal of Applied Ecology* 7: 145-149.
15. Devasagayam, M. M. and Ebenezar, E.G. (1996). Allelopathic effect of eucalyptus on arable crops. *Journal of Ecotoxicology and Environmental Monitoring* **6** :73-75.
16. Duhan, J.S. and Lakshminarayana, K. (1995). Allelopathic effect of *Acacia nilotica* on cereal and legume crops grown in fields. *Allelopathy Journal* **2** : 93-98.
17. Duhan, J.S., Sharma, P.K. and Lakshminarayana, K. (1994). Allelopathic effect of *Acacia nilotica* on nodulation and nitrogen fixation by Rhizobium (cowpea). *Allelopathy Journal* **1** :47 - 52.
18. Fogg, G.E. (1947). *Proceedings of Royal Society Ser. B*. London. **134** : 503 - 522.
19. Gaynar, D.G. and Jadhav, B.B. (1992). Allelopathic effects of *Terminalia tomentosa* Roth. on germination of rice and cowpea. *Indian Journal of Plant Physiology* **35** : 288-291.
20. Gliessman, S.R. and Muller, C.H. (1978). The allelopathic mechanisms of dominance in bracken (*Pteridium aquilinum*) in southern California. *Journal of Chemistry and Ecology* **4** : 337-362.

21. Goel, U., Nathawat, G.S. and Goel, U. (1990). Relative effects of *Prosopis juliflora* Swartz and *Prosopis cineraria* (L.) Druce on seed germination and seedling growth. *Acta Botanica* – India **18** : 76-79.
22. Igboanugo, A.B.I. (1988). Effects of eucalyptus trees on yield of *Vigna unguiculata* L. Walp., *Zea mays* L. and *Sorghum bicolor* L. *Agriculture, Ecosystem and Environment* **24** : 453-458.
23. Inderjit, Moral, R. Del, and Del, Moral, R. (1997). Is separating resource competition from allelopathy realistic? *Botanical Review* **63** : 221-230.
24. Jadhav, B.B. and Gaynar, D.G. (1992). Allelopathic effects of *Acacia auriculiformis* A. Cunn. on germination of rice and cowpea. *Indian Journal of Plant Physiology* **35** : 86-89.
25. Jadhav, B.B. and Gaynar, D.G. (1994). Effect of *Tectona grandis* (L.) leaf leachates on rice and cowpea. *Allelopathy Journal* **1** : 66-69.
26. Jadhav, B.B. and Gayanar, D.G. (1995). Effect of *Casuarina equisetifolia* J.R. leaf litter leachates on germination and seedling growth of rice and cowpea. *Allelopathy Journal* **2** : 105-108.
27. King, K.F.S. and Chandler, M.T. (1998). *The Waste Lands.* Nairobi, Kenya: International Council for Research in Agroforestry .
28. Kohli, R.K., Singh, D. and Verma, R.C. (1990). Influence of eucalyptus shelterbelt on winter season agroecosystems. *Agriculture, Ecosystems and Environment* **33**: 23-31.
29. Konar, J. and Kushari, D.P. (1995). Effect of *Eucalyptus globulus* leachates on the growth and diosgenin content of *Costus Speciosus. Allelopathy Journal* **2** : 215-218.
30. Koul, V.K. (1990). The effect of soil beneath *Leucaena leucocephala* and its decomposed leaves on germination of rice. *Leucaena Research Reports* **11**: 54-55.
31. Kumar, R., Kumar, S. and Singh, D. (1992). Impact of *Acacia nilotica* on soil physical and chemical properties. *Nitrogen Fixing Tree Research Reports* **10** : 108-109.
32. Kumari, S. and Devi, S. (1994). Influence of forest litter degradation on soil dehydrogenase activity. *Current Research* **23** : 5-7.
33. Molisch, H. (1937). *Der Enfusslinear Pflanze Aufdie Andere - Allelopathie*, GDR: Fisher, Jena. pp. 130.
34. Muller, C.H. (1974). Allelopathy in the environmental complex, in Handbook of Vegetation Science, Part VI. Vegetation and Environment (Eds., B.R. Strain and W.D. Billings), pp.37-85.The Hague: W. Junk.
35. Noor, M., Salem, U., Khan, M.A., Noor, M. and Salam, U. (1995). Allelopathic effects of *Prosopis juliflora* Swartz. *Journal of Arid Environment* **31** : 83-90.
36. O' Connel (1986). Effect of legume understorey on decomposition and nutrient content of eucalyptus forest litter. *Plant and Soil* **92** : 235-248.
37. Padhy, B. and Khan, P.A. (1996). Physiological and biochemical effects of allelopathic substances of eucalyptus leaves on rice. Abstract No. E-6 . *See Reference No.*2, pp. 170.
38. Padhy, B, Khan, P.A., Acharya, B. and Buxipatra, N.P. (1992). Allelopathic effects of eucalyptus leaves on seed germination and seedlings growth of finger millet . *See Reference No.*4, pp. 102-104.
39. Patil, B.P. (1994). Effects of Glyricidia maculata L. extracts on field crops. Allelopathy Journal 1 : 118-120.
40. Permual, R. and Francis H.J. (1991). Soil fertility improvement using integrated nutrient management in agroforestry. In *Agroforestry, Extension and Training* (Eds., G. Perumal, J. Oiver, and R. Annamalai),pp. 39-41. Coimbatore, India: Tamil Nadu Agricultural University.
41. Phlomina, N.S. and Srivasuki, K.P. (1996). Allelopathic studies on agroforestry species : Effect of leaf leachates on seed germination of crop plants. *Indian Journal of Forestry* **19** : 45-53.
42. Prawoto, A.A. (1997). A study of allelopathy effect of *Cassia siamea* and *Adenanthera microsperma* to cocoa. *Pelita Perkebunan* **13** (1) :16-23.
43. Puri, S. (1992). The allelopathic effects of *Eucalyptus tereticornis* in an agroforestry system. *See Reference No.* 4, pp. 101.
44. Rafique, M.(1995). Allelopathic effects of *Eucalyptus camaldulensis* on wheat variety "Inqualab - 91". *Pakistan Journal of Forestry* **45** : 21-24.
45. Ramamoorthy, M. and Paliwal, K. (1993). Allelopathic compounds in leaves of *Glyricidia sepium* (Jacq.) Kunth ex Walp. and its effect on *Sorghum vulgare* L. *Journal of Chemical Ecology* **19** : 1691-1701.
46. Rao, O.P., Saxena, A.K. and Singh, B.P. (1994). Allelopathic effects of certain agroforestry tree species on the germination of wheat, paddy and gram. *Annals of Forestry* **2** : 60-64.
47. Rastogi, R.P. and Mehrotra, B.M. (1991). In *Compendium of Indian Medicinal Plants* (Vol.II),pp 832, Lucknow :Central Drug Research Institute.
48. Rice, E.L. (1984). *Allelopathy* .IInd edition. New York: Academic Press.
49. Rizvi, S.J.H., Sinha, R.C. and Rizvi, V. (1990). Implications of mimosine allelopathy in Agroforestry. *Proceedings XIX IUFRO World Congress on Forestry*, Montreal, Canada. **2** : 22-27.
50. Sajjan, A.S., Hiremath, S.M. and Badanur, V.P. (1997). Allelopathic effect of eucalyptus on getminatioon of seedling characteristics in some crops. *Annals of Plant Physiology* **11** : 54-57.

51. Sankhla, N., Baxi, D. and Chatterjee, W.N. (1965). Eco-physiological studies on arid zone plants. 1. Phytotoxic effects of aqueous extracts of mesquite (*Prosopis juliflora* D.C). *Current Science* **34**: 612-614.
52. Saxena, A. and Sharma, A.K. (1996). Allelopathic potential of *Acacia tortilis* in agroforestry systems of arid region. *Allelopathy Journal* **3** : 81-84.
53. Saxena, A., Singh, D.V. and Joshi N.L. (1996). Allelopathy in Agroecosystems. *Field Crop Abstracts* **49** : 891-899.
54. Sharma, K.K. (1992). Wheat cultivation in association with *Acacia nilotica* (L.) Wild ex. Del. field bund plantation-a case study. *Agroforestry Systems* **17** :881-884.
55. Shivanna, L.R. and Prasanna, K.T. (1992). Allelopathic effects of eucalyptus ; an assessment on the response of agricultural crops. *Myforest* **28** : 131-137.
56. Singh, D., and Kohli, R.K. (1992). Impact of *Eucalyptus tereticornis* Sm. shelterbelts on crops. *Agroforestry Systems* **20** : 253-266.
57. Singh, D., Kohli, R.K. and Saxena, D.B. (1991). Effect of eucalyptus oil on germination and growth of *Phaseolus aureus* Roxb. *Plant and Soil* **137** : 223-227.
58. Singh, H.K. (1993). Allelopathic effects of tree species on crop plants. In *Proceedings and Recommendations of National Workshop on Farm Forestry Management*, (Ed., T.H. Badu), pp. 18-22 Bhopal : Indian Institute of Forest Management.
59. Singh, P.N., Gupta, S., Azmi, S. and Singh, G. (1992). Allelopathic effects of *Eucalyptus citriodora* Hook leaf litter leachate on germination and seedling growth of wheat, chickpea and toria. *See Reference No.*4, pp. 105-107.
60. Siqueira, J.O., Nair, M.G., Hammerschmidt, R. and Safir, G.R. (1991). Significance of phenolic compounds in plant soil microbial systems. *Critical Reviews in Plant Sciences* **10** :63-128.
61. Sivagurunathan, M., Devi, S.G. and Ramasamy, K. (1997). Allelopathic compounds in *Eucalyptus* spp. *Allelopathy Journal* **4** : 313-320.
62. Srivastava, S., Tripathi, A. K. and Jain, A. (1996). Phytochemical investigation of some forest tree species for their allelopathic potential . *Indian Journal of Ecology* **23** : 21-28.
63. Sundramoorthy, S. and Kalra, A. (1991). Allelopathic potential of wheat (*Triticum aestivum*) straw on selected weed species. *Weed Science* **30** :495-497.
64. Sundramoorthy, S., Kalra, N. and Chawan, D.D. (1995). Allelopathy and *Prosopis juliflora* provenance Israel in semi-arid agrforestry systems. *Indian Journal of Forestry* **18** : 214-220.
65. Surendran, (1991). Recent technology development in agroforestry. In *Agroforestry, Extension and Training* (Eds., G.Perumal, J.Oliver and R. Annamalai,). Pp. 12-16. Coimbatore, India: Tamil Nadu Agricultural University, India.
66. Suresh, K.K. and VinayRai, R.S. (1987). Studies on the allelopathic effects of some agroforestry trees on crops. *International Tree Crop Journal* **4**: 109-115.
67. Swaminathan, C. (1996). Effect of leachates of multipurpose trees on germination and seedling growth of maize, pigeonpea and sesame. *Allelopathy Journal* **3** : 77-80.
68. Swaminathan, C., VinayRai, R.S. and Suresh, K.K. (1989). Allelopathic proclivities of *Acacia nilotica* (L.). *Journal of Tropical Forest Science* **2** : 56-60.
69. Tian, G. And Kang, B.T.(1994). Evaluation of phytotoxic effects of *Gliricidia sepium* (Jacq) Walp. pruning on maize and cowpea seedlings. *Agroforestry Systems* **26**:3:249-254.
70. Tripathi, S., Tripathi, A. and Banerjee, S.K. (1996). Comparative study of chemical nature and role of leaf and root leachates on productivity. *Advances in Forestry Research in India* **14** : 183-194.
71. Tripathi, S., Tripathi, A., Kori, D.C. and Tiwari, S. (1998). Effect of tree leaves aqueous extracts on germination and seedling growth of soybean. *Allelopathy Journal* **5** : 75-82.
72. Tukey, H.B. Jr.(1970). The leaching of substances from plants. *Annual Review of Plant Physiology* **21** : 305-324.
73. Waller, G.R. (1989). Allelochemical action of some natural products. In *Phytochemical Ecology : Allelochemicals, Mycotoxin, Insect Pheromones and Allomones*. (Eds., C.H. Chou and G.R. Waller), pp. 129-154. Taipei, Taiwan : Institute of Botany, Academia Sinica.
74. Whittaker, R.H. (1970). The biochemical ecology of higher plants. In *Chemical Ecology* (Eds.,E. Sondheimer and J.B. Simeone), pp 43-70. New York: Academic Press.

15

Tree allelopathy in agroforestry

M. PARAMATHMA*, J. A. AMAL and M. RAJKUMAR
Forest College and Research Institute
Tamil Nadu Agricultural University, Mettupalayam 641 301, India

Abstract

Introduction of trees in agro-ecosystems have helped us to withstand the present pace of deforestation. The farmers have planted trees in good agricultural lands and have grown crops in association with trees. In these agroforestry systems, the yields of crops may be reduced due to competition and allelopathic interactions with trees. Allelopathic studies have revealed that ideal tree-crop combinations with mutual stimulatory effects optimise the yields of both components. The paper describes the stimulatory and inhibitory interactions of few agroforestry trees.

Key words: Agroforestry, allelochemicals ,allelopathy, crops, trees.

1. Introduction

To overcome the harmful effects of increased deforestation, massive tree planting programmes have become necessity. This is feasible, only when tree cultivation is coupled with agriculture. Hence, the agroforestry has been understood as a sustainable land use system, to meet the food and wood requirements of the people and to protect our earth from environmental hazards.

Research programmes focusing on the sustainable agroforestry development are being practised World over. Allelopathy is a new and potential field and has been recognized as thrust area of research in agroforestry. Allelopathy is referred to the addition of toxins from one or more plant species in the association and their inhibitory/ stimulatory effects on receiver plants. All tree species are rich in secondary metabolites (allelochemicals), which influences the pattern and distribution of adjoining flora and also regulates the nitrification, nitrogen fixation and ecosystem balance. Suppressive effects of allelochemicals have been reported for indigenous trees used in agroforestry systems in Himalayas (4,23). The possible allelopathic interactions between tree species and crop plants in agroforestry systems have received little attention. Evaluation of allelopathic potential of the trees on crop plant (effects on seed germination, growth and metabolism) is necessary to understand the interactions between crop plants and the tree component in agroforestry systems. Hence, there is a strong need to screen those tree

* Correspondence author

S.S. Narwal et al. (eds.), Allelopathy in Ecological Agriculture and Forestry, 229–235.

species, which are intended as agroforestry components against their negative allelopathic interactions with agricultural crops. Keeping in view such factors, this paper reviews the salient findings of allelopathic interactions of trees in agroforestry systems.

2. Allelochemicals

The allelochemicals in trees are stored in the cells in a bound form such as water soluble glycosides, polymers including tannins, lignins and salts. These are released upon cleavage through enzymatic reactions or due to environmental stress. Such chemicals may be released through volatalization, leaf or stem leachates, root exudates and decomposition of plant residues (36, 68).

I. Volatalization: Allelochemicals may volatalize from the tree leaves and may be absorbed from the atmosphere by the crop plants. The trees rich in terpenoids, release them continuously into the atmosphere. Since volatalization is enhanced in hot weather, hence this phenomenon is common in arid regions. Volatile terpenes present in *Eucalyptus* exerts maximum allelopathic potential of the tree (38). *Eucalyptus* oil vapours get obsorbed on the soil surface, thus adversely affected the germination, shoot and root growth of surface feeding plants.

II. Leaching: Leaching is the removal of water soluble substances from plants during rain, dew, mist, snow and fog (60). Leached substances may reach the soil beneath, or aerial parts of ground vegetation beneath the trees intercepts those chemicals or absorb them through roots. These substances cause adverse effects in receiver species. At times, the phytotoxic allelochemicals of leaves and barks of the tree dissolved in the leachates may cause injury to the donor plants.

III. Root exudates: A variety of compounds are exuded from the roots, which may influence the growth of microorganisms and associated plants. The rhizosphere of some trees is rich in allelochemicals which are injurious to underneath or neighbourhood vegetation (26). For example, the allelochemicals found in the *Eucalyptus* rhizosphere are injurious to the nearby vegetation (45).

IV. Decomposition of residues: The allelochemicals may be added to the rhizosphere through death and decay of fallen plant parts, mainly the leaves and crop residues. In agroforestry, this is the largest source of the allelochemicals. Among the various tree tested, the allelochemicals in the leaves of *Leucaena leucocephala* were found highly phytotoxic (26). Such released allelochemicals may inhibit photosynthesis and or respiratory metabolism.

2.1. MECHANISMS OF ALLELOPATHY

The mechanisms of action of allelochemicals present in trees on the underneath agricultural crops is not clearly understood. Daizy and Kohli (12) reported that allelochemicals of *Eucalyptus* reduced the chlorophyll content of leaves, the cellular respiration rate and the water content of target plants as a response of allelopathy. The contents of protein, RNA and carbohydrates and enzymes (protease, polyphenol oxidase, peroxidase, α - amylase and β - amylase) activities are badly affected (26).

3. Allelopathic effects of trees

3.1. *EUCALYPTUS*

Leachates of fresh leaves, litter, root, bark and seeds of *Eucalyptus citriodora*, *E.globulus* and *E.tereticornis* are found to contain caffeic, coumaric, gallic, gentisic, hydroxybenzoic, syringic and vanillic acids and catechol (50). These chemicals inhibited the growth of blackgram, greengram, rice and tomato. Blaise *et al.* (6) reported that the slow decomposition of *E.tereticornis* roots released the phenolics slowly, thereby, affecting the plant growth at a later phase. The root growth of test plants viz., wheat, maize and cowpea was affected more than the shoot, probably because their roots remain in direct contact with allelochemicals. *Eucalyptus tereticornis* also reduced the yields in many crops (Table-1).

*Table.1.*Inhibitory effect of *Eucalyptus tereticornis* on yield in crop plants

Crops	References
Kidneybean, wheat	6, 30, 43
Barley, potato	14, 15
Pigeonpea, soybean, cowpea	59
Pearlmillect, fingermillet	2, 35, 46
Castor, sunflower	24, 55
Sesbania, sunhemp	11, 25
Jute	1
Groundnut, linseed, rapeseed	26, 52
Egyptian clover, lentil	35, 34
Cauliflower, mustard	3, 25
Radish, fenugreek	5, 31
Brinjal, coriander	13, 47
Maize, sorghum, oats	16, 56
Horsegram, blackgram	14, 47
Desmodium, *Panicum maximum*, ginger grass	35, 37

Similar inhibitory effects of other eucalyptus species on agricultural crops have also been observed. Sorghum and cowpea are relatively tolerant to allelochemicals of *Eucalyptus* (32,37). Bhaskar *et al.* (3) advocated that where the problem of allelopathic effects of *Eucalyptus* have been observed in agroforestry systems, the leaf litter may be skooped, piled and then burnt. This reduced the allelopathic potential and increase the plant growth through nutrient recycling as well.

3.2. LEGUMINOUS TREES

3.2.1. *Leucaena leucocephala*

The presence of mimosine in leucaena increase its allelopathic potential. Its inhibitory effects have been observed in many crops viz., Greengram, rye, wheat, maize, pea, mustard, sorghum, sunflower, cowpea, rice, kidney bean, lettuce, turnip, radish, carrot and *Bidens pilosa* (8,9,22,28,29,40,51,55,58, 61).

3.2.2. *Acacias*

Swaminathan (56) reported that among the test crops, sesame was found most sensitive to the presence of tannins in *Acacia nilotica.* Allelopathic effect of *A.nilotica* has been visually observed on pearlmillet and clusterbean (18). Other annuals, whose growth has been found to be inhibited are: tomato, egg plant, okra, chillies, sunflower, sorghum, cotton, mustard, lentil, fieldbean, clusterbean, cowpea and wheat (13,42,43,48,49,57). The allelopathic effects of *A.nilotica* on *Rhizobia* (17) inhibited the nodulation and nitrogen fixation in cowpea.

The inhibition in plant growth and germination of pearlmillet has been observed, perhaps due to the presence of chemical substances in soil, root, leaf and inflorescence extracts of *A.tortilis* (41). They suggested that phytotoxic effects could be more pronounced in low or erratic rainfall areas, where phytotoxic substances may accumulate in top soil over the time, as the rainfall is insufficient to remove them by runoff or leaching. This tree specie affects sesame, clusterbean, greengram and mothbean (53,54).

3.2.3. *Prosopis spp*

The allelopathic effect of *Prosopis juliflora* leaf litter has been proved in pot culture experiments (7). However sowing of blackgram and *P. juliflora* at 1 week after leaf litter incorporation of *P. juliflora* and sowing of sorghum after 2 weeks are the suitable agronomic management strategies, against the allelopathic effect of its leaf litter. Chouhan *et al.* (10) found that under silvipastoral system, the grasses associated with *P.juliflora* did not show its effect during the early stages of tree growth, however, at later stages it exerted maximum inhibitory allelopathic effect. Till now, the beneficial effects in crops have been recorded with *P. cineraria* only (13).

3.3. MISCELLANEOUS TREES

Swaminathan (56) proved that the bark extracts of *Ailanthus excelsa, Bauhinia* spp., *Casuarina equisetifolia, Dalbergia sissoo, Tectona grandis, Acacia nilotica* and *A. leucophloea* inhibited the germination of maize, pigeonpea and sesame. The magnitude of reduction in germination was maximum in maize from the extracts of *Ailanthus excelsa* and minimum in pigeonpea from the extracts of *Bauhinia* spp. Patil (33) indicated the strong negative influences of aqueous extracts of *Gliricidia maculata* on germination and growth of fingermillet, sorghum and mustard. Survival and growth of weed species viz., *Boeviria diffusa* and *Phyllanthus niruri* and corn (*Zea maize*) were inhibited by leaf extracts of *Albizia amara* (1,19).

Jadhav and Gaynar (20, 21) reported that leaf leachates of *Tectona grandis* and *Casuarina equisetifolia* significantly affected germination and dry matter production in rice and cowpea. The inhibition was more in rice than in cowpea.

4. Conclusions and future areas of researh

Although beneficial effects of agroforestry trees on crops are always preferred, but most of the tree species exert an adverse effect on the growth of understorey vegetation, however, the effect varies with trees and crops spp. Allelopathy studies may identify the best tree-crop combinations for agroforestry systems. In high rainfall areas or under managed agroforestry system, the residual allelopathic effects are not visible and have not been studied. But in rainfed or low rainfall areas such harmful effects are pronounced. Allelopathic studies in agroforestry are limited to Asian countries. Hence, in view of great scope of agroforestry systems, in future allelopathy research may be done to identify better tree-crop combinations for these systems.

4.1. FUTURE AREAS OF RESEARCH

I. While studying the effect of allelochemicals on crop growth, the cumulative effects of all environmental factors often confuse the scientists. Hence, the isolation, identification and characterization of allelochemicals from the tree parts and their influence on crops in bioassays need to be worked out.

II The results of *in-vitro* studies, may be verified in field studies with the same combination of crop components.

III. To determine the compatability of agroforestry tree species with field crops, screening of field crops and their varieties need to be done region-wise.

IV. The promising germplasm strains of each tree may be screened for crop compatability.

V. The effect of microbial population under tree canopy may be studied on crop growth. If the microflora composition is influenced notably by allelochemicals, the condition needs to be further clarified.

VI. Multipurpose, nitrogen fixing trees may be evaluated for their allelopathic potential on crop yield and those trees which are stimulatory or least harmfull to crop yield may be recommended for agroforestry.

VII. Allelopathic effects of trees on crops may be evaluated for each agro-climatic zone.

VIII. To provide initiative to farmers to take up successful agroforestry practices, the tree-crop combinations may be first worked out. Crops most tolerant to different tree species may be screened for their allelopathic effects and may be recommended for cultivation in agroforestry. In-depth allelopathic studies may reveal the best tree-crop combinations for different agro-climatic zones.

References

1. Bansal, G.L. (1988). Allelopathic effects of aqueous extracts of stem and leaves of tree species on the germination of some crops and weeds. In *Trends in Tree Science*, pp. 118 –123. Solan, India. Indian Society of Tree Scientists.

2. Basu, P.R., Kapoor,K. S., Nath, S. and Banerjee, S. K. (1987). An assessment on the response of agricultural crops growing near *E.tereticornis*. *Indian Journal of Forestry* **10**: 267-271.

3. Bhaskar, V., Arali, A. and Shankaralingappa, B.C. (1992). Alleviation of allelophathic effects of *E.tereticornis* through litter burning. *Proceedings First National Symposium on Allelopathy in Agroecosystems*. (Eds., P.Tauro and S.S.Narwal), pp. 118-119. Hisar, India : Indian Society of Allelopathy.

4. Bhatt, B.P. and Todaria, N.P. (1990). Studies on the allelopathic effects of some agroforestry tree crops of Garhwal Himalaya. *Agroforestry systems* **12**: 251 -255.
5. Bisla, S.S., Nandal, D.P.S. and Narwal, S.S. (1992). Influence of aqueous leaf extracts of *Eucalyptus* and poplar on the germination and seedling growth of winter crops. *See Reference No.*3. pp.95-97.
6. Blaise, D. Smith, V. and Fowden, R.(1997). Effects of *Eucalyptus* on wheat, maize and cowpea. *Allelopathy Journal* **4**: 341-344.
7. Chellamuthu, V., Balasubramanian, T.N., Rajarajan, A. and Palaniappan, S.P. (1997). Allelopathic influence of *Prosopis juliflora* (Swartz) DC on field crops. *Allelopathy Journal* **4**: 291-302.
8. Chou, C.H. (1983). Allelopathy in agroecosystems in Taiwan. In *Allelochemicals and Pheromones.* (Eds., C.H. Chou and G.R. Waller), pp. 27 –64. Taipei, Taiwan : Institute of Botany, Academia Sinica.
9. Chou, C.H. and Kuo,Y.L. (1986). Allelopathic exclusion of understorey by *Leucaena leucocephala* (Lam.) de wit. *Journal of Chemical Ecology* **12**: 1431-1448.
10. Chouhan, G.S., Mathur, A.N., Bhandari,M.M.C. and Jat, P.K. (1992). Allelopathic effect of some tree species on associated grasses under silvipastrol system. *See Reference No.*3. pp. 130-131.
11. Craig, I.A., and Saenalo,M. (1988). Effects of paddy-bund planted *Eucalyptus* trees on the performance of field crops. *Working Paper No.*1. The Phar, Khon Kaen, Thailand : Nerad Project.
12. Daizy, R., and Kohi, R.K. (1991). Fresh water is not an appropriate relation unit for chlorophyll content. Experience from experiments on effects of herbicides and allelopathic substances. *Photosynthetica* **25**: 655-657.
13. Dalal, M.R., Dahiya, D.S., Sarmah,M.K., and Narwal,S.S. (1992). Suppression effects of arid zone trees on plant stand and growth of crops. *See Reference No.*3. pp. 132-35.
14. Del Moral, R. and Muller, C.H. (1970). The allelopathic effect of *E.camaldulensis. American Midland Naturalist* **83**: 254-282.
15. Del Moral, R. and Muller, C.H. (1969). Fog drip: a mechanism of toxin transport from *E.globulus. Bulletin of Torrey Botany Club* **96**: 467-475.
16. Devasagayam, M.M. and Ebenezar, E.G. (1996). Allelopathic effect of Eucalyptus on arable crops. *Journal of Ecotoxicology and Environmental Monitoring* **6**: 73-75.
17. Duhan, J.S., Sharma, P.K. and Lakshminarayana, K. (1994). Allelopathic effect of *Acacia nilotica* on nodulation and nitrogen fixation by *Rhizobium* (Cowpea). *Allelopathy Journal* **1**: 47-52.
18. Duhan, J.S. and Lakshminarayana, K. (1995). Allelopathic effect of *Acacia nilotica* on cereal and legume crops grown in fields. *Allelopathy Journal* **2**: 93-98.
19. Eyini, M., Maheswari, A.U., Chandra, T. and Jayakumar, M. (1996). Allelopathic effects of leguminous plants leaf extracts on some weeds and corn. *Allelopathy Journal* **3**: 85 - 88.
20. Jadhav, B.B. and Gaynar, D.G. (1994). Effect of *Tectona grandis* (L.) leaf leachates on rice and cowpea. *Allelopathy Journal* **1**: 66-69.
21. Jadhav, B.B. and Gaynar, D.G. (1995). Effect of *Casuarina equisetifolia* J.R. leaf litter leachates on germination and seedling growth of rice and cowpea. *Allelopathy Journal* **2**: 105-108.
22. Joshi, P.C. and Prakash, O. (1992). Allelopathic effects of litter extract of some tree species on germination and seedling growth of agricultural crops. *See Refence No.*3. pp. 127 - 128.
23. Kaletha, M.S., Bhatt, B.P. and Todaria, N.P. (1996). Tree-crop interactions in traditional agroforestry systems of Garhwal Himalaya. 1. Phytotoxic effects of farm trees on food crops. *Allelopathy Journal* **3**: 247 - 250.
24. Kohli, R.K. (1987). Eucalyptus - an antisocial tree for social forestry. In *Social Forestry for Rural Development*, (Eds., P.K. Khosla and R.K. Kohli), pp. 235 – 241, Solan, India : Indian Society of Tree Scientist.
25. Kohli, R.K. (1990). *Allelopathic Properties of Eucalyptus.* Project Report MAB DoEn Project, pp. 199.
26. Kohli, R.K., Kaur, K., Chaudhary, P., Kumari, A. and Saxena, D.B. (1987). Negative aspects of *Eucalyptus* farming. In *Agroforestry for Rural Needs*, (Eds., P.K. Khosla and D.K. Khurana), pp. 225 – 241, Solan, India : Indian Society of Tree Scientists.
27. Kohli, R.K., Singh, D. and Verma, R.C. (1990). Influence of *Eucalyptus* shelterbelt on winter season agroecosystems. *Agricultural Ecosystems and Environment* **33**: 23-31.
28. Koul, V.K., Raina, A., Khanna, Y.P., Tickoo, M.L. and Singh, H. (1991). Evaluation of allelopathic influence of certain farm grown tree species on rice (*Oryza sativa* L. cv PC - 19). *Indian Journal of Forestry* **14**: 54-57.
29. Kuo, Y.L., Chou, C.H. and Hu, T.W. (1983). Allelopathic potential of *Leucaena leucocephala. In Allelochemicals and Pheromones*. (Eds., C.H. Chou and G.R. Waller), pp.107-117. Taipei : Taiwan, Institute of Botany, Academic Sinica.
30. Melkania, N.P. (1984). Influence of leaf leachates of certain woody species on agricultural crops. *Indian Journal of Ecology* **11**: 82-86.

31. Nandal, D.P.S., Bisla, S.S. and Narwal, S.S. (1992). Allelopathic influence of *Eucalyptus* and poplar leaf extracts on the germination and seeding growth of winter vegetables. *See Reference No.* 3. pp. 98-100.
32. Narwal, S.S. and Sarmah, M.K. (1992). Suppression effect of *E. tereticornis* on the field crops. *See Reference No.* 3. pp.111-113.
33. Patil, B.P. (1994). Effects of *Glyricidia maculata* L. extracts on field crop. *Allelopathy Journal* **1**: 118-120.
34. Puri, S. (1992). The allelopathic effects of *E. tereticornis* in an agroforestry system. *See Reference No.* 3. pp.101.
35. Puri, S. and Khara, A. (1991). Allelopathic effects of *Eucalyptus* on *Phaseolus vulgaris* seedlings. *International Tree Crops Journal* **6**: 287-293.
36. Putnam, A.R. (1985). Volatalization of leachates. *Weed physiology* (Ed., S.O.Duke). Vol.**1** : 131-135, Boca Raton , Florida, USA : CRC Press.
37. Rao, N.S. and Reddy, P.C. (1984). Studies on the inhibitory effects of *Eucalyptus hybrid* leaf extracts on the germination of certain food crops. *Indian Forester* **110:** 218-222.
38. Rice, E.L. (1984). *Allelopathy.* Second edition. London : Academic press, pp.422.
39. Rishi, A. and Dhillion, M.S. (1997). Allelopathic influence of *L. leucocephala* on growth of oats. *Indian Forester.* **123**: 831 -833.
40. Rizvi, S.J.H., Sinha,R.C. and Rizvi, V. (1990). Implications of mimosine allelopathy in agroforestry. *Proceedings 19th World Congress on Forestry* **2**: 22-27.
41. Saxena, A. and Sharma, A.K. (1996). Allelopathic potential of *Acacia tortilis* in agroforestry systems of arid regions. *Allelopathy Journal* **3**: 81-84.
42. Sharma, K.K. (1992). Wheat cultivation in association with *Acacia nilotica* (L.) wild ex Del. field bund plantation - a case study. *Agroforestry systems* **17**: 43-51.
43. Sharma, K.M.S., Dhillon, M.S. and Dhingra, K.K. (1967). Presence of germination inhibitors in the leaf leachates of some farm grown trees. *Indian Forester* **113:** 816-820.
44. Shivanna, L.K., Prassanna, K.T. and Mumtaz, J. (1992). Inhibitory allelopathic effects of Eucalyptus: An assessment on the response of agricultural crops. *See Reference No.*3. pp. 108 -110.
45. Singh, D. and Kohli, R.K. (1992a). Reasons of poor underfloor vegetation of *Eucalyptus. See Reference No.*3. pp.114-117.
46. Singh, D.and Kohli, R.K. (1992b). Impact of *Eucalytpus tereticornis* S.M. shelterbelts on crops. *Agroforestry systems* **20:** 253-266.
47. Singh, P.N., Gupta, S., Azmi, S. and Singh, G. (1992). Allelopathic effects of *Eucalyptus citriodora* Hook, leaf litter leachate on germination and seedling growth of wheat, chickpea and *toria. See Reference No.*3. pp.105-107.
48. Singh, R.P. and Nandal, D.P.S. (1993). Allelopathic effects of aqueous leaf extracts of important agroforestry tree species on some fodder crops. *Forage Research* **19**: 59-61.
49. Singh, R.P. and Nandal, D.P.S. (1993). Allelopathic effects of *Eucalyptus* and *Leucaena* leaf litter on the germination and seedling growth of some fodder crops. *Forage Research* **19:** 13-16.
50. Sivagurunathan, M., Sumitra Devi, G. and Ramasamy, K. (1997). Allelopathic compounds in *Eucalyptus* spp. *Allelopathy Journal* **4**: 312-320.
51. Smith, I.K., and Fowden, C.J. (1966). A study of mimosine toxicity in plants. *Journal of Experimental Botany* **17**: 750-761.
52. Srinivasan, K., Ramasan, M. and Shanta, R. (1990). Tolerance of pulse crops to allelochemicals of tree species. *Indian Journal of Pulses Research* **3**: 40-44.
53. Sundramoorthy, S., and Kalra, A. (1991). Allelopathic potential of *Acacia tortilis* plantation in Indian desert. *Annals of Arid Zone* **30**: 259-266.
54. Sundramoorthy, S., Kalra, N. and Sen, D.N. (1992). Allelopathic potential of *Acacia tortilis* on seed germination and seedling growth of some legumes. *See Reference No.*3. pp.123-124.
55. Suresh, K.K. and Vinay, Rai, R.S. (1987). Studies on the allelopathic effects of some agroforestry tree crops. *International Tree Crops Journal* **4**: 109-115.
56. Swaminathan, C. (1996). Effect of bark leachates of multipurpose trees on germination and seedling growth of maize, pigeonpea and sesame. *Allelopathy Journal* **3**: 77-80.
57. Swaminathan, C., Vinay, Rai. R.S. and Suresh,K.K. (1989). Allelopathic potentials of *Acacia nilotica* (L) Wills ex. Del. *Journal of Tropical Forest Science* **2**: 56-60.
58. Tawata, S. and Hongo, F. (1987). Mimosine allelopathy of leucaena. *Leucaena Research Report* **8**: 40-41.
59. Tomer, G.S. and Srivastava, S.K. (1986). Preliminary studies of rice cultivation in association with trees. *In Agroforestry Systems : A New Challenge*, (Eds., P.K.Khosla and S.Puri,), pp. 207-211. Solan, India : Indian Society of Tree Scientists.
60. Tukey, Jr., H.B. (1970). The leaching of substrates from plants. *Annual Review of Plant Physiology* **21**: 305 - 324.
61. Wilson, W.F. and Bell, E.A. (1979). Amino acids and related compounds as inhibitors of lettuce growth. *Phytochemistry* **18:** 1883-1884.

16

Toxic allelochemicals in leguminous forages and their ecological significance

S. S. PARIHAR
Division of Grassland and Silvipasture
Indian Grassland and Fodder Research Institute, Jhansi-284 003, India

Abstract

Numerous allelochemicals belonging to various groups of secondary metabolites viz. non-protein amino acids, cyanogens, phenolic compounds and alkaloids reported from promising leguminous forages and involved in plant - animal and plant-plant interactions are reviewed. Their ecological significance has also been discussed.

Key words : Alkaloids, amino acids, cyanogens, ecological significance, forages, grasses, legumes, pastures, phenolics, toxic allelochemicals.

1. Introduction

Leguminous forages are superior to grasses and other fodder species in protein and mineral content. Since many tropical pastures are deficient in indigenous legumes, therefore, a search is continuing for new species suitable for introduction as forages. Many leguminous forages (*Leucaena leucocephala, Leucaena escuenta, Indigofera spicata, Canavalia ensiformis, Sesbania* spp., *Gliricidia sepium, Robinia pseudoacacia, Dioclea megacarpa, Dichrostachys* spp., *Acacia* spp. *and Albizzia* spp.) are quite promising. However, the presence of toxic allelochemicals (secondary metabolites) in these species has restricted their full exploitation as forage species. Livestock poisoning by these leguminous forages is result of management errors, range conditions, or type of animal, rather than the presence of toxic substances alone. Further, the toxicity levels in the forages can vary depending upon the environmental factors such as site, temperature, precipitation and light. For example, cyanide in plants varies diurnally or during the drought.

Whittaker (42) proposed the term "allelochemic" and defined it, as a non-nutritional chemical produced by one plant species, which effects the growth, health, behaviour or population biology of another species. A variety of allelochemicals involved in biochemical interactions among living organisms are reported from several plant species, including those

S.S. Narwal et al. (eds.), Allelopathy in Ecological Agriculture and Forestry, 237–243.

of high forage value. These belong to a few major groups of secondary compounds including non-protein amino acids, cyanogens, phenolic compounds (phenolic acids and tannins), glucosinolates and alkaloids. Allelochemicals do not play a primary role in the biochemical activities such as plant growth, development and reproduction. They are of prime importance for the plant which synthesise them, in maintaining the relationships that exist between one plant and another and between plant and other living organism with which they grow. In natural plant communities, allelochemicals are part of the checks and balances that maintain relative stability. Many responses to allelochemicals are, therefore, subtle having developed during the evolution of the plant community (21).

In this communication, review of allelochemicals pertaining to leguminous forages, showing toxicity symptoms in herbivores and allelopathic effects on associated plant species is made. Ecological significance of these compounds to plant-animal and plant-plant interactions is also discussed.

2. Major groups of toxic allelochemicals

2.1. NON-PROTEIN AMINO ACIDS

The toxic non-protein amino acids are not normal components of proteins but are found in many leguminous forages. Presence of these toxic amino acids, is likely to undermine efforts at exploiting many of these plants as feed resources. Important non-protein amino-acids from forage legumes (Table 1) have been reviewed (6, 7, 11, 35, 37). The seed is generally the most concentrated source of these compounds (8).

Table 1. Toxic non-protein amino acids present in leguminous forages

Toxic amino acid	Legume species
Canavanine	*Canavalia ensiformis* (Linn.) DC., *Glyricidia sepium* (Jacq.) Munth ex Walp, *Robinia pseudoacacia* Linn., *Dioclea megacarpa* Ducke, *Indigofera spicata* Forassk, *Sesbania sesban* (Linn.) Merrill
Indospicine	*Indigofera spicata* Forssk
S-methyl-methionine	*Canavalia ensiformis* (Linn.) DC.
Mimosine	*Leucaena leucocephala* (Lamk.) de Wit, *Pithecellobium angulatum* (Grah.) Benth.
Djenkolic acid	*Leucaena esculenta* (Moc & Sesse) Benth, *Pithecellobium lobatum*
Dichrostachinic acid	*Dichrostachys* spp.
Albizzine	*Albizzia* spp.
n-acetyldjenkolic acid	*Acacia* spp.
BCNA	*Vicia sativa* Linn.
ODAP	*Lathyrus sativus* Linn.
L-DOPA	*Mucuna spp.*
DABA	Lathyrus sylvestris Linn.

BCNA (β-cyanoalanine), ODAP (β-N-Oxalyl-a,β-diaminopropionic acid), L-DOPA (L-3,4- dihydroxy phenylalenine), DABA (a, y-diamonobutyric acid).

One of the most ubiquitous toxic amino acids is canavanine found in *Canavalia ensiformis*, *Glyricidia sepium*, *Robinia pseudoacacia*, *Dioclea megacarpa* and *Sesbania sesban*. During the metabolic process in herbivores, canavanine releases the carcinogenic compound canaline, which has been implicated in the development of systemic lupus erythematosus (SLE), an auto-immune disease (15). The diverse distribution of toxic non-protein amino acids is further illustrated by the presence of canavanine and indospicine in *Indigofera spicata* (Table 1).

Another important toxic amino acid is mimosine, present in *L. leucocephala* and *P. angulatum*. Concentration of mimosine in *L. leucocephala* usually ranges between 10-25 g kg $^{-1}$ dry matter, while even higher concentrations may occur in seed. The deleterious properties of mimosine (Table 2) are extensive and include disruption of reproductive processes, teratogenic effects, loss of wool and even death (29, 30). Similar effects may be induced by feeding *L. leucocephala* seeds or foliage to ruminants. Thus, the defleecing effect in sheep may be precipitated by the administration of crystalline mimosine or by feeding of *Leucaena* forage.

Table 2. Effect of mimosine on sheep

Method of administration	Effect
Single intravenous injection (20-350 g kg^{-1} BW*); or intravenous infusion (10-20 g kg-1 BW)	No ill effect
Intravenous infusion (8 g over two days; 77-96 mg kg $^{-1}$ BW)	Reduced wool strength, defleecing
Intravenous infusion (24 g over 4 days; 147 mg kg $^{-1}$ BW)	Reduced wool growth and strength, defleecing, loss of appetite, organ damage, death
Single oral dose (450-600 mg kg $^{-1}$ BW)	Shedding of fleece

*BW = body weight, source: 29, 30

The toxicity of *Leucaena* in ruminants depends upon the rate and extent of mimosine degradation by the rumen microflora. During the breakdown of mimosine, 3, 4-DHP is synthesised, which is a potent goitrogen, a property not shared by mimosine. However, 3, 4 - DHP may also be produced in the leaf of the legume due to post-harvest enzymatic action. Ruminants consuming *Leucaena* macerate the leaves and add alkaline saliva, providing the optimum conditions favouring autolysis of mimosine. Therefore, significant portion of mimosine may undergo degradation to 3, 4-DHP even before reaching the rumen. Some rumen microbes are also capable of degrading 3, 4-DHP to non-toxic products (19). Ruminants of some tropical regions are believed to possess bacteria that are required for 3, 4-DHP breakdown which may account for the absence of *Leucaena* toxicity in these regions.

The lathyrogenic group of non-protein amino acids mainly consist of β-cyanoalanine (BCNA) and β-N-oxalyl-a, β-diaminopropionic acid (ODAP) in species of *Vicia* and *Lathyrus* (Table 1). *Lathyrus sativus* is a rich source of ODAP, the most potent neurotoxic amino acid, responsible for neo-lathyrism in humans, caused by consumption of *Lathyrus sativus* (25). Deleterious properties of non-protein amino acids cover a wide range (15,36-38).

Mammals, possess a significant capacity to detoxify these toxic allelochemicals. The ability of ruminants to detoxify mimosine is subject of geographical differences in rumen ecology (19). In parts of the tropics, cattle and goats are capable of detoxifying mimosine. Recent biotechnical advances have allowed the transfer of 3,4-DHP (the toxic product of mimosine) degrading bacteria from Hawaiian to Australian cattle.

Detoxification of crop plants containing non-protein amino acids is currently under investigation and plant breeding techniques are used to eliminate these toxic substances. However, it should be recognised that the toxic amino acids may confer insect and fungal resistance, removal may result in agronomic disadvantages. For example, the infestation by psyllid insect has extensively damaged the *Leucaena* crop in some parts of the tropics and attempts to develop low mimosine lines may further increase this problem. Removal of toxic allelochemicals with solvent or aqueous solutions is an effective method. Extraction of seeds of *C. ensiformis* with dilute $KHCO_3$ solution at 80 °C reduces canavanine concentration to negligible levels.

2.2. CYANOGENS

Cyanogenic glycosides are toxic to livestock, because they can be hydrolysed enzymatically to release hydrogen cyanide (HCN). In the intact plant, the hydrolytic enzyme and glycoside remain separate, but if the plant tissue is damaged, the two are brought together and HCN is liberated (18, 23). Hydrogen cyanide is toxic to many organisms and the lethal dose for humans is estimated at 0.5 to 3.5 mg/kg body weight. Ruminants are more susceptible to HCN poisoning than non-ruminants because micro-organisms in the rumen induce greater enzymatic breakdown of cyanogens (20). The ability to produce HCN is widespread in the plant kingdom, but it is more in taxa belonging to *Leguminosae* and *Gramineae*. Important leguminous forages containing cyanogenic glycoside and responsible for toxicity to livestock are *Canavalia ensiformis, Prosopis juliflora, Vicia sativa, Acacia* spp., *Phaseolus* spp. and *Trifolium* spp. (2, 17). Forages containing toxic cyanogenic glycoside will loose most of their cyanogen content when converted into silage and exposure to air may give off large quantities of free HCN. Drying or hay making, chaffing and aeration also reduces cyanogen content (28).

2.3. PHENOLIC COMPOUNDS

Phenolic compounds constitute a wide and diverse group of allelochemicals found in forages. The most common are tannins, a polyphenolic compound, represented in 54% of all Angiospermic genera studied (36). Examples of common leguminous feed containing condensed tannins are common bean (*Phaseolus vulgaris*), faba bean (*Vicia sativa*) and field bean or pea (*Pisum* spp.). Many browse species/fodder trees viz. *Acacia* spp., *Albizia stipulata, Bauhenia variegata, Prosopis cineraria, Leucaena leucocephala,* are also rich in tannins (3, 9, 24). Tannins precipitate the proteins and this reduces palatability and adversely affects availability of digested nutrients (13). However, sometime polyphenolic compounds are considered beneficial in ruminant nutrition, because of the suppressive effect on bloat and also the prevention of excessive degradation of high quality leaf protein in rumen. Detoxification of condensed tannins is possible by heat treatment. Most alkaline treatments using dilute ammonia, potassium carbonate or calcium oxide will reduce tannin content in feeds.

2.4. ALKALOIDS

Alkaloids are basic nitrogenous compounds widely found in nature. The toxic alkaloids reported from forage legumes include quinolizidine alkaloids present in *Lupinus* spp. and pyrrolizidine alkaloids of *Crotalaria* and *Sesbania* spp. (4).

3. Ecological significance

It is now widely accepted that the most secondary metabolites from plants act as protective agents against herbivores and microbes. These plants can be broadly grouped into two categories: those which escape predation by the compounds that are highly toxic and those which rely more on feeding deterrents rather than highly active toxins e.g. grasses and browse species (22).In the first group, legume seeds are large and provide rich nutrition to herbivores and if there are no means of protection, the seeds would be overeaten. The main protective role of toxic amino acids in forage legume seeds, seems to be against insect feeding. For example, the occurrence of L-DOPA in *Mucuna* seeds at 6-9 % dry weight with its relative non-toxicity to mammals indicate a defensive role against insects and microbes (14). In the other group, plants species may be selected due to herbivore impact where the species can survive moderate grazing. In such cases, several non-specific deterrents which deter the herbivore not to graze too heavily on any one species, may have a collective action to favour survival. These allelochemicals can be grouped into two classes: digestibility reducing allelochemicals (DRA) and toxic allelochemicals. The later are mainly nitrogen based. The DRA are the 'poor plants' toxins, since in nitrogen deficient areas, nitrogen rich toxins would be an additional drain on a resource that is already in short supply (5).

Numerous allelochemicals are also reported in plant-plant interactions - allelopathy. Mimosine produced by *L. leucocephala* is implicated in allelopathic interactions in various forestry and agroforestry systems (12, 26, 33). Allelopathic potentials of *L. leucocephala* and *G. sepium* are being utilised to control weeds (1, 16). Cyanogens and phenolic compounds are also involved in allelopathic interactions in several instances (10, 31, 32, 40, 41). Some fodder trees viz., *Acacia nilotica*, *A. tortilis* and *A. catechu* that contain tannins and other allelochemicals are also reported in allelopathic interactions (12, 27, 34, 39). Despite the negative attributes of non-protein amino acids distributed in forage legumes, there is increasing evidence that these toxic properties may be harnessed for therapeutic and other economic purposes. For example, mimosine is endowed with anti-tumour activity and its use as a chemical defleecing agent has also been considered. Also administration of DOPA as a palliative in the management of Parkinson's disease is widely recognised.

Acknowledgement

The author is grateful to Dr. P. S. Tomer, Director, Indian Grassland and Fodder Research Institute, Jhansi for constant encouragement.. I am also grateful to Dr. M.J. Reigosa, University of Vigo, Spain for inviting me to this lecture.

References

1. Anoka, U.A., Akobundu, I.O. and Okonkwo, S.N.C. (1991). Effect of *Gliricidia sepium* and *Leucaena leucocephala* on growth and development of *Imperata cylindrica. Agroforestry System* **16:** 1-12.
2. Bhadoria, B.K. and Gupta, R.K. (1981). A note on HCN content in *Acacia leucophloea. Current Science* **50:** 689-690.
3. Bhadoria, B.K., and Upadhyay, V.S. (1994). Tannin - An antiquality constituent in the leaves of *Acacia catechu. Proceedings of 81st Indian Science Congress Part IV.* pp. 40.
4. Cordell, G.A. (1981). *Introduction to Alkaloids.* New York: John Wiley & Sons.
5. Crawford, R.M.M. (1989). *Studies in Plant Survival.* New York: Blackwell Scientific Publications.
6. D'Mello, J.P.F. (1992). Chemical constraints to the use of tropical legume in animal nutrition. *Animal Feed Science and Technology* **38:** 237-261.
7. D'Mello, J.P.F., Duffus, C.M. and Duffus, J.H. (1991). *Toxic Substances in Crop Plants.* Cambridge: The Royal Society of Chemistry.
8. Dorsaz, A.C., Hostettmann, M. and Hostettmann, G.K.(1988). Molluscicidal saponins from *Sesbania sesban. Planta Medica* **54:** 80-83.
9. Ernst, W.H.O., Kulters, A.T., Nelissen, H.J.M. and Tolsma, D.J. (1991). Seasonal variations in phenolics in several Savannah tree species in Botswana, *Acta Botanica Neolandia* **40:** 63-74.
10. Fisher, N.H. (1991). Plant terpenoids as allelopathic agent. In *Ecological Chemistry and Biochemistry of Plant Terpenoids.* (Eds., J.B. Harborne and F.A. Tomes), pp. 377-398. Oxford: Clarendon Press.
11. Fowden, L. (1981). Non-protein amino acids. In *The Biochemistry of Plants.* (Ed., E.E. Conn), Vol. **7:** 215-247. New York: Academic Press.
12. Gill, A.S. and Parihar, S.S. (1996). Allelopathy in agroforestry system. In *Allelopathy: Field Observation and Methodology.* (Eds. S.S. Narwal and P. Tauro), pp. 121-126. Jodhpur, India: Scientific Publishers.
13. Griffiths, D.W., (1991). Condensed tannins. In *Toxic Substances in Crop Plants.* (Eds., D'Mello, J.P.F, C.M. Duffus and J.H. Duffus), pp. 206-228. Cambridge: The Royal Society of Chemistry.
14. Harborne, J.B (1993). *Introduction to Ecological Biochemistry.* 4th edn. New York: Academic Press.
15. Hegarty, M.P. (1986). Toxic amino acids in foods of animal and man. *Proceedings of Nutrition Society of Australia* **11:** 73-81.
16. Jama, B., Getahum, A. and Ngugi, D.N (1991). Shedding effect of alley cropped *L. leucocephala* on weed biomass and maize yield at Mtwapa, Coast Province, Kenya. *Agroforestry Systems* **13:** 1-11.
17. Jones, D.A. (1988). *Cyanide Compounds in Biology.* Ciba Foundation Symp. 140, Chichester : Wiley.
18. Jones, D.A., Keymer, R.J. and Ellis, W.M. (1978). Cyanogenesis in plants and animal feeding. In *Biochemical Aspects of Plant and Animal Co-evolution.* (Ed., J.B. Harborne), pp. 21-32. New York: Academic press.
19. Jones, R.J. and Lowry J.B. (1984). Australian goats detoxify the goitrogen 3-hydroxy-4 (IH) pyridone (DHP) after rumen infusion from an Indonesian goat. *Experentia* **40:** 1435.
20. Kingsbury, J.M. (1964). *Poisonous Plants of United States and Canada.* Englewood Cliffs, N.J: Prentice Hall.
21. Lovett., J.V. and Levitt, J. (1981). Allelochemicals in future agriculture. In *Biological Husbandry - A Scientific Approach to Organic Farming.* (Ed., B. Stonehouse), pp. 169-180. London: Butterworths.
22. Mac Naughton, S.J. (1983). Physiological and ecological implications of herbivory. In *Encyclopaedia of Plant Physiology.* Vol. **12 C:** 657-677, Berlin: Springer - Verlag.

23. Nahrstedt, A., (1987). Recent development in chemistry, distribution and biology of cyanogenic glycoside. In *Biologically Active Natural Products.* (Eds., K. Hostettmann and P.J. Lea), pp. 213-234. Oxford: Clarendon Press.
24. Negi, S.S., and Makkar, H.P.S. (1989). Influence of method of calculation and length of period of rumen fermentation on the effective degradability of dry matter and nitrogen in some tree forages. *Animal Feed Science and Technology* **26:** 309-322.
25. Padmanaban, G. (1980). Lathyrogens. In *Toxic Constituents of Plant Foodstuffs.* (Ed., I.E. Liener), pp. 239-258. New York: Academic Press.
26. Parihar, S. S. (1990). Prospects of *Leucaena leucocephala* in agroforestry systems - An allelopathic point of view. In *Silvipastoral Systems in India.* (Eds., P.S. Pathak and P. Singh.*)*, pp. 121-125. Jhansi, India: Range Management Society of India, Indian Grassland and Fodder Research Institute.
27. Parihar, S. S. (1994). Allelopathic MPTS and their prospects in agroforestry systems in India. In *Agroforestry Systems for Degraded Lands.* (Eds., Panjab Singh, P. S. Pathak and M.M. Roy), Vol. **I** : 371-378. New Delhi: Oxford & IBH Pub. Co. Pvt. Ltd.
28. Prasad, J. (1987). *Animal Health Problems Associated with Forage Based Diet.* Jhansi, India: Indian Grassland and Fodder Research Institute.
29. Ries, P.J., Tunks, D.A. and Chapman, R.E. (1975a). Effect of mimosine - A potential chemical defleecing agent on wool growth and the skin of the sheep. *Australian Journal of Biological Science* **28:** 69-84.
30. Ries, P.J., Tunks, D.A. and Hegarty, M.P. (1975b). Fate of mimosine administered orally to sheep and its effectiveness as a defleecing agent. *Australian Journal Biological Science* **28:** 495-501.
31. Rice, E.L. (1984) *Allelopathy.* II Edition, New York: Academic Press.
32. Rice, E.L. (1987) Allelopathy : An overview. In *Allelochemicals: Role in Agriculture and Forestry.* (Ed., G.R. Waller), pp. 8-22. Washington DC: American Chemical Society.
33. Rizvi, S.J.H. and Rizvi, V. (1987). Increasing crop productivity in India. In *Allelochemicals: Role in Agriculture and Forestry.* (Ed., G.R. Waller), pp.69-75. Washington DC: American Chemical Society.
34. Rizvi, S.J.H. and Rizvi, V.(1992). *Allelopathy: Basic and Applied Aspects.* London: Chapman and Hall.
35. Rosenthal, G.A. (1982). *Plant Non-Protein Amino and Imino Acids.* New York: Academic Press.
36. Rosenthal, G.A. and Janzen, D.H. (1979). *Herbivores - Their Interactions with Secondary Plant Metabolites.* New York: Academic Press.
37. Roy, D.N. (1981). Toxic amino acids and proteins from lathyrus plants and other leguminous species : A literature review, *Nutrition Abstract Reviews* **51:** 691-710.
38. Seawright, A.A., Hegarty, M.P., James, L.F. and Keeler, R.F. (1985). *Plant Toxicology.* Yeerongpilly, Australia :Queensland Poisonous Plant Committee.
39. Tauro, P. and Narwal, S.S. (1992). *Proceedings 1st National Symposium Allelopathy in Agroecosystems (Agriculture and Forestry).* Hisar, India: Indian Society of Allelopathy, Haryana Agricultural University.
40. Thompson, A.C. (1985). *The Chemistry of Allelopathy.* Washington DC: American Chemical Society.
41. Waller, G.R. (1989). Allelochemical action of some natural products. In *Phytochemical Ecology: Allelochemics, Mycotoxins and Insect Pheromones and Allomones* (Eds., C.H. Chou and G.R. Waller.), Monograph, No. **9** : 155-165. Taipei, R.O.C.: Academia Sinica.
42. Whittaker, R.H., (1970). The biochemical ecology of higher plants. In *Chemical Ecology* (Eds., E. Sondheimer and J.B. Simeone), pp. 47-70. New York: Academic Press.

SECTION V

APPENDICES

INTERNATIONAL ORGANISING COMMITTEE

Prof. R.S. Paroda, Director General (ICAR)	Chairman
Prof. M. Mahadeveppa, Vice Chancellor	Co-Chairman
Prof. C.J. Itnal, Director of Research	Organising Secretary
Prof. S.S. Narwal, National Fellow (Allelopathy)	Co-Orgainising Secretary

Members

Prof. G.S. Dhaliwal, Ludhiana
Prof. S.C. Dutt, Calcutta
Prof. P. Tauro, Mangalore
Prof. R.H. Dilday, USA
Prof. R.E. Hoagland, USA
Prof. M.J. Reigosa, Spain

LOCAL ORGANISING COMMITTEE
(University of Agricultural Sciences, Dharwad –580 001, INDIA)

Dr. M. Mahadeveppa, Vice-Chancellor	Chairman
Dr. C. J. Itnal, Director of Research	Co-Chairman

Members

Dr. H.M. Manjunathiah, Professor
Dr. K. Giri Raj, Senior Scientist (Oilseeds)
Dr. R.V. Hiremath, A.D.R. (HQ)
Dr. Hanchinal, Senior Breeder (Wheat)
Dr. B.S. Nadagoudar, Senior Scientist (Agroforestry)
Dr. I.G. Hiremath, Senior Scientist (Sorghum)
Dr. M.B. Chetti, Professor (Crop Physiology)
Dr. M.R. Kachapur, Coordinator (Extension)
Dr. S.L. Lingappa, Prof. & Head (Entemology)
Dr. J.H. Kulkarni, Prof. & Head (Microbiology)
Dr. (Ms). S. Chimad, Technical Officer

III International Congress
Allelopathy in Ecological Agriculture and Forestry
August 18-21, 1998.

University of Agricultural Sciences, Dharwad- 580 007, India

TECHNICAL PROGRAMME

Date (Day)	Time (h)		
Aug. 17,	16.00-20.00	**REGISTRATION**	
Aug. 18,	08.00-09.30	**REGISTRATION**	
(Tuesday)	09.30-11.00	**INAUGURAL CEREMONY**	
		Chief Guest	: Minister (Cooperation), Karnataka Government
		Chairman	: Prof. M.M. Mahadeveppa
		Co-Chairman	: Prof. C.J. Itnal
	11.00-11.30	**TEA BREAK**	

SYMPOSIA 1. CROP ALLELOPATHY

Chairman : Prof. R.H. Dilday, USA
Co-Chairman : Prof. M. Olofsdotter, Philippines
Rapporteur : Prof. V.K. Garg, India

Date	Time	Abst.No.	Speaker(country) Title of Lecture
Aug. 18 Tuesday	11.30-12.30	1.1	Prof. R.H. DILDAY(USA). Allelopathy as a weed management strategy **(Key Note)**
	12.30- 13.30	1.3	K. OKUNO (China). Variability of allelopathic activity in rice (*Oryza sativa* L.) with water soluble extracts **(Invited)**
	13.00-14.30		**LUNCH BREAK**
	14.30-15.00	1.4	S.M. HASSAN(Egypt). Improving weed management in rice (*Oryza sativa* L.) with water soluble extracts **(Invited)**

15.00- 15.30	1.5	J.D.MATTICE (USA). Chemical aspects of rice allelopathy for weed control **(Invited)**
15.30- 16.00	1.6	T.L. LAVY (USA). Laboratory and greenhouse studies of rice allelopathy to ducksalad **(Invited)**
16.00-16.30		**TEA BREAK**
16.30-17.00	1.7	W.X.LIN (China). Allelopathic potential and its possible modes of action on barnyard grass (*Echinochloa crus-galli*) **(Invited)**
17.00-18.00	1.2	K. MOLDENHAUER (USA). Developing rice (*Oryza sativa* L.) varieties with allelopathic traits **(Plenary)**
18.00-18.15		**Chairman Remarks**

SYMPOSIA 2. CROPPING SYSTEMS

Chairman : Prof. S.M. Hassan (Egypt)
Co-Chairman : Prof. G. Kazinczi (Hungary)
Rapporteur : Prof. M.K. Porwal

18.15-19.15	2.1	S.S. NARWAL. Allelopathic interactions in cropping systems **(Key Note)**
19.15-19.45	2.3	G.DHARAMARAJ. Sunflower allelopathy **(Invited)**
19.45-20.00	2.17	G. KAZINCZI (Hungary). Allelopathic effect of common milkweed (*Ascelpias syriaca* L.) on crops and weeds **(Contributory)**
20.00-20.15	2.19	M.K. PORWAL Allelopathic effects of *Avena fatua* and *Phalaris minor* on wheat **(Contributory)**
20.15-20.30	2.14	S.S. NARWAL. Influence of weeds on crops **(Contributory)**
20.30-20.45	2.24	G.L.BANSAL. Allelopathic potential of *Eupatorium odoratum* on *Lantana camara* **(Contributory)**
20.45-21.00		C.J. ITNAL

	21.00-21.15		**Chairman Remarks**
	21.15-22.15		**DINNER**

CONCURRENT (SIMULTANEOUS) SYMPOSIA

SYMPOSIA 4. ALLELOCHEMICALS AS HERBICIDES

(Venue: University Auditorium)

Chairman : Prof. F.A. Macias (Spain)
Co-Chairman : Prof. S.J. Cutler (USA)
Rapporteur : Prof. M.A.B. Mallik (USA)

Aug. 19 Wednesday	09.00-10.00	4.1	S.J. CUTLER(USA). Plant and microbial compounds as herbicides **(Key Note)**
	10.00-10.30	4.2	F.A. MACIAS (Spain). Chemistry and biological activity of sesquiterpene lectones as models of natural herbicides **(Invited)**
	10.30-11.00	4.5	C.G. KNUDESEN (USA). Discovery of the Triketone class of p- Hydroxyphenylpyruvate Dioxygenase (HPPD)-inhibiting herbicides and their relationship to naturally occurring β-Triketones **(Invited)**
	10.30-11.30		**TEA BREAK**
	11.30-12.00	4.6	F. SUGAWARA (Japan). Phytotoxins from pathogens as potential herbicides **(Invited)**
	12.00-12.30	4.7	S.J.CUTLER (USA). Agrochemicals and pharmaceuticals : Bridging the gap **(Invited)**
	12.30-13.00	4.8	M.A.B. MALLIK (USA). Metabolites from soil microorganisms with herbicidal potential
	13.00-14.30		**LUNCH BREAK**
	14.30-15.00	4.9	M.J.REIGOSA (Spain). Effect of allelochemicals and herbicides on physiological processes **(Invited)**
	15.00-15.30	4.13	A.M.ZOBEL (Canada). Phytochemistry of medicinal plants under stress conditions **(Invited)**

	15.30-16.30	4.4	M. OLOFSDOTTER. Rice allelochemicals a: herbicides **(Plenary)**
	16.30-16.45		**Chairman Remarks**
	16.45-17.15		**TEA BREAK**
	17.30-18.30		General Body Meeting, ISAL
	18.30-20.30		**Cultural Programme**
	20.30-21.30		**DINNER**

SYMPOSIA 3. AGROFORESTRY/FORESTRY

(Venue: College of agriculture Auditorium)

Chairman : Prof. M.J. Reigosa (Spain)
Co-Chairman : Prof. M.S.M. Rawat
Rapporteur : Prof. S.S. Bisla

Aug.19 Wednesday	09.00-10.00	3.1	M.J. REIGOSA (Spain). Allelopathic interactions i forests **(Key Note)**
	10.00-10.30	3.3	A.S. GILL. Allelopathic interactions in agroforestr systems **(Invited)**
	10.30-11.00	3.4	J. JOHN. Prospects of allelopathic research in mult storey cropping systems **(Invited)**
	11.00-11.30		**TEA BREAK**
	11.30-11.45	3.5	C.S. HUNSAL. Scope of allelopathy research i agroforetry systems of South India **(Contributory)**
	11.45-12.00	3.7	S.S.BISLA. Allelopathic potential of nee (*Azadirachta indica* A.Juss) on germination ar seedling growth of summer crops **(Contributory)**
	12.00-12.15	3.25	L. GONZALEZ (Spain). Microbial changes in the sc under *Eucalyptus globulus, Pinus radiata* and *Acac melonoxylon* in Galicia (Spain) **(Contributoy)**
	12.15-12.30	3.28	S.S.PARIHAR. Toxic allelochemicals in forage legume Their ecological significance **(Contributory)**

12.30-13.30	3.2	M.S.M. RAWAT. Chemical interactions in agroforestry systems **(Plenary)**
13.30-14.30		**LUNCH BREAK**

SYMPOSIA 5. PLANT PROTECTION

(Venue : College of Agriculture Auditorium)

Chairman : Prof. R.C. Saxena (Kenya)
Co-Chairman : Prof. N.C. Sukul
Rapporteur : Prof. (Ms) V. Mozumder

14.30-15.30	5.1	R.C. SAXENA. Neem for sustainable management of plant parastic nematodes affecting banana and tomato **(Key Note)**
15.30-15.45	5.2	D. SINGH. Intercropping for insect pest management **(Contributory)**
15.45-16.00	5.3	K.C. BHAGAT. Role of behaviour modifying chemicals (BMC) in integrated pest management **(Contributory)**

TEA SERVED DURING THE LECTURES

16.00-16.15	5.8	S.N. HOLIHOSUR. Allelopathic control of castor semilooper **(Contributory)**
16.15-16.30	5.12	V . MOZUMDER. Neem in nematode management **(Contributory)**
16.30-16.45	5.19	N.C. SUKUL. Antagonistic effect of *Acacia auriculiformis* and *Artemisia cina* against root-knot nematodes **(Contributory)**
16.45-17.00	5.21	A.A. YAZDI (Iran). Effect of resistant catch crops on population of sugarbeet nematode *(Heterodera schachtii* Schmidt) in Khorassan province of Iran **(Contributory)**
17.00-17.15	5.23	M.F. RASTEGAR (Iran). Investigating the allelopathic activity of *Trichoderma* to control

			Fusarium oxysporium f.sp. cumini **(Contributory)**
	17.15-17.30		**Chairman Remarks**
	17.30-18.30		General Body Meetings, ISAL
	18.30-20.30		**Cultural Programme**
	20.30-21.30		**DINNER**

SYMPOSIA 6. ECOLOGICAL AGRICULTURE

Chairman : Prof. C.J. Itnal
Co-Chairman : Prof. S.S. Narwal
Rapporteur : Prof. (Ms) P.Challa

Aug. 20 Thursday	9.00-10.00	6.1	S.S. NARWAL. Allelopathy in ecological agriculture **(Key Note)**
	10.00-10.30	6.2	S. FACKNATH (Mauritius). Allelopathic interactions in soil **(Invited)**
	10.30-11.00	6.3	B.LALLJEE (Mauritius). Allelopathic interactions in soil **(Invited)**
	11.00-11.15	6.6	P. CHALLA. Allelopathic potential of mango leaves for weed management in rose (*Rosa hydrida* Cv Happiness) basins **(Contributory)**
	11.15-11.30		**TEA BREAK**

SYMPOSIA 7. PLENARY (Recommendations Made)

Chairman : Prof. M. Mahadevepha
Co-Chairman : Prof. S.S. Narwal
Rapporteur : Prof. C.S. Hunshal

	11.30-13.00		Presentation of report by Chairmans of Oral and Poster Symposia No. 1-6
	13.00-13.30		**Final recommendations made**
	13.00-14.30		**LUNCH BREAK**
	14.30-19.30		Visit to Research Farm, Dharwad/Hublicity.
Aug. 21, Friday	07.00-21.00		Excursion Trip to Goa

POSTER PRESENTATION

(Venue: Symposia Sites)

Display Date	Symposia No. and Title	Chairman
August 18	1. Crop Allelopathy	Prof. J.D. Mattice, USA
August 18	2. Cropping Systems	Prof. G. Dharamraj, India
August 19	3. Agroforestry	Prof. A.S. Gill, India
August 19	4. Allelochemicals	Prof. C.G. Khundsen, USA
August 19	5. Plant Protection	Prof. D. Singh, India
August 19	6. Ecological Agriculture	Prof. S. Facknath , Mauritus

LIST OF PARTICIPANTS

III INTERNATIONAL CONGRESS

ALLELOPATHY IN ECOLOGICAL AGRICULTURE AND FORESTRY

A. FOREIGN PARTICIPANTS

1. Dr. M. Vicherkova, Masaryk University, Brno, CZECH REPUBLIC
2. Dr. S.M. Hassan, Rice Research and Training Centre, Sakha, Kafr El Sheikh, EGYPT.
3. Dr. K. Ebana, National Institute of Agrobiology, Kannondai 2-1-2, Tsukuba, Ibaraki, JAPAN.
4. Dr. M. Takahashi, JAPAN.
5. Dr. M. Olofsdotter, International Rice Research Institute, P.O. Box 933, 1099 Manila, PHILLIPPINES.
6. Dr. I.T. Breazu Department of Breeding, Grassland Research Institute, 5, Cuculi Street, 2200 Brasov, ROMANIA.
7. Dr. F.A. Macias, Department de Quimica Organica, Facultad de Ciencias, Universidad de Cadiz, Apdo. 40, 11520-Puerto Real, Cadiz, SPAIN.
8. Dr. M.J. Reigosa, Deptt. Biologia Vegetale Ciencia, Faculatad de Ciencias, Universidad de Vigo, Apdo. 874, 36200 Vigo, SPAIN.
9. Dr. Luis Gonzalez, Deptt. Biologia Vegetale Ciencia Facultad de Ciencias, Universidade de Vigo, Apdo. 874, 36200 Vigo, SPAIN.
10. Dr. R.H. Dilday, USDA-ARS, National Rice Germplasm Centre, Stuttgart, AR-72160-0135, USA.
11. Dr. J.D. Mattice, Deptt. of Agronomy, University of Arkansas, Fayetteville, AR 72704, USA.
12. Dr. Terry Lavy, Deptt. of Agronomy, University of Arkansas, Fayetteville, AR 72704, USA.
13. Dr. K. Moldenhauer, Deptt. of Agronomy, University of Arkansas, Fayetteville, AR 72704, USA.
14. Dr. S.J. Cutler, Mercer University, Atlanta, GA 30341, USA.
15. Dr. C.G. Knudsen, Zeneca Agri. Products 12005, 47th Street, Richmond, CA 948089, USA.
16. Dr. Rowshan Mallik, Agricultural Research, Langston University, Langston, OK 73050, USA.
17. Dr. M.A.B. Mallik, Agricultural Research, Langston University, Langston, OK 73050, USA.
18. Dr. R.C. Saxena, International Centre of Insect Physiology, P.O. Box 30, Mbita, KENYA.

B. INDIAN PARTICIPANTS

19. Dr. M. L. Chetty, Department of Botany, S. V. University, Tirupati - 517 502, Andhra Pradesh

20. Dr. A. Kashinath, Organic Chemistry Division- I, Indian Institute of Chemical Technology, Hyderabad - 500 007
21. Dr. B. Venkatiah, Organic Chemistry Division- I, Indian Institute of Chemical Technology, Hyderabad - 500 007
22. Dr. R. R. Venkata Raju, Department of Botany, Sri Krishnadevaraya University, Anantpur - 515 003, Andhra Pradesh
23. Dr.Vijaylakshmi Mojumder, Division of Nematology, Indian Agricultural Research Institute, New Delhi - 110 012
24 Dr. Sushil Kumar, Agricultural Experimental Station, Gujarat Agricultural University, Paria - 396 145, District Valsad, Gujarat
25. Dr. S. S. Bisla, Department of Agroforestry, CCS Haryana Agricultural University, Hisar 125 004
26. Dr. S. S. Narwal, Department of Agronomy, CCS Haryana Agricultural University, Hisar - 125 004
27. Dr. S. B. Puri, Department of Horticulture, CCS Haryana Agricultural University, Hisar - 125 004
28. Dr. S. R. Dhawan, Department of Botany, Government Postgraduate College, Jind - 126 102, Haryana.
29. Dr. G. L. Bansal, Department of Plant Physiology, Himachal Pradesh Krishi Vishvavidyalaya, Palampur - 176 062, Himachal Pradesh
30. Dr. S. R. Ambika, Department of Botany, Jnana Bharati, Bangalore University, Bangalore 560 056, Karnataka, India.
31. Ms. S. Somanatha, Department of Botany, Jnana Bharati, Bangalore University, Bangalore 560 056, Karnataka, India
32. Dr. Parbha Challa, Indian Institute of Horticultural Research, Hessaraghatta, Bangalore - 560 089
33. Dr. S. N. Holihosur, Department of Zoology, Karnataka University, Dharwad - 580 004, Karnataka
34. Dr. C. S. Hunsal, Department of Agronomy, University of Agricutural Sciences, Dharwad - 580 005, Karnataka
35. Dr. S. B. Kalaighati, University of Agricultural Sciences, Dharwad - 585 005, Karnataka
36. Dr. B. S. Nadgoudar, Department of Agroforestry, University of Agricutural Sciences, Dharwad - 580 005, Karnataka
37. Dr. H. V. Nanjappa, Department of Agronomy, University of Agricultural Sciences, Bangalore - 560 065, Karnataka.
38. Dr. B. K. Singhal, Laboratory of Mulberry Physiology, Central Sericultural Research Institute, Mysore - 570 008, Karnataka
39. Dr. Jacob John, Division of Agronomy, Rubber Research Institute of India, Pudupally, Kottayam - 686 009, Kerala
40. Dr. P. M. Shafi, Department of Chemistry, Calicut University, Calicut - 673635, Kerala
41. Dr. P. Oudhia, Department of Agronomy, Indira Gandhi Agricultural University, Raipur - 492 001, Madhya Pradesh

42. Mr. S. S. Chitwadgi, IFS, Bharat Forestry Consultancy, 156/A, Indrapuri, Bhopal - 462 021, Madhya Pradesh
43. Dr. (Ms.) K. B. Pawar, Department of Botany, Shivaji University, Kolhapur - 416 004, Maharashtra.
44. Dr. B. Padhy, Department of Botany, Berhampur University, Berhampur - 760 007, Orissa
45. Dr. M. Singh, Division of Perennial Cropping System, Central Arid Zone Research Institute, Jodhpur - 342 003, Rajasthan
46. Dr. M. K. Porwal, Rajasthan Agricultural University, Agricultural Research Station, Banswara - 327 001, Rajasthan
47. Dr. G. Dharamaraj, Department of Crop Physiology, Tamil Nadu Agricultural University, Coimbatore - 641 003, Tamil Nadu
48. Dr. M. Eyini, Research Department of Botany, Thiagarajah Autonomous College, Madurai - 625 020, Tamil Nadu.
49. Dr. M. Jayakumar, Research Department of Botany, VIINSN College, Virudhunagar - 626001, Tamil Nadu.
50. Dr. M. Manikandan, Research Department of Botany, VHNSN College, Virudhunagar - 626001, Tamil Nadu.
51. Dr. K. Arumugam, Department of Botany, Annamalai University, Annamalainagar - 608 001, Tamil Nadu.
52. Dr. M. Rajangam, Post Graduate Centre in Botany, A.P.A. College of Arts and Culture, Palani - 624 601, Tamil Nadu
53. Dr. A. D. Ravindra, Department of Biology, Rural Institute, Gandhigram - 624 302, Kerala
54. Dr. K. Umapathi, Department of Agronomy, Tamil Nadu Agricultural University, Coimbatore - 641 003, Tamil Nadu.
55. Dr. R. K. Bhadoria, Grasslands and Silvipasture Division, Indian Grassland and Fodder Research Institute, Jhansi 284 003, U. P.
56. Dr. A. S. Gill, Indian Grassland and Fodder Research Institute, Jhansi 284 003, U. P.
57. Dr. Banwari Lal, Division of Crop Production, Indian Grassland and Fodder Research Institute, Jhansi 284 003, U. P.
58. Dr. S. S. Parihar, Grassland Management Division, Indian Grassland and Fodder Research Institute, Jhansi 284 003, U. P.
59. Dr. M. S. M. Rawat, Department of Chemistry, H. N. B. Garhwal University, Srinagar (Garhwal) - 246 174, U. P.
60. Dr. Dwijendra Singh, Entomology Division, Central Institute of Medicinal and Aromatic Plants, Lucknow - 226 015, U. P.
61. Ms. Poonam Mehrotra, Department of Botany, Kumaon University, Nainital - 263 001, U. P.
62. Ms. Beena Joshi, Department of Botany, Kumaon University, Nainital - 263 001, U. P.
63. Dr. J. Konar, Department of Botany, University of Burdwan, Golapbag, Burdwan-713 104, West Bengal

64. Dr. B. Nandi, S. P. Sinha Babu, Department of Zoology, Visva-Bharati University, Santiniketan - 731 235, West Bengal
65. Dr. S. P. Sinha Babu, Department of Zoology, Visva-Bharati University, Santiniketan - 731 235, West Bengal.
66. Dr. N. C. Sukul, Department of Zoology, Visva-Bharati University, Santiniketan - 731 235, West Bengal.
67. N. Kari, Department of Zoology, Karnataka University, Dharwad-580 001, Karnanataka.
68. R. Kulkrani, Department of Zoology, Karnataka University, Dharwad -580 001, Karnanataka.
69. M. Manikandan, V.H.N.S.N. College, Virudhunagar, Tamil Nadu.
70. Dr. S.S. Ali, Indian, Institute of Pulses Research, Kanpur, U.P.
71. Dr. H.T. Channal, Department of Agronomy, University of Agricultural Sciences, Dharwad-580 005, Karnataka.
72. Dr. S.B. Danawadagi, University of Agricultural Sciences, Dharwad – 580 005, Karnataka.
73. Dr. S.M. Mutanal, Department of Agroforestry, University of Agricultural Sciences, Dharwad – 580 005, Karnataka.
74. Dr. S.B. Devarnavadagi, Department of Agroforestry, University of Agricultural Sciences, Dharwad – 580 005, Karnataka.
75. Dr. H.S. Murthi, Department of Microbiology, College of Fishries, Mangalore – 575 002, Karnataka.
76. Dr. J. Parbhakaram, Tamil Nadu Agricultural University, Coimbatore – 641 003, Tamil Nadu.
77. A.M. Ramesh, University of Agricultural Sciences, Dharwad – 5800 005, Karnataka
78. C.K. Doddagauder, University of Agricultural Sciences, Dharwad – 5800 005, Karnataka
79. Seema Singh, University of Agricultural Sciences, Dharwad – 5800 005, Karnataka
80. B.M.K. Pandit, University of Agricultural Sciences, Dharwad – 5800 005, Karnataka
81. Sunil Desai, University of Agricultural Sciences, Dharwad – 5800 005, Karnataka

SECTION VI

INDEXES

SUBJECT INDEX

AUTHOR INDEX

MIX
Papier aus verantwortungsvollen Quellen
Paper from responsible sources
FSC® C105338

If you have any concerns about our products,
you can contact us on
ProductSafety@springernature.com

In case Publisher is established outside the EU,
the EU authorized representative is:
Springer Nature Customer Service Center GmbH
Europaplatz 3, 69115 Heidelberg, Germany

Printed by Libri Plureos GmbH
in Hamburg, Germany